TOWARD A COMPETITIVE TELECOMMUNICATION INDUSTRY

Selected Papers from the

1994 Telecommunications

Policy Research Conference

TELECOMMUNICATIONS
A Series of Volumes Edited
by Christopher H. Sterling

Bracken/Sterling • Telecommunications Research Resources: An Annotated Guide

Brock • Toward a Competitive Telecommunication Industry: Selected Papers from the 1994 TPRC

Lehr • Quality and Reliability of Telecommunications Infrastructure

Mody/Straubhaar/Bauer • Telecommunications Politics: Ownership and Control of the Information Highway in Developing Countries

Teske • American Regulatory Federalism and Telecommunications Infrastructure

Williams/Pavlik • The People's Right to Know: Media, Democracy, and the Information Highway

TOWARD A COMPETITIVE TELECOMMUNICATION INDUSTRY

Selected Papers from the 1994 Telecommunications Policy Research Conference

Gerald W. Brock
George Washington University

NEW YORK AND LONDON

First Published by
Lawrence Erlbaum Associates, Inc., Publishers
10 Industrial Avenue
Mahwah, New Jersey 07430

Transferred to Digital Printing 2009 by Routledge
270 Madison Ave, New York NY 10016
2 Park Square, Milton Park, Abingdon, Oxon, OX14 4RN

Cover design by Jennifer Sterling

Library of Congress Cataloging-in-Publication Data

Toward a competitive telecommunication industry : selected papers from the 1994 Telecommunications Policy Research Conference / edited by Gerald W. Brock.
p. cm.
Includes bibliographical references and index.
ISBN 0-8058-2030-2 (alk. paper).—ISBN 0-8058-2031-0
1. Telecommunication policy—United States—Congresses. 2. Telecommunication policy—Congresses. I. Brock, Gerald W. II. Telecommunications Policy Research Conference (1994)
HE7781.T68 1995
384'.0973—dc20 95-19204
CIP

Publisher's Note
The publisher has gone to great lengths to ensure the quality of this reprint but points out that some imperfections in the original may be apparent.

Contents

Authors

Chris Avery
Harvard/Kennedy School of Government

Johannes M. Bauer
Michigan State University

Marjory S. Blumenthal
National Research Council

Timothy J. Brennan
University of Maryland

Gerald W. Brock
George Washington University

Richard A. Cawley
European Commission

Richard L. Cimerman
Public Service Commission of Maryland

Robert Corn-Revere
Hogan & Hartson, Washington, DC

Nicholas Economides
Stezy School of Business, New York University

John Haring
Strategic Policy Research, Inc.

Robert G. Harris
University of California, Berkeley

Charles Himmelberg
Stezy School of Business, New York University

Heather Hudson
University of San Francisco

Bridger Mitchell
Charles River Associates

Milton Mueller
Rutgers University

Werner Neu
Wissenschaftliches Institut für Kommunikations Dienste (WIK)

Karl-Heinz Neumann
Wissenschaftliches Institut für Kommunikations Dienste (WIK)

Paul Resnick
Massachusetts Institute of Technology

Jeffrey Rohlfs
Strategic Policy Research, Inc.

Gregory L. Rosston
Federal Communications Commission

L. Keta Ruiz
The World Bank

Pamela Samuelson
University of Pittsburgh

Harmeet Sawhney
Indiana University

Harry M. Shooshan III
Strategic Policy Research, Inc.

Padmanabhan Srinagesh
Bellcore

David J. Teece
University of California, Berkeley

Ingo Vogelsang
Boston University

Geoffrey J. Waldau
Public Service Commission of Maryland

Richard Zeckhauser
Harvard/Kennedy School of Government

Acknowledgments

For many years, the Telecommunications Policy Research Conferences (TPRCs) have produced high quality research and commentary on a wide range of telecommunication policy issues. Although the conference papers have resulted in numerous publications in many different journals, there has not been a routine publication channel for the research presented at the conference. Through the efforts of Christopher Sterling, the series editor of the LEA Telecommunications Series, Hollis Heimbouch of LEA, and John Haring and Bridger Mitchell in their roles as former chairpersons of the TPRC Board of Directors, an arrangement was made for LEA to publish a series of volumes based on research presented at the TPRC. This is the first volume in that series and consists of revised versions of selected papers presented at the 22nd annual TPRC, held in October 1994.

I am grateful to the many members of the organizing committee and session moderators who served as referees to help choose the highest quality papers for inclusion in this volume.

The 22nd annual TPRC, and therefore the publication of this volume, was made possible through generous grants from the John and Mary R. Markle Foundation and from the National Science Foundation.

Gerald W. Brock

Acknowledgments

[illegible]

[illegible]

Foreword

Pamela Samuelson
University of Pittsburgh

On behalf of the organizing committee for the 22nd Annual Telecommunications Policy Research Conference (TPRC), I am pleased to introduce you to the printed edition of selected papers presented at this conference.

1994 was an unusually rich year for telecommunications policy initiatives and research. The Federal Communications Commission (FCC) had undertaken some important new projects, such as a plan for auctioning the spectrum for personal communication services. The U.S. Congress was considering a major revision to the Communications Act of 1934 that would have dramatically altered the telecommunication regulatory landscape. State regulatory agencies were experimenting with a number of new approaches to traditional telecommunication regulation. On an international scale, countries that had previously maintained telecommunication services as state monopolies were considering opening up these services to competition. In the United States, a number of major telecommunication firms had announced mergers with other major players in the communications field, raising antitrust concerns in the minds of many.

Seemingly dwarfing all of these important telecommunication initiatives was the Clinton administration's National Information Infrastructure (NII) initiative and its formation of the NII Task Force, as well as a number of policy working groups (such as the one on privacy and the one on intellectual property rights) under the Task Force's jurisdiction and an Advisory Council of industry and policy leaders to provide input on the NII projects.

The plethora of important telecommunication policy initiatives in 1994 made the job of the organizing committee for TPRC easier in one sense and harder in another. The easy part was brainstorming about the kinds of sessions that the organizing committee thought should be offered at the TPRC in 1994. The hard part was deciding which of the many desirable sessions could actually fit within the constraints of the 2-day schedule and limiting the number of presenters and commentators so that there was adequate time for audience participation and general discussion. Also difficult was scheduling the paper sessions in multiple tracks.

There were inevitably overlaps in sessions that organizing committee members and many conference attendees wanted to attend at any given hour. This is all the more reason why we are fortunate now to have a printed volume containing many of the papers presented at the conference. Overall, the organizing committee was pleased at the large number of excellent papers the conference yielded and at the broad appeal of the plenary and paper sessions to a diverse set of conference attendees.

The organizing committee is grateful to the session moderators for their efforts to attract top-notch research papers for presentation at the conference, to ensure that the paper authors provided the conference staff with copies far enough in advance of the conference so that they could be photocopied and made available at the conference, and for working with the session authors in the preparation of this volume.

I also wish to think the other members of the organizing committee, without whose work there would have been no conference: Martin Cave of Brunel University, Al Hammond of New York Law School, Alex Hills of Carnegie Mellon University, Brian Kahin of Harvard University, Robin Prager of the Sloan School at MIT, David Reed from the FCC, David Sappington from the University of Florida, and David Waterman of Indiana University. It was a pleasure working with them.

Our thanks also go to Dawn Higgins and Lori Rodriguez for the outstanding administrative work that underlay this and previous conferences, and to Gerald Brock for undertaking the substantial effort of organizing and editing the volume that you are now able to enjoy.

I

INTRODUCTORY MATERIAL

1

Introduction

Gerald W. Brock
George Washington University

For many years after the United States began its experiments with competition in the telecommunication industry, other countries retained their government-owned monopoly form of doing business. Even in the United States, there has long been the assumption that competition is an adjunct to the regulated monopoly marketplace, and that competitors are dependent on connecting to the regulated monopoly services of the local exchange companies. All of that is changing as competition comes to the local market in the United States and the competitive model is adopted in other countries as well.

However, the current telecommunication market is still far from the textbook model of perfect competition. For services dependent on spectrum usage, licensing procedures and the limited availability of the spectrum has tended to limit the number of competitors. For example, the duopoly structure of cellular communication was created by spectrum restrictions. The network externality makes interconnection arrangements and prices critical to competitive feasibility. The existing set of laws, rules, and expectations regarding communication networks limits the freedom of companies to enter new markets. The critical importance of communications for defense and economic development causes most countries, including the United States, to limit the freedom of foreign-owned companies to offer a full range of communication services. The various chapters in this volume focus on the policy issues created by the difficult transformation of telecommunication services around the world from a monopoly-oriented, government-dominated structure into a more market-oriented structure.

L. Keta Ruiz's paper won first place in the annual TPRC student paper contest. Ruiz's study of the cellular duopoly provides an important theoretical and empirical contribution to understanding that market. The chapter develops a supergame framework for understanding tacit duopoly cooperation with varying levels of homogeneity. The theoretical results are then clarified with an econometric analysis of the United States cellular industry that concludes that capacity constraints raise prices but neither regulation nor multimarket interaction produces a statistically significant effect on cellular telephone prices. Ruiz's chapter uses more mathematical and economic tools than other chapters in this volume. Readers who are unfamiliar with those techniques should begin with the Economides and Himmelberg chapter rather than the Ruiz chapter.

Economides and Himmelberg analyze the role of network externalities in telecommunication services. Network externalities occur in telecommunication services because the value of a particular service (such as telephone, facsimile, or electronic mail) to a particular subscriber is dependent not only on the characteristics of that subscriber but also on the number and identity of other people connected to the network for that service. The subscriber with the only telephone in the world, or the only facsimile machine, or the only e-mail account, gains no value from that service. Economides and Himmelberg show that with network externalities, the demand curve may have an inverted U shape rather than falling over its entire range, and that perfect competition does not necessarily lead to an efficient outcome. The authors clarify their theoretical analysis with an empirical examination of the growth of the market for facsimile machines. They conclude that the very rapid growth of the facsimile market during the late 1980s was strongly influenced by network externalities.

Network externalities are also discussed in several other chapters in this volume because they are such a critical aspect of communication competition. Because of network externalities, new entrants normally seek interconnection agreements with incumbent firms in order to benefit from the existing customer base of the incumbent firms. The terms and conditions of interconnection among competing firms have played a central role in the development of competition in communications, and continue to pose important questions for the further development of competition.

Part II contains three chapters examining the role of interconnection arrangements and prices for the development of local telephone competition. When the local operating companies were divested from AT&T at the beginning of 1984, it was assumed that local competition was a natural monopoly service. A fundamental goal of the divestiture was to separate the providers of potentially competitive services (long distance, information services, and manufacturing) from the providers of the natural monopoly local exchange service. The structural separation was designed to prevent the anticompetitive use of the natural monopoly services to restrict competition in the potentially competitive services. We now observe increasing competition in the local exchange and corresponding policy controversies regarding the appropriate conditions for entry and interconnection of competitors with the incumbent local exchange carriers (LECs), and regarding the necessary conditions for removing the existing restrictions on LEC provision of long-distance service.

The Harris, Rosston, and Teece chapter sets out one vision of the future competitive local service industry structure. They describe the growth in competition for high-density communications in the center of major cities, and evaluate the potential for greater future competition from wireless systems and upgraded cable-TV capacity. They also expect existing LECs to improve their facilities and gain the capability to provide broadband services in competition with cable-TV companies. Thus they forecast a world in which there will be extensive competition with many new services and no clear dividing line between telephone-type services and cable-TV-type services. The policy question is how the LECs can compete in such a world without either being unfairly hobbled by past restrictions created for a monopoly environment or making anticompetitive use of their established position.

The Harris, Rosston, and Teece answer is the Ameritech plan, a proposal for substantial unbundling in exchange for freedom from the current restrictions imposed on services provided by LECs. The plan is a more radical version of the FCC's Open Network Architecture (ONA) approach developed during the 1980s. The basic concept is to create particular units of service that are useful to customers and competitors and to offer them on a nondiscriminatory basis. That is an alternative vision to the structural separation concept embedded in the divestiture. Just as ONA was used by the FCC to justify eliminating its structural separation rules in the Computer III proceeding, so the unbundling of the Ameritech plan is designed to provide a basis for eliminating the divestiture decree structural prohibitions on LEC entry into long-distance service. The authors observe that there is considerable controversy over the plan and suggest that it should be used as a controlled experiment. They advocate implementing the plan only in the Ameritech territory with a temporary waiver of the prohibition on LEC provision of long-distance service until the performance of the plan can be evaluated.

A second perspective on local competition in a different institutional context is provided by the Mitchell, Neu, Neumann, and Vogelsang chapter. Whereas the Harris, Rosston, and Teece chapter is specifically concerned with the United States restrictions on LEC provision of long-distance service, the Mitchell et al. work was prepared by an international group of experts as a part of the European Commission's (EC) efforts to develop a framework for competitive telecommunications. The EC does not have the institutional history and structure of the United States, but still faces the same problems of interconnection in order to make a viable network of networks in a more competitive future.

Mitchell et al. provide an assessment of the existing theoretical models related to interconnection charges, clarifying the implicit assumptions made and providing insight into why various authors advocate inconsistent recommendations. Mitchell, et al. suggest that the basic standard for interconnection should be the marginal cost or average incremental cost of providing the connection, with a possible markup above incremental cost when special cause is shown by the incumbent. They discuss several practical problems of measuring cost, and conclude that interconnection charges should generally be based on charges determined from the capital cost of providing the peak load capacity. They argue that interconnection charges should not be used to provide for universal service obligations.

The chapter is an unusually complete and clear framework for evaluating prices and regulations for interconnected networks. It helps define the relevant questions and provides a set of possible answers. Even those who disagree with the specific recommendations provided should find the framework useful for adding clarity to a complex and confusing subject. Because of the similarity of the interconnection economics issues in telephone systems around the world, the chapter should assist not only the EC, for whom it was written, but a variety of other jurisdictions in approaching the problem of interconnecting sets of networks.

The Cimerman and Waldau chapter provides an account of the efforts of the Maryland Public Service Commission (PSC) to deal with the 1993 application of Metropolitan Fiber Systems-Intelenet (MFS) to provide and resell local exchange and interexchange telephone service to business customers in Maryland. The incumbent carrier, Bell Atlantic, argued that interconnection rates should be set at incremental cost plus foregone contribution (opportunity cost). The Maryland PSC staff argued that costs should be set near long-run incremental cost with a minimal contribution to supporting the incumbent's shared and common cost. The staff specifically argued for separating the universal service support funding from the interconnection costs. MFS proposed "bill and keep" for terminating calls (each company bills local usage charges to its customers and keeps the resulting revenues), equivalent to a terminating charge of zero. The Maryland Commission adopted a terminating charge of $.061 per call for MFS to pay Bell Atlantic for terminating calls, a compromise between the high charge implied by the Bell Atlantic analysis and the low charges proposed by MFS and the commission staff. The Cimerman and Waldau chapter is particularly valuable for grounding a theoretical discussion in a specific case.

Part III contains three chapters dealing with competitive issues in international telecommunication services. The first two examine the issue of interconnection charges from the perspective of international communication settlement rates for jointly provided service. They provide sharply different viewpoints on the significance of the current difference between international settlement rates and the cost of terminating international traffic. Most international calls are supplied through the joint efforts of a company in the originating country and a company in the terminating country. Countries generally do not allow a foreign carrier to terminate calls in that country and therefore it is not possible for a single company to carry an international call. Normally, the company in the originating country specifies the charge for the call and collects from the customer. The originating company then pays the terminating company a fee (the settlement rate) determined separately for each country pair route by negotiation. Because costs of international communication have been dropping much faster than settlement rates, the settlement rates are generally well above the cost of terminating the call. The excessive settlement rates have been a particular concern in the United States because of a large net outflow of traffic and a correspondingly large net flow of payments to foreign telecommunication operators. High settlement rates are generally a benefit to countries with net inflow of traffic and a cost to countries such as the United States with a net outflow of traffic.

The Haring, Rohlfs, and Shooshan chapter attempts to quantify the costs to the United States from foreign restrictions on United States companies and from above-cost settlement rates. They argue that the current prices for international calls are far above cost primarily because of impediments in the market. According to their analysis, the removal of market imperfections would cause the price of calls originating in the United States to decline by 63% (to 37% of current levels), and the price of calls terminating in the United States to decline by 88%. The dramatic price declines would greatly stimulate demand and increase United States Gross Domestic Product by a cumulative amount between $120 billion and $210 billion over a 10-year period, according to the authors' forecasts, and would substantially improve the United States trade balance and create large numbers of new jobs. The authors advocate vigorous bargaining by the United States government in order to remove foreign impediments to United States companies, and argue that without United States government pressure, the current impediments will remain in place and limit the benefits obtainable from international switched telecommunication services.

Richard Cawley's chapter examines the same set of issues with a very different perspective on both the current situation and the prospects for the future. He focuses more narrowly on transatlantic communication, but notes that the United States and Western Europe account for 70% of the worldwide international public switched traffic. Cawley acknowledges that current settlement rates are well above cost, but argues that far too much attention is given to the rates. Cawley emphasizes the opportunities for working around the high settlement rates through three different methods. International simple resale (resale of interconnected private lines to provide the equivalent of switched service) is available on the United States–Canada and United States–U.K. route and is likely to spread to other routes. International resale undermines the role of high settlement rates as a barrier to entry because the rates only apply to switched traffic, not to the private lines used to produce the equivalent of switched service. The expansion of virtual private networks also reduces the role of settlement rates by providing an alternative to the public switched network on which settlement rates are computed. New international alliances have the potential to further undermine the role of settlement rates. Whereas Haring, Rohlfs, and Shooshan see the current imbalances as significant and increasing over time without remedial government action, Cawley views the current imbalances as less significant and expects them to decrease over time through market forces without government intervention.

Johannes Bauer's chapter examines the significance of changing the model of international telecommunication services from a cooperative effort of two or more nationally based companies to a service of one company or consortium. He provides considerable information on the development of direct foreign investment and alliances among companies that provide the potential capability of future international service without a hand-off of the communication to a national terminating company. Bauer also provides a theoretical perspective and antitrust analysis of the trend toward greater foreign investment. He concludes that asymmetric regulatory conditions could prevent direct foreign investment from having the normally expected beneficial effects and suggests that improvements should be made in the regulatory framework for international communication.

Part IV examines universal service concerns created by increasing competition and advancing technology. The United States achieved effectively universal service in basic telephone many years ago, although some segments of the population remain underserved. Most other industrialized countries now have almost universal service availability, but the poorer countries of the world continue to have very limited telephone service. Universal service concerns have been used to oppose competition in telecommunication services because of the fear that competition would force prices toward cost and eliminate subsidies. Universal service concerns have also been used to justify high access charges so that competitors of a dominant firm contribute toward the universal service obligations of incumbent firms.

New technologies are related to universal service concerns because of the controversy over what should qualify as desirable universal service. Universal service has generally been understood in the past to mean access to a telephone. With changing technology, the question is how to update that definition. Should it remain the same? Should universal service obligations be extended to new forms of technology? Does everyone have a right to Internet services or to wideband digital services? The answers to the universal service questions will have a critical effect on the shape of the future telecommunication industry. Treating a service as a necessary component of universal service implies considerable government control along with possible subsidies and entry restrictions.

The three chapters in Part IV examine universal service from quite different perspectives. Harmeet Sawhney places the search for universal telecommunication service in the broader context of attempts to reach other universal objectives such as universal electrification, universal education, and universal suffrage. He reviews the lofty rhetoric of such efforts, which ranges from improved efficiency to grandiose expectations for the transformation of society. Sawhney then examines the more practical concerns that led to the realization of universal education and other accomplishments. Rather than attempts to implement fundamental rights, each was the result of particular interest groups and political jurisdictions attempting to solve particular problems. For example, Sawhney traces the development of free public schools in the 19th century to efforts to assimilate the large number of immigrants rather than to beliefs about the value of education to a democracy. Sawhney applies those lessons to the current universal service debate to conclude that the effort to develop a precise definition for future universal service is misguided. He suggests that a better alternative is to allow competition among both service providers and political jurisdictions with an expectation that successful innovations will be copied and unsuccessful ones will wither away.

Milton Mueller challenges the "public finance" perspective on universal service. The public finance approach assumes that universal service is at least in part a function of the subsidies developed during the regulated monopoly period in the telephone industry. Because competition tends to erode away those subsidies, the public finance solution to universal service under competition is to establish the subsidies in a way that is compatible with competition, either through an explicit contribution element built into access charges or a tax on service providers that is used to provide the necessary subsidies. Mueller suggests that the public finance approach should be replaced with an "appropriability" analysis that considers telephone access analogous to intellectual property. According to Mueller's analysis, the value of telephone access increases with the number of connections, and therefore an incumbent carrier ought to charge more for connection to a network with widespread access. There should be no general right to nondiscriminatory interconnection charges and to resale. Rather, incumbent carriers should make a distinction in charges between subscribers who desire service for themselves and those who expect to resell it.

Mueller's prescription produces results similar to the regulatory approach of allowing incumbent carriers to add an explicit contribution charge on to the actual cost of providing interconnection, but the foundations of the two methods are quite different. Whereas the contribution charge approach assumes that the incumbent carrier has monopoly power along with universal service obligations and that interconnection conditions are under the control of the regulator, the Mueller approach assumes a more market-oriented structure in which companies do not have either interconnection obligations or full market power. According to Mueller, attempts to charge unreasonably high interconnection fees will lead to broader competition without interconnection. Mueller supports his analysis with examples from the early period of telephone competition without regulation or interconnection, and from current cases of overnight delivery services that strive to offer near universal service without any interconnection obligation.

Heather Hudson's chapter examines universal service in the context of less developed countries. It is a follow-up to the Maitland Commission report of 1984 that set a goal of worldwide universal telephone service. Hudson presents numerous statistics on the development of telephone service in the poorer countries of the world. She finds that although telephone growth is faster than GNP growth, the rate of telephone development is slower than the rate of television growth and still lags far behind the industrialized countries. Only a relatively small number of countries of the world have effectively universal basic telephone service. Most countries have traditionally provided telephone service through a government monopoly, and even customers willing to pay the full price of service have often not been able to obtain service. Hudson reviews the many reforms and changes underway, including openings to foreign investment in telephone systems, that hold promise for improving the rate of telephone development.

Whereas Parts II, III, and IV are largely concerned with telephone policy problems, Part V considers some of the same economic issues in the context of the Internet "network of networks." The unregulated Internet has grown extremely rapidly and provides an alternative model to the regulated structure of the telephone industry. Interconnection agreements and arrangements for using the Internet are determined by private negotiation. The free and unstructured Internet provides an important set of services in its own right, and also provides insight into how other communication services might evolve with greater freedom from regulation.

Padmanabhan Srinagesh's chapter examines the Internet cost structure and interconnection agreements. Internet service providers do not have a prescribed accounting system or requirement to submit the regulatory reports that telephone companies submit. There is consequently much less information on Internet costs and quantity of services than there is for telephone companies. Although the cost data provided by Srinagesh is somewhat limited, it helps clarify an area on which little information is available.

The original Internet structure was created to link educational and governmental researchers, with no commercial component. As the number of individuals connected to the Internet has grown, many commercial access companies have been started. Their interconnection agreements have been developed by negotiation among the companies involved without mandated interconnection or governmental supervision. In general, the current agreements are a form of "sender keep all." That is, each network agrees to accept traffic from others without charge in exchange for the right to send its traffic to other networks. That method is in sharp contrast to the telephone methods in which it is typical for one carrier to pay the other a prescribed fee for terminating traffic (as long-distance carriers pay to local exchange carriers) or to share revenues on a prescribed basis (as in the international settlements system). The "sender keep all" method is particularly simple and eliminates the need for many kinds of measurement and accounting mechanisms that are required under the various alternative compensation schemes. Srinagesh describes the actual methods in use and analyzes the incentives created by the existing interconnection methods for the Internet. He notes that the interconnection methods are not entirely consistent with the current charging schemes and allow opportunities for profitable arbitrage by resellers.

Marjory Blumenthal's chapter considers the lessons of the internet for creating the National Information Infrastructure (NII). The NII has been discussed widely, but without any clear definition of what it would be or how it differs from services available today. Blumenthal provides criteria for creating the NII that incorporate lessons from the structure of the Internet. The Internet's openness and ability to evolve with combined public and private investments have been viewed as key aspects of an NII structure. Some therefore see the future NII as a grown-up universally available Internet with improved access tools.

Blumenthal introduces a concept known as the *Open Data Network* as a basis for evaluating proposals for the future NII. She suggests the NII should be characterized by "openness, independence from specific technologies, large-scale and decentralized organization, heterogeneity ... and flexible or extensible basic service" (p. 278). She then reviews a number of lessons to be drawn from the Internet in achieving the goals of an open data network.

The chapter by Resnick, Zeckhauser, and Avery tackles another aspect of the Internet problem: how to filter the vast amount of information available. The existing Internet and expected future NII greatly ease the transmission of data. However, they do nothing to enhance the individual user's ability to comprehend, sort, and utilize that data. One dubious benefit of the Internet is to create the ability for every individual subscriber to be an originator of bulk volume junk mail. The limited commercial use of the Internet has tended to minimize that problem in the past, but the large and rapidly growing number of users make an attractive advertising target. If users begin receiving vast numbers of advertising messages, they will need some method of filtering the information available to decide what needs to be seriously considered. Resnick, Zeckhauser, and Avery examine the potential role for "electronic brokers" that would filter the information. Using the specific case of product evaluations, the authors consider the issues of privacy and censorship that would occur in an electronic brokerage service, and also consider the incentives to provide electronic brokerage.

The three chapters together make a substantial contribution not only to understanding the existing Internet, but also to the future enhanced NII. The problems of creating an open network that can allow many different types of technologies to be used together, the problem of understanding the cost arrangements and connecting a network of networks, and the problem of filtering massive quantities of information to separate the useful from the irrelevant are all critical to creating an NII.

Part VI concludes the book with an examination of applying the First Amendment to changing communication technology. At first sight, it may seem strange to include a section on the First Amendment in a book primarily concerned with competition, interconnection, and universal service. However, the First Amendment issues provide an important social context to the remaining discussions. As technology changes, it does not recreate the entire social order. New technology and new ways of accomplishing tasks have to be fit into the existing legal and political framework. The First Amendment issues are critical in their own right and also are important as a reminder of the broad context in which technological changes take place. The two papers in this section take quite different approaches to clarifying the role of the First Amendment in the emerging and converging communication technologies.

Timothy Brennan's chapter attempts to use economic analysis to evaluate the reasons for the First Amendment. His innovative work draws on the existing broad economic literature of information economics. That literature shows that information asymmetry may lead to adverse selection or moral hazard. Brennan applies that insight to the concept of First Amendment protection of freedom of the press. He suggests that freedom of the press may break down the information asymmetry between candidates and voters and improve the efficiency of the "electoral market."

Brennan's chapter and Resnick et al.'s chapter are both concerned with information as a commodity and how to use that information to enhance other processes. Both recognize that information may be misrepresented to benefit those who control it. Resnick et al.'s solution is an honest broker who develops a reputation for providing accurate information and is disinterested in the substantive commerce. Brennan's comparable solution is a press that reports accurately on the accomplishments of politicians.

Robert Corn-Revere's chapter provides a legal analysis of the history of First Amendment interpretations in the presence of technological change. He traces the tortured history of First Amendment litigation in broadcasting and other industries and concludes that free speech rights are gradually added to new technologies, rather than being automatically granted to them when they perform the older functions. Thus television broadcasting was initially not included in First Amendment protections even though its news provided a clear substitute for newspapers that were. Over time, the protections for television have been increased.

The development of different bodies of law for common carriers, cable-TV systems, and broadcasters creates difficulties when converging technologies allow any of them to provide the same service. Corn-Revere concludes that the past approach risks diluting the significance of First Amendment protections. As new technologies supplant old, the First Amendment is diminished if it only applies to the old technologies. Some have even advocated using the broadcast restrictions to limit the freedom of newspapers if they use spectrum at any stage of their production (such as transmitting their copy to remote plants for printing). In other words, the divided approach allows the possibility for changing technology to undermine the First Amendment freedoms. Corn-Revere argues that approach is bad policy and should be rejected in favor of a strong First Amendment approach that is independent of the technology by which communication is carried out.

2

Pricing Strategies and Regulatory Effects in the U.S. Cellular Telecommunications Duopolies

L. Keta Ruiz
The World Bank

Cellular telecommunication in the United States is a case of duopoly with no entry as defined by the Federal Communications Commission (FCC). Each player in each market uses a portion of radio spectrum to produce cellular telecommunication services. Scarcity and rivalry of radio spectrum is the main reason why the FCC granted only two licenses for cellular operation at each market. The main characteristics of this industry are defined later and are summarized as follows: (a) cellular telecommunication has no close substitutes and is a nearly homogeneous service in each market and nationwide; (b) some players meet as competitors in several markets; and (c) cellular services are regulated in some states, whereas in others they are not. These characteristics compelled me to make a theoretical and empirical assessment of this young industry, which started in 1982.

The theoretical part of the analysis is developed in the section on theoretical analysis. Theoretical models of tacit cooperation that best represent the characteristics of the U.S. cellular telecommunications duopolies are developed in a game theoretic framework. Duopolists are assumed to compete in a nonhomogeneous goods market, where the strategic variable is price. The production technology corresponds to constant marginal costs and consumers' preferences are represented by a quadratic utility function. Then the conditions needed for tacit cooperative equilibria and the possible pricing outcomes of the duopolies are derived. In addition to the basic nonhomogeneous goods model, regulation is imposed to see its implications on the cooperative equilibria and on the pricing strategies.

Empirical models are derived and estimated in another section. Based on the theoretical predictions of duopoly behavior, a series of hypotheses are empirically tested on duopoly pricing strategies in response to stated restrictions. Econometric techniques are used to find the determinants of prices of these duopolies and a reduced equation of price is estimated to overcome the lack of data.

Finally, the last section provides a set of conclusions. Some of the conclusions open a door to further research because they do not provide compelling arguments for policy recommendations. Thus, the main contribution of this dissertation falls in the area of positive analysis of the cellular telecommunications duopolies. Further data is necessary to provide a precise set of recommendations.

CELLULAR TELECOMMUNICATIONS

Cellular Telecommunications Technology

Cellular telecommunication was developed as a solution to congestion, difficulty to subscribe to the service due to long waiting lines, and problems of low quality of the conventional mobile telephone service. Mobile telephone service was first provided by AT&T in St. Louis in 1946.[1] This technology is based on a frequency modulated (FM) radio link between a single large transmitter that provides coverage to large areas, usually 40 to 80 square miles, and the mobile unit. Innovations in telecommunications permitted improvements in mobile radio communications, which were introduced in the mid-1960s. This more advanced service was known as Improved Mobile Telephone Service (IMTS), which included features such as automatic trunking, direct dialing, and full duplex service.[2] These features, combined with the reduced radio spectrum needed per channel, increased the capacity of the system. But capacity remained a problem that cellular sought to solve. Cellular telecommunication is based on the use of low power transmitters that cover small geographic areas, usually 2 miles in diameter, called *cells*. The low power allows frequency reuse because there is no interference in nonadjacent cells that use the same frequency, and therefore provides more capacity. Cellular telecommunication requires the use of radio spectrum because it utilizes radio links from the mobile units to the cell-sites.[3]

New developments in cellular telecommunications include data transmission, voice encryption, voice mail, hands-free talk, speech recognition voice dialer, and the coming digital cellular radio.

Although the cellular idea was worked out in the 1960s by AT&T, and the FCC first requested comments on frequency allocations in 1974, the final report and order was issued in 1981. In June 1982, a very tedious and complicated process of licensing started. Finally, the first cellular service started in Chicago in 1983 by the then about-to-be-divested AT&T.

---

1. For a history and evolution of both mobile and cellular telecommunications see Calhoun (1988).

2. *Automatic trunking* is an arrangement by which the receiver searches for a vacant radio channel in the group of channels available to the unit, instead of the caller searching for one manually. *Direct dialing* avoids the need for an operator to connect the calls. *Full duplex* transmission permits both parties to talk simultaneously instead of alternating.

3. A detailed explanation of the engineering characteristics can be found in Lee (1988).

Market Structure

Radio spectrum for cellular use in the United States is allocated to carriers by the FCC. In 1975, the FCC allocated 115 MHz of spectrum of the 900 MHz frequency range to mobile communications (FCC, 1975). The FCC allocated 40 MHz of these to cellular communications, 30 MHz to conventional mobile and dispatch services, and 45 MHz was left on reserve. The FCC also proposed to allow one cellular carrier per market. The first authorizations for trials were given to AT&T for a trial in the Chicago area and to American Radio Telephone Service (ARTS) in the Washington/Baltimore/Northern Virginia area in 1977, although service would not be provided until 1983.

After a long rule-making process the FCC determined that cellular service would be provided by two facilities-based carriers in each of the "markets" or Cellular Geographic Service Areas (CGSAs), which more or less encompass the major Metropolitan Statistical Areas (MSAs) as defined in the 1980 United States Census.[4] The decision to allow two carriers was based on the desire of the FCC to introduce facilities based competition, while still letting the providers be efficient, because AT&T sustained that sharing spectrum with other licensees would significantly increase costs. The FCC maintained that the benefits of diversity of technology, service, and price were not outweighed by the benefits of economies of scale and increased efficiency of a single carrier system versus the two-carrier system, which became known as *wireline carrier* and *nonwireline carrier* (FCC, 1981).

The wireline carrier is a separate subsidiary of one of the local telephone companies in the MSA, either a Regional Bell Operating Company (RBOC) or an independent telephone company. A block of frequencies, usually known as "Block B," was assigned to this carrier. The nonwireline carrier is usually a firm or group of firms from the land mobile radio industry, private investors, or one of the RBOCs or independent telephone companies that do not provide service in the area. These carriers were assigned "Block A" frequencies. Because there is this geographic carving of the serving areas, "roaming"[5] agreements had to be established among adjacent carriers and carriers with frequent travelers (e.g., New York and Los Angeles), as well as security measures to prevent fraud.

In terms of substitutes, cellular phone competes with IMTS, dispatch services, paging, and the landline telephone network, but none of these services are close substitutes.[6]

There has been a rapid growth of cellular subscribers, due to reduction of equipment costs, even though service prices have not fallen very much. Cellular service

4. In this study I analyze cellular services in the MSAs because the Rural Statistical Area (RSA) licenses have been awarded only recently, and there is little information about them.

5. *Roaming* is the handling of a call initiated by a cellular caller who belongs to other CGSA.

6. The FCC is in the rule-making process of a new service called Personal Communications Service (PCN or PCS), which seems to be a close substitute for cellular service, and is expected to be more affordable than cellular.

TABLE 2.1.
U.S. Cellular Industry Figures

Year	*Number of Subscribers*	*Capital Investment in millions*	*Service Rev. in millions*	*Number of Workers*	*Number of Systems*	*Number of Cell-sites*	*Bill $ per month*	*Call Length (minutes)*
1984	91,600	355	178	1,404	32	346	NA	NA
1985	340,213	911	482	2,727	102	913	NA	NA
1986	681,825	1,437	823	4,334	166	1,531	NA	NA
1987	1,230,855	2,235	1,152	7,147	312	2,305	96.83	2.33
1988	2,069,4411	3,274	1,960	11,400	517	3,209	98.02	2.26
1989	3,508,944	4,480	3,341	15,927	584	4,169	89.3	2.48
1990	5,283,055	6,282	4,549	21,328	751	5,616	80.9	2.20
1991	7,557,148	8,672	5,709	26,327	1,252	7,847	72.74	2.38
1992	11,032,753	11,262	7,823	34,348	1,506	10,307	68.68	2.58
1993	16,009,461	13,946	10,892	39,775	1,529	12,805	61.48	2.41

Note: Figures are as of December of corresponding year.
Source: Cellular Telecommunications Industry Association (1994).

prices are still high, compared to the local landline network.[7] With the introduction of digital technologies, the prospects of growth are better since capacity will be increased, though initial prices of equipment, will probably be high. Table 2.1 gives a summary of some statistics of the U.S. cellular industry.

Regulation

Federal regulation has been limited to the issuing of licenses and the supervision of transfer of them,[8] among other functions. The FCC also authorizes the carriers to make changes and expansions of their systems. It reviews complaints against carriers. The FCC can regulate interstate cellular rates, but has not been active in doing so.

State regulation, on the other hand, has been more active. Table A.1 in Appendix A summarizes the status of regulation of cellular telecommunications by State Public Utilities Commissions (SPUCs) in the United States. State regulation can be at the wholesale or at the retail level. It can govern three aspects of cellular operations. The SPUC can require a certificate of public convenience and necessity (CPCN) whenever the cellular carrier starts operations or makes expansions to its system. The second aspect regulated by some public utilities is the rates charged for cellular services. In some states, each time a new tariff is introduced or a change of an existing tariff is proposed, it has to be approved after a review process. This review process usually entails the filing of cost support data and, if changes are considerable, a public hearing. In other states, the filing of tariffs is only for informational purposes. The third aspect regulated by some SPUCs is to

7. For example, for the landline network average residential monthly rates for unlimited local calling ranged from $8.14 to $27.11, as of October 1990 (see Mitchell & Vogelsang, 1991, p. 159). On the other hand, an average monthly bill for cellular service, although it might include long distance as well as roamer charges, was $80.90 (see CTIA, 1992).

8. Resale of the license by the original operator is allowed, as long as the buyer is not the operator of the other license in the same market.

require cellular companies to follow a uniform system of accounting (USOA). This is done to facilitate audits in case of complaints. Some states do not regulate at all.

As can be seen in the table in Appendix A, 23 of the 50 states and the District of Columbia require the filing of tariffs at the wholesale level, and 14 at the retail level. Most of these states require from 10 to 60 days before any change in rates, sometimes only in an upward direction, becomes effective.

There are several factors that worry some state regulators about the competitiveness of cellular communications:

1. Although resale is allowed and there can be downstream competition, the number of firms in each market is fixed and small (only two); so the cellular providers are likely to recognize their interdependence and can coordinate prices.
2. The homogeneity of the service and the similarity of the costs. Technology is somewhat standard and both carriers offer similar quality and coverage.[9]
3. There are few close substitutes to cellular services.
4. Players in one market compete in other markets as well. There can be a live-and-let-live attitude because they face each other in several markets.[10]

THEORETICAL ANALYSIS

Literature on Regulation, Tacit Collusion, and Supergames

The objective of this research is to see what effects product differentiation and regulation have on pricing of cellular telecommunications in the U.S. metropolitan areas. I take the definition of economic regulation by Joskow and Rose (1989). According to this definition "economic regulation" is both direct legislation and administrative regulation of prices and entry into specific industries or markets, as opposed to "social regulation" that deals with government intervention in markets analyzing, for example, environmental, health, and safety issues.

Most regulation analysis deals with natural monopoly industries in which there is only one regulated entity, or with contestable markets in which there is potential entry. For example, in the Salant and Woroch (1992) model, using a game theoretic framework, the relationship is between a regulator and a public utility. In contrast, cellular telecommunications presents a special case of regulation, in which there is a small fixed number of players, and where the theory of tacit collusion suitably comes to facilitate an analysis of this market.

9. Recently, some carriers have tried to differentiate their services by providing nationwide coverage. With this service, cellular customers are able to make and receive calls in any system in the United States without any burdensome codes or dialing. McCaw Cellular has been the first one to do so.

10. In the study by the United States General Accounting Office (1992), they mention that public utility officials of the six most populous states told them that cellular is not an essential service, and the industry is sufficiently competitive, so traditional utility regulation was not necessary.

Chamberlin (1929) was the first to recognize the possibility of tacit collusion in a concentrated market with homogeneous products and where firms were aware of their interdependence. Since then, there have been several models of tacit collusion and tests for it. A good review of this literature was done by Jacquemirn and Slade (1989), who also examined policy issues related to collusion. The theory of supergames and repeated games[11] provides a useful framework with which to analyze tacit collusion. Under certain conditions with repeated interaction, players of a noncooperative supergame can attain cooperative outcomes without entering into binding agreements. Friedman (1990) gave a detailed account of the equilibria of the various classes of supergames. Tirole (1988) gave a concise review of the literature of supergames applied to oligopoly theory.

Model

In this section, a supergame framework is used for the analysis of tacit cooperation. The first subsection details the basic model, a repeated game, with product differentiation. The cellular market, as mentioned earlier, is a somewhat homogenous market, but still there is some product differentiation that is worth modeling. In addition to this, building a more general model that allows for product differentiation opens the possibility for use of this model to explain performance in other industries. In the last subsection, a repeated game with regulation is presented to see the effects of regulation on the tacit cooperation equilibrium.[12]

I use the supergame framework to model the cellular telecommunications market. A supergame is a sequence of one-shot games, finite or infinite in number, played by a fixed set of players. The set of available moves and the payoff functions can change from period to period. A special case of supergames is a repeated game, in which the same game is played at each period (neither the set of available moves nor the payoff functions change).

A Repeated Game

The Market and Payoffs. This is a market where there are two firms denoted by the subscript i, $i = 1, 2$. For convenience, they are assumed to behave as if they were in the market forever. They produce a differentiated good, and each firm's output is x_i. Both firms have the same constant marginal cost c, up to capacity. For simplicity let $c = 0$.[13] The firms sell to a very large number of consumers whose

11. For a definition of supergames, see next section.

12. For this section, I have drawn from Deneckere (1983) and Ross (1992), who made a detailed analysis of duopoly supergames with product differentiation.

13. I can generalize the results for a constant marginal cost function by considering $a = a' - c$, where a, the intercept of the inverse linear demand (see footnote 14) includes the constant marginal cost $c \neq 0$.

demand is represented by the functions $x_i = x_i\ (p)$ and $x_j = x_j\ (p)$; $p = (p_i, p_j)$; $i,j = 1,2$. Let us assume inverse linear demands given by:[14]

$$p_i = a - bx_i - dx_j$$

The ratio $z = d/b$ gives the degree of product differentiation. Let $d \leq b$, that is, own effects are at least as large as cross effects, therefore $z \in [0,1]$. When $z = 1$ the goods sold by i and j are perfect substitutes. If $z = 0$ they have independent demands.

Let $\prod_t^i (p_{1t}, p_{2t})$ be the profits of firm i at time t. Each firm maximizes the present discounted value of its profits:

$$\sum_{t=0}^{\infty} \delta^t \, \Pi_t^i(p_{1t}, p_{2t})$$

where δ is the discount factor; $\delta = 1/(1 + r)$, and $r \geq 0$ is the discount rate.[15]

The Game. Let Γ be the game in which, at each date t, the two firms choose strategies Φ_{it}, $\Phi_{it} : h_t \rightarrow p_{it}$ simultaneously, dependent on the entire history:

$$h_t \equiv (p_{10}, p_{20}; \ldots ; p_{1t-1}, p_{2t-1})$$

where (p_{1s}, p_{2s}), $s = 0, \ldots, t{-}1$ are the actual prices chosen at time s.[16]

The fact that the firms choose their prices simultaneously allows for deviations of a given strategy, without retaliation by the affected firm for one period, because they cannot react instantaneously to any deviation.

The assumption that the firms' choices depend on h_t implies perfect recall or complete memory, that is, the firms accumulate information as the game progresses. Therefore, their choices at any period are a function of the other firm's previous actions.

---

14. The implicit assumption here is that utility is quadratic. The demand of good i is given by:

$$x_i = \frac{a}{b+d} - \frac{b}{b^2-d^2}p_i + \frac{d}{b^2-d^2}p_j; \qquad for\ x_i > 0, \quad i,j \ 1,2$$

The intercept $a/(b + d)$ can be interpreted as all other factors that affect the demand of good i in addition to the prices of goods i and j. Income, and commute time are among these other factors for the case of cellular services.

15. I can allow δ to change from period to period, but for simplicity, I keep it constant.

16. Note that the strategic variable here is price. The other elements of demand, a, b, d are considered as given parameters. Therefore, z, the degree of product differentiation, is exogenous. Firms can strategically choose how close to locate from each other, but this case is not analyzed here. Cooperative behavior in terms of pricing is difficult to achieve by firms, and cooperating in the degree of product differentiation makes cooperation even more difficult.

Equilibria. Equilibrium strategies will form a "perfect equilibrium" if for any history h_t at date t, firm i's strategy from date t on maximizes the present discounted value of its profits given firm j's strategy from that date on.

This game has multiple equilibria. One of them is the repeated Bertrand equilibrium, with each firm charging the Bertrand price (p_{it}^B):[17]

$$p_{it}^{B} = \frac{a(b-d)}{2b-d}$$

and earning profits of:

$$\pi_i^B = \frac{a^2 b(b-d)}{(2b-d)^2(b+d)}$$

Another equilibrium is given by the following strategy:

$$p_{it}(h_t) = p^m \quad \text{if } p_{is} = p^m \text{ for all } i \text{ and } s = 0, \ldots t\ 1$$
$$p_{it}(h_t) = p^B \quad \text{otherwise}$$

Where $p^m = a/2$, the collusion price and p^B is the Bertrand price.[18]

Equation 6 is known as a trigger strategy because firm i will use the action $p_{it} = p^m$ as long as firm j uses the same action. However, if firm j deviates from this strategy of charging p^m, as soon as this deviation is detected by firm i, it will use the "grim" strategy, which is to charge the Bertrand price p^B from then on.[19] Any deviation by firm j from the cooperative path will "trigger" a punishment by firm i. If firm j deviates at time T, while firm i charges $p^m = a^2/2$, the price that maximizes its one-time profits, the chiseling price, will be given by:

$$p_{jT}^{ch} = \frac{a(2b-d)}{4b}$$

where *ch* is for chiseling.

This price is calculated by replacing $p_i = p^m = a/2$ in the Bertrand reaction function, given by:[20]

17. See Ruiz (1994) for the deduction of this price.

18. See Ruiz (1994) for the deduction of these prices.

19. As opposed to a finite reversion strategy in which the punishment lasts for only a finite number of periods.

20. The Bertrand reaction function, $p_j = [a(b-d) + dp_i]/2b$, gives the change in price of firm j when the price of firm i changes.

$$p_j = \frac{a(b - d) + dp_i}{2b}$$

Demand is given by:

$$x_T^j(p^m, p^{ch}) = \frac{a(2b - d)}{4(b^2 - d^2)}$$

and therefore profits are:

$$\Pi_T^j(p^m, p^{ch}) = \frac{a^2(2b - d)^2}{16b(b^2 - d^2)}$$

Note that the profits $\Pi \to \infty$ when $b = d$. This is so because when one of the firms cheats, the maximization of its profits drives the output of the other to negative values. Therefore we have to find the price p_{jT}^{ch2} and profits Π_T^j (p^m, p^{ch2}) subject to the restriction that the output of the other firm is nonnegative.

In this case the price of firm j will be given by:

$$p_{jT}^{ch2} = \frac{a(2d - b)}{2d}$$

And its profits are given by:

$$\Pi_T^j(p^m, p^{ch2}) = \frac{a^2(2d - b)}{4d^2}$$

Therefore, the price of firm j if it chisels at time T is given by:[21]

$$p_{jT}^{ch} = \frac{a(2b - d)}{4b} \quad if \quad z \leq 0.7321$$

$$p_{jT}^{ch} = \frac{a(2d - b)}{2d} \quad otherwise$$

and its profits at time T will be given by:

$$\Pi_T^j(p^m, p^{ch}) = \frac{a^2(2b - d)^2}{16b(b^2 - d^2)} \quad if \quad z \leq 0.7321$$

$$\Pi_T^j(p^m, p^{ch}) = \frac{a^2(2d - b)}{4d^2} \quad otherwise$$

I concentrate the analysis on trigger strategies, although these are not the only ones that can sustain tacit cooperation, they have a characteristic that makes them important in the framework where collusion is illegal. Trigger strategies are very simple and easy to communicate, and therefore they are easily self-enforcing.

Proposition 1 shows that for a large enough δ the trigger strategies given by Equation 6 form a noncooperative equilibrium.

Proposition 1. Let Γ be the repeated game described in the previous subsection. The trigger strategy given by Equation 6 is a perfect equilibrium of Γ with both firms pricing at p^m forever if:[22]

$$\delta \geq f(z) = \frac{4 - 4z + z^2}{8 - 8z + z^2} \quad \text{if } z \leq 0.7321$$

$$\delta \geq f(z) = \frac{(-2 + z)^2(1 - z - z^2)}{4 - 8z + z^2 + 3z^3 - 2z^4} \quad \text{otherwise}$$

For a proof, see Ruiz (1994).

Therefore, if the trigger strategy was to charge p^m, this strategy is a perfect equilibrium in which the firms charge this price forever. Because p^m is the collusion price, then, without cooperation, the duopolists have engaged in a tacitly cooperative behavior.

It is interesting to analyze how the condition for the cooperative equilibrium to be sustainable varies nonmonotonically with the degree of product differentiation z.[23] There has long been the belief in oligopoly theory (see Stigler, 1966) that cartels are more sustainable when products are more homogeneous. But from the results, when the degree of homogeneity rises from completely independent goods ($z = 0$) to less heterogeneous goods ($z = 0.7321$),[24] the conditions for the cooper-

21. Firm j will be indifferent between charging any of these two chiseling prices if they give it the same amount of profits, therefore we can find the level of $z = d/b$ for which each of the prices and profits are valid by solving:

$$\Pi_T^j(p^m, p^{ch}) = \Pi_T^j(p^m, p^{ch2})$$

then

$$\frac{a^2(2b - d)^2}{16b(b^2 - d^2)} = \frac{a^2(2d - b)}{4d^2}$$

$$(2b - d)^2 d^2 - 4b(b^2 - d^2)(2d - b) = 0$$

$$d^4 + 4bd^3 - 8b^3 d + 4b^4 = 0$$

because $z = d/b$ and $b \neq 0$, dividing the previous equation by b^4, and writing it in terms of z:

$$z^4 + 4z^3 - 8z + 4 = 0$$

which has as one solution (the other solutions are -0.7321, 2.7321, and -2.7321, which are not of interest because they fall outside the range of values z can take ($0 \leq z \leq 1$)):

$$z = d/b = 0.7321$$

22. For perfect substitutes, that is, when $z = 1$, the condition is that $\delta \geq 1/2$.

23. This result was first found by Deneckere (1983).

24. The value of $z = 0.7321$, which is going to appear in the conditions, divides the profit functions for the chiseling case. It is the value at which the chiseler is indifferent between maximizing its unconstrained profits when it chisels and its profits subject to the output of the other firm being nonnegative.

ative equilibrium to arise are more stringent. Therefore, product homogeneity makes cooperation less sustainable, exactly the opposite of what the belief was. Although when products get more homogeneous ($z > 0.7321$) up to the point of perfect substitutability ($z = 1$), the condition for cooperation turns less stringent. Therefore the belief that more homogeneity leads to more sustainability of tacit collusion is true only for high product homogeneity. These differences are due to the fact that profits from chiseling are nonmonotonic with respect to z.

A Repeated Game With Regulation

I am going to work in the framework of the model described in the preceding section. In this case, I would like to analyze the effects of different kinds of regulation on the cooperative equilibrium. I am going to analyze two types of regulation:

- Firms have to file price changes or new pricing plans to a regulatory agency. This is either to lower or raise a price or to offer a special "high user" plan. The price change becomes common knowledge from the filing and it has to go through an approval process that takes some time.
- The regulatory agency sets a price cap for a given period of time. Changes in the price of a service, or a basket of services within certain ranges do not have to be filed or go through any special process during the given period of time. When this period of time is over, the regulatory agency will set another price cap for another period of time, and so on.

Firms Have to File New Prices or Price Changes

This kind of regulatory regime can have three effects:

1. Rates become common knowledge. Therefore, firms use the regulatory agency as a cartel board through which information is disseminated. They signal the other duopolist of what their pricing plans are. This is not an uncommon case. For example, one of the principal functions of the Joint Executive Committee Railroad Cartel analyzed by Porter (1983a, 1983b) was information gathering and dissemination to member firms.

1. *Price cuts.* The Joint Executive Committee kept weekly accounts to inform members of the developments in the industry. This function was very important because the information was used to see if any member firm was chiseling, and if so, to punish it. In the case of the kind of regulation I am analyzing, any price change or new price has to be filed with the regulatory agency. Members of the duopoly are therefore not able to make price cuts because whenever they make a price change, they have to file it with the regulatory agency. In this case regulation increases the possibility of tacit cooperation because it is acting as the "cartel board."
2. *Price increases.* Because explicit arrangements are banned by law, how can a duopolist raise its price without losing its market share and trying to make the other follow suit? They can do it through advertisement; this has

been claimed to be used by the airlines. They can do it through the regulatory agency by filing price changes. In this case, regulation increases the possibility of tacit cooperation because as in the case of price cuts, it is acting as the "cartel board," where pricing strategies are posted.

2. It introduces a lag to the price changing process. This lag consists of how fast the other duopolist can react to the price change. Because no rates can be changed without a filing and approval process, if it takes 1 week for the other firm to prepare an application for this price change, the other firm will be ahead in whatever rate cut it makes. Let σ be the time that it takes a firm to file new prices or price changes.[25]

In this case even though price changes are observed immediately after the new price is filed, because a firm cannot retaliate immediately, the trigger price strategy will be given by:

$$p_{it}(h_t) = p^m \quad \text{if} \quad p_{is} = p^m \quad \forall\ i \text{ and } s = 0, \ldots, t-\sigma$$
$$p_{it}(h_t) = p^B \quad \text{otherwise}$$

where σ is the time it takes a firm to file new prices.

The condition for the cooperative equilibrium to be sustainable is given by:[26]

$$\delta \geq f(z,\sigma) = \left(\frac{4 - 4z + z^2}{8 - 8z + z^2}\right)^{\frac{1}{\sigma}} \quad \text{if } z \leq 0.7321$$
$$\delta \geq f(z,\sigma) = \left(\frac{(-2 + z)^2(1 - z - z^2)}{4 - 8z + z^2 + 3z^3 - 2z^4}\right)^{\frac{1}{\sigma}} \quad \text{otherwise}$$

In this case regulation might be a cause of cooperation breakdown. However if firms are aware of this, then they can always have a price change application ready so that $\sigma \to 1$, and therefore, we are under the case of perfect information. Consequently, if firms are behaving in an implicitly cooperative way, they can always do that and therefore regulation will have no effect because any price cut will be matched by the other duopolist, and because of that they will not engage in chiseling.

3. Firms might want to cut prices to test the market, but regulation can introduce price rigidities. Firms are afraid of cutting their prices because they will have to file a tariff to do so, and if the price cut was not profitable, then they have to file another tariff to raise it again, which might not be approved by the regulatory agency. Therefore firms might behave more conservatively and have higher prices due to regulatory constraints. So even if cooperation is not present, regulation will have a negative effect because, due to rigidities, firms will have higher prices.

25. Note that the time the regulatory agency takes to review and approve the rates does not matter because it is the same for both firms.

26. For derivation see Ruiz (1994).

On the other hand, these price rigidities imposed by the regulatory agency can also affect the cooperative equilibrium because the threat, for the trigger strategy to work, has to be credible. If firms are unable to cut their prices due to regulation, or can cut them only to a certain point, the conditions for the cooperative equilibrium will be more restrictive. There is also the case in which they cannot chisel due to the same restrictions.

In this case, let us suppose that the punishment price cannot be less than the Bertrand price, p^B, that is, the trigger strategy will be given by:[27]

$$p_{it}(h_t)\ p^m \quad \text{if } p_{is}\ p^m \quad \forall\ i \text{ and } s\ 0, \ldots, t\ 1$$

$$p_{it}(h_t)\ \omega_1 p^B \quad \text{otherwise}, \quad 1 < \omega_1 < \frac{p^m}{p^B}$$

And the chiseling price can only be above the optimum chiseling price:

$$p_{jT}^{chl}\ \ \omega_2 p_{jT}^{ch}\ \ \frac{\omega_2 a(2b\ \ d)}{4b} \quad \text{if } z \le 0.7321$$

$$p_{jT}^{chl}\ \ \omega_2 p_{jT}^{ch}\ \ \frac{\omega_2 a(2d\ \ b)}{2d} \quad \text{otherwise}$$

Where $1 < \omega_2 < p^m/p^{ch}$.

Proposition 2 states the conditions under which the trigger strategy given by Equation 18 form a noncooperative equilibrium.

Proposition 2. Let Γ be the repeated game described in the previous section but with regulation that restricts price cuts. The trigger strategy given by Equation 18 is a perfect equilibrium of Γ with both firms pricing at p^m forever if:

$$\delta \ge f(z,\omega_1,\omega_2)\ \frac{4(\ 2\ \ z)^2(\ 1\ \ z)\ \ (\ 2\ \ z)^4(2\ \ \omega_2)\omega_2}{16(\ 1\ \ z)^2(\ 2\ \ \omega_1)\omega_1\ \ (\ 2\ \ z)^4(2\ \ \omega_2)\omega_2} \quad \text{if } z \le 0.7321$$

$$\delta \ge f(z,\omega_1,\omega_2)\ \frac{(\ 2\ \ z)^2(\ z^2\ \ \omega_2\ \ z\omega_2\ \ 2z^2\omega_2)}{4(\ 1\ \ z)z^2(2\ \ \omega_1)\omega_1\ \ (2\ \ z)^2(1\ \ z)(\ 1\ \ 2z)\omega_2} \quad \text{otherwise}$$

For a proof, see Ruiz (1994).

The variable ω_1 can be interpreted as how restrictive regulation is with respect to the punishment price. The higher ω_1 the more restrictive the punishment price is. The value of $\omega_1 = 1$ means that the punishment price is p^B, whereas a larger ω_1 means that the punishment price is getting closer to the collusion price and there-

27. The case when the punishment price is less than the Bertrand is not considered because here only price cuts are analyzed.

fore punishment is less credible. The variable ω_2 can be interpreted as how restrictive is regulation is with respect to the chiseling price. The value of $\omega_2 = 1$ means that the chiseling price is equal to p_{jT}^{ch}, that is, there are no restrictions on chiseling. On the other hand, a larger ω_2 means that the chiseling price is close to the collusive price. The higher ω_2 the more restrictive regulation is with respect to the chiseling price. Both ω_1 and ω_2 restrict the amount that prices may fall either when firms chisel or punish.

When we analyze these propositions, the more restrictive regulation is with respect to the punishment price (higher ω_1) the more sustainable the cooperative equilibrium is. This is because the returns from chiseling plus punishment are smaller. This is not true for ω_2 for the case of more homogeneous goods. The more restrictive regulation is with respect to the chiseling price, the less sustainable cooperation is. This is due to the fact that for higher chiseling prices, the returns (from chiseling plus punishment) get higher. Overall, more restrictive regulation facilitates cooperation, especially for less homogeneous goods.

Regulation That Sets a Price Cap

In this case p_i can be changed within certain ranges. Let us say if $\lambda_1 p_i \leq p_i \leq \lambda_2 p_i$; $\lambda_1 < 1, \lambda_2 > 1$, then the firms can change rates without any tariff filing or announcement. I assume that the initial rates are known at the beginning, and that firms are tacitly colluding. So I analyze if these "price caps" have any effect on the cooperative equilibrium. Because they are tacitly cooperating, and they will not set their price above p^m, the upper bound will not matter.[28]

On the other hand, in the case of the punishment price, if firms can always file a new price to the regulatory agency and have it changed, then threats of retaliation against chiseling become credible. Therefore the price caps do not matter in relation to the punishment price. So the lower bound is important only from the standpoint of the chiseling price because firms would like to be able to cut prices without the other realizing it. There is another variable that matters in this case, information lags, so I take into account two cases: when there is perfect monitoring and when there are information lags.

Either under perfect or imperfect monitoring, when $\lambda_1 p_i \leq p_i^{ch}$, there is no effect of regulation, as the firms can cut their prices up to p_i^{ch} without the other being able to react. In this case the conditions for the cooperative equilibrium to be sustainable are given by Proposition 1 in the case of perfect monitoring, and by Equation 17 in the case of information lags.

Therefore the only case worth analyzing is when $\lambda_1 p_i > p_i^{ch}$, that is, when the chiseling price, due to the price cap, cannot be the optimum one, but is higher. Under this condition, the chiseling price will be given by:

---

28. Unless there are demand changes that will make them willing to raise their prices. I assume that demand does not change, so I do not study this case.

$$p_{jT}^{ch2} \quad \omega_2 p_{jT}^{ch} \quad \lambda_1 p_i$$

There is perfect monitoring. Because regulation only matters in terms of the chiseling price, then we are under a special case analyzed previously, that is, when there are rigidities to price cuts. We only have to set $\omega_1 = 1$, to find the condition so that the cooperative equilibrium is sustainable. This condition is then given by:

$$\delta \geq f(z,\omega_2) \quad \frac{4(\ 2\ z)^2(\ 1\ z)\ \ (\ 2\ z)^4(2\ \ \omega_2)\omega_2}{16(\ 1\ z)^2\ \ (\ 2\ z)^4(2\ \ \omega_2)\omega_2} \qquad \text{if } z \leq 0.7321$$

$$\delta \geq f(z,\omega_2) \quad \frac{(\ 2\ z)^2(\ z^2\ \ \omega_2\ \ z\omega_2\ \ 2z^2\omega_2)}{4z^2\ \ 4z^3\ \ 4\omega_2\ \ 8z\omega_2\ \ 3z^2\omega_2\ \ 7z^3\omega_2\ \ 2z^4\omega_2} \qquad \text{otherwise}$$

From this proposition, the condition for cooperation to be sustainable varies nonmonotonically with respect to the level of product differentiation and the degree of regulation.

Prices are observed with τ lags. In this case any price cut is observed with a delay of τ periods. We can show[29] that the condition for the cooperative equilibrium to be sustainable is:[30]

$$\delta \geq f(z,\omega_2,\tau) = \left(\frac{4(\ 2\ z)^2(\ 1\ z)\ \ (\ 2\ z)^4(2\ \ \omega_2)\omega_2}{16(\ 1\ z)^2\ \ (\ 2\ z)^4(2\ \ \omega_2)\omega_2}\right)^{\frac{1}{\tau}} \qquad \text{if } z \leq 0.7321$$

$$\delta \geq f(z,\omega_2,\tau) = \left(\frac{(\ 2\ z)^2(\ z^2\ \ \omega_2\ \ z\omega_2\ \ 2z^2\omega_2)}{4z^2\ \ 4z^3\ \ 4\omega_2\ \ 8z\omega_2\ \ 3z^2\omega_2\ \ 7z^3\omega_2\ \ 2z^4\omega_2}\right)^{\frac{1}{\tau}} \qquad \text{otherwise}$$

The results and conditions of the theoretical analysis, however, leave us with uncertainty on the effect of the variables on the pricing behavior of cellular carriers because most of the theoretical results give nonmonotonic relationships. Therefore, it is the empirical evidence examined in the following section that will help find more specific relationships for the case of cellular telecommunications services.

AN EMPIRICAL MODEL

Methodology

Econometric techniques are used to find out which variables determine the price of these duopolies. A reduced form equation of price is estimated to overcome the

29. The proof is in Ruiz (1994).

30. Note that when $\tau = 1$ we are in the case of perfect monitoring detailed earlier.

lack of data on quantities and costs. Hypotheses on product heterogeneity, capacity constraints, regulation, multimarket interaction, and others are tested. The analysis consists of seeing if these variables induce the duopolies to price higher or lower. In order to see if the duopolies are earning extraordinary profits, the margin of price over costs should ideally be measured. However, it is not possible to do so because of lack of data.

There are several empirical studies on cooperation based on supergame strategies. A summary of the ones most related to the study can be found in Ruiz (1994). The empirical analysis I make tries to capture different price strategies for a cross section. Although similar tests to the ones tried here have been done before, the data used in this study are special because there is no entry allowed into the U.S. cellular telecommunications industry. One important test I make is regulation. The empirical studies reviewed do not examine this issue because the markets examined by them are unregulated. I lack data to test for cooperative behavior per se, but I am able to identify whether variables believed to have effects on cooperative pricing effectively raise or lower prices.

A Reduced Form Equation for Price

I have taken the repeated game framework because repeated interaction is an important feature of the market under analysis. Trigger price strategies offer the convenience of being easy to communicate, an important characteristic if collusion is illegal, as is the case because U.S. laws ban it. As seen in the section on theoretical analysis, any price–quantity solution between the Bertrand and the cooperative one can constitute a perfect equilibrium for certain discount rates. This imposes the difficulty of which equilibria to consider. The objective of the empirical analysis is to test several hypotheses to uncover the pricing strategies of the duopolies, based on the theoretical predictions.

Let us take the grim strategy when prices are observed with a lag of τ periods:

$$\begin{aligned} &p_{it}(h_t) \; p^m \quad \textit{if } p_{is} \; p^m \; \forall i \textit{ and } s \; 0, \ldots, t \; \tau \\ &p_{it}(h_t) \; p^B \quad \textit{otherwise} \end{aligned}$$

As seen in the previous section, for certain conditions of δ, $p_{it} = p^m$ is a tacitly cooperative outcome. If those conditions are not met, then the noncooperative equilibrium will be reached with $p = p^B$.

If we look at a price in cellular market k, p_k:

$$\begin{aligned} &p_k \; p_k^B \quad \textit{if there is no cooperation} \\ &p_k \; p_k^c \quad \textit{otherwise} \end{aligned}$$

where p_k^B is the Bertrand price in market k, and p_k^c is the collusion price in this market.[31]

Let us take the highest collusion price, that is, p_k^m. So we have:

$$p_k \quad p_k^c \quad p_k^m \quad c_k \mu_k$$

where c_k is the marginal cost in market k, $\mu_k = 1/(1 - 1/\eta_k)$ and η_k is the elasticity of demand in market k at p_k^m.

Taking logarithms of Equation 26:

$$ln \; p_k \quad ln \; c_k \quad ln \; \mu_k$$

To estimate $ln \; p_k$, we need equations for $ln \; c_k$ and $ln \; \mu_k$. Let:

$$\ln c_k \quad \alpha_0 \quad \alpha_1 y_k \quad \epsilon_1$$

where the αs are the parameters, $\mathbf{y}_k$ is the vector of cost driver variables and ε_1 is an econometric error.[32] And,

$$\ln \mu_k \quad \beta_0 \quad \beta_1 x_k \quad \beta_2 STRAT \; \beta_3 REG \; \epsilon_2$$

where the βs are the parameters, x_k is the level of demand, *STRAT* is the vector of strategic variables, *REG* is a regulatory variable and ε_2 is an econometric error. Here the assumption is that any pricing above marginal cost depends on demand and will be affected by the strategic variables and the regulatory regime. Note that regulation is considered exogenous here.[33]

Because demand is an endogenous variable, the quantity demanded in market k, x_k, is given by:

$$x_k \quad \gamma_0 \quad \gamma_1 \ln p_k \quad \gamma_2 z_k \quad \epsilon_3$$

where the γs are the parameters, p_k is the price in market k, z_k is the vector of demand drivers, and ε_3 is an econometric error. Substitute Equations 28 through 30 into Equation 27, to get the reduced form equation for price:

$$\ln p_k \quad \rho_0 \quad \rho_1 y_k \quad \rho_2 z_k \quad \rho_3 STRAT_k \quad \rho_4 REG_k \quad \nu_k$$

31. A similar analysis, but for the effects of regulation on intrastate landline telephone service, was done by Kaestner and Kahn (1990).

32. We can introduce the level of output in this equation to take into account variations of cost due to different levels of output, for example, economies of scale.

33. It might be the case that regulation is endogenous, but I do not analyze this situation here.

where

$$\rho_0 \quad \frac{1}{1 \quad \beta_1\gamma_1}(\alpha_0 \quad \beta_0 \quad \beta_1\gamma_0)$$

$$\rho_1 \quad \frac{1}{1 \quad \beta_1\gamma_1}(\alpha_1)$$

$$\rho_2 \quad \frac{1}{1 \quad \beta_1\gamma_1}(\beta_1\gamma_2)$$

$$\rho_3 \quad \frac{1}{1 \quad \beta_1\gamma_1}(\beta_2)$$

$$\rho_4 \quad \frac{1}{1 \quad \beta_1\gamma_1}(\beta_3)$$

$$\nu_k \quad \frac{1}{1 \quad \beta_1\gamma_1}(\epsilon_1 \quad \epsilon_2 \quad \beta_1\epsilon_3)$$

This reduced form equation for price is the one that will be used to test the various hypotheses. It offers the advantage that neither cost nor demand are necessary to estimate the effect of the exogenous variables in price. The only data needed for each market k are: prices (p_k), cost drivers (y_k), demand drivers (z_k), strategic variables ($STRAT_k$) and regulation (REG_k). With these data the parameters ρ_0, ρ_1, ρ_2, ρ_3, ρ_4 can be estimated and hypotheses on effects of product heterogeneity, capacity constraints, regulation, and strategic variables can be tested.

Hypotheses to Test

I test hypotheses based on the theoretical predictions developed earlier. Unfortunately I cannot test dynamic hypotheses because I only have a cross section of data. Therefore, I am only able to test static hypotheses. The hypotheses I test are the following.

- *Does product differentiation affect pricing?* The theoretical model showed that, depending on the degree of product differentiation, the conditions for the collusive equilibrium to arise might be more stringent. Because cellular telecommunications service presents some product differentiation (e.g., through a wider coverage and a higher quality of service), the estimations will try to find the sign and significance of the effect of this variable on pricing.
- *Does regulation affect pricing?* Does it increase or reduce the prices charged by the duopolies? The theoretical model shows a nonmonotonic relationship between regulation and prices, therefore the estimations will try to find the sign and significance of the effect of regulation on pricing.

- *Do capacity constraints matter on pricing?* Do they increase prices or reduce them?
- *Does multimarket interaction influence pricing?* Are prices higher or lower for firms that meet in several markets? This is an issue that has not been theoretically modeled here but because the available data allows testing of hypotheses with multimarket interaction, I show what effect this variable has on cellular pricing.
- *Does brand recognition affect pricing?* For example, the cellular carrier affiliated with the local landline telephone company might be able to charge a premium because it might be better known than the other cellular carrier. This is an issue that has not been theoretically modeled here, but I empirically test it because the available data allows testing of its effects on cellular pricing.

Data

The data are described in the following. Table 2.3 gives summary statistics for all the variables used in the empirical analysis.

Price Data

I collected prices for the cellular markets for the 306 metropolitan areas as of October 1991.[34] There are several pricing plans. Carriers report as many as 14 retail price plans, and 7 wholesale price plans. On average there are five retail plans and one wholesale plan per carrier. Charges usually comprise a monthly service charge, a per-minute charge, and charges for several special features such as voice mail, call forwarding, roaming, and so on.

The price variables are an index of retail prices that reflects the price paid by a customer that uses the cellular phone for 200 minutes per month,[35] of which 80% are at peak times, with an average length of call of 2.5 minutes. The first one, INDEXMIN, corresponds to the least expensive plan that does not require signing a contract.[36] The second one, INDEXCON, corresponds to the least expensive plan that requires signing a contract of 1 year or less. The third one, INDEX, corresponds to the least expensive plan. It might require signing a contract or not, and the contract might exceed 1 year of commitment on behalf of the customer.

I calculate these three indices per each cellular carrier (one for the wireline and one for the nonwireline), because in some markets they differ. The average of the absolute value of the difference (see Table 2.2) is more than $7, which accounts for an 8% difference between the mean prices of each of the carriers. I have tested

34. The list of variables and data sources is in Appendix B.

35. 200 minutes per month is the average nationwide usage per customer as reported by the CTIA for the date of analysis.

36. Most cellular carriers have opted to offer "special" plans. These consist of signing up for the service for 6 months or more and getting a handset at very low prices. If the contract is breached the customer is subject to penalty.

TABLE 2.2
***t*-test for Price Differences Between Carriers**

Variable	*N*	*Mean*	*Std Dev*	*Std Error*	*T*	*Prob > |T|*
INDEXMIN	230	7.25	10.13	0.66823	10.84491	0.0001
INDEXCON	230	7.18	10.08	0.66488	10.80318	0.0001
INDEX	226	7.51	11.27	0.74934	10.02537	0.0001

for the significance of these differences. Table 2.2 shows the means test for the absolute value of the difference between the price of each carrier. The null hypothesis of no difference between indices can be rejected with better than 1% of significance.[37]

Cost Data

As cost drivers I have included:

- The number of cell-sites per carrier (CELL). Each carrier has laid out cell-sites to provide cellular service. The number of cell-sites determines the cost in operation and maintenance (O & M) that the cellular carrier has to incur. The average number of cell-sites is 8.13, with some carriers having only 1 cell-site and others having as many as 130. This variable can also serve as a "proxy" of product differentiation because the number of cell-sites allows the cellular carrier to offer certain coverage and capacity, which at the same time translates into the quality of the service provided by the carrier.
- The population density of the metropolitan area served by the cellular carrier (DENSITY). The more dense a metropolitan area is, the fewer cell-sites and connections it will need, because the population is concentrated.
- The density of cells per square mile (DENCELL). The number of cell-sites per square mile will also affect cost because fewer cells per square mile imply longer connections and larger O & M costs. They imply larger costs because there is more equipment to maintain and more O & M personnel is needed given that it is more difficult to share them if cell-sites are more spread.
- The population per cell (POPCELL). This a capacity variable. The larger the number of people per cell, the more expensive it will be to serve the area, up to a certain point (capacity limit), above which no more customers can be served. This variable is used to test for capacity constraints.
- The cost of high capacity lines or trunks in the area the cellular carriers serve (MONDS1, MONDS3, NONRDS1, NONRDS3). The cellular net-

37. I did not include the price of the cellular phone and other equipment necessary to have cellular service because for the cross section analysis, there seem to be no significant differences in phone prices across the country.

TABLE 2.3
List of Variables

Variable	*M*	*SD*	*No. of Obsrvations*
INDEXMIN: price index (contract = 0)	101.19	19.77	452
INDEXCON: price index (contract ≤ 1)	94.15	18.14	460
INDEX: price index	93.98	18.15	460
CELL: number of cell-sites	8.14	14.45	573
DENSITY: MSA population density	345.22	425.84	594
DENCELL: density of cells	0.00496	0.00958	560
POPCELL: MSA population per cell	98,864	66,062	564
MONDS1: monthly DS1 charge	718.12	130.79	598
MONDS3: monthly DS3 charge	8,964	3,355	594
NONRDS1: nonrecurring DS1 charge	814.56	238.92	598
NONRDS3: nonrecurring DS3 charge	1775	1762	594
MONTHOPE: months carrier has been operating	66.07	17.62	500
WAGES: annual wages of telecom workers	29,947	4,039	594
BLOCK: one if carrier is nonwireline	0.5	0.5	610
AREA: area of the MSA	1,922	2,335	594
COMMUTE: minutes of daily commute	41.73	6.41	596
POP89: MSA total population 1989	6,564	13,609	598
PERIN89: MSA total personal income 1989	12.3e + 6	29.2e + 6	596
INCOMEPC: MSA annual income per capita	16,323	3,149	596
CON: length of contract for indexcon	0.52	0.5	460
CONTRACT: length of contract for index	0.63	0.67	460
REGULATION: if carrier has to file tariff change	0.43	0.5	600
MULTIM: number multimarket interactions	0.86	2.54	610

work is connected through these high capacity lines, therefore the cost will depend on the cost of these connections.

- The number of months the cellular carrier has been operating (MONTHOPE). This variable affects cost because a longer operating cellular carrier might reduce costs due to learning. It might also be that a recently established cellular carrier is using penetrating pricing. A third way this variable might affect pricing is through collusion: The longer the firm is operating, the more it knows about the other firm.
- The annual wages of local telecommunications workers (WAGES). Wages are part of the costs. The higher the wages in an area, the higher the costs of the cellular carrier in that area will be.

- The area in square miles of metropolitan area where the cellular franchise is (AREA). The larger the area of the cellular franchise, the higher total cost the cellular carrier will incur.
- A dummy variable that indicates if the carrier is a nonwireline carrier or not (BLOCK), that is, if it is not affiliated with the local exchange company that serves the area.[38] It takes the value of 1 if it is a nonwireline carrier, and 0 if it is a wireline carrier. This variable can be used as a "proxy" of brand recognition.

An important variable in the cost of cellular carriers is marketing. It constitutes a big share of expenses (40% in some cases). Data on marketing expenditures were not collected; however, its effect might have been captured by other variables such as wages, given that marketing has a very high component of wages, and the nonwage components (newspaper or television advertising rates) might be also correlated with wages.

Demand Data

As demand drivers I have considered:

- The number of minutes of daily commute in the area served by the cellular carrier (COMMUTE). The longer the commute time, higher the demand for cellular.
- The population of the area (POP89). The larger the population in the area, the higher the demand for cellular.
- Total personal income in the corresponding area (PERIN89). The higher the personal income in the area, the higher the demand for cellular.
- The income per capita in the corresponding area (INCOMEPC). The higher the income per capita in the area, the higher the demand for cellular.

Strategic and Regulatory Data

As strategic variables I included:

- The length of contract for the corresponding price index (CON, CONTRACT). Cellular carriers will trade lower prices for cellular service in exchange for a contract so that the customer must be a subscriber of the service for the length of the contract.
- As regulation variable (REGULATION) a dummy is used, which takes the value of 1 if the state requires the carrier to file a tariff whenever there is a price change, and 0 otherwise.
- Multimarket interaction (MULTIM) takes the value of 0, if the cellular carriers in that franchise do not meet with each other in any other market, 1 if they meet in one other market, 2 if they meet in two other markets, and so forth. It is expected that cellular carriers that compete against each other in more markets will have higher prices, because this multimarket interaction

38. In some cases the wireline carrier is a subsidiary of a local landline exchange company that serves only a small part of the territory, but that has been awarded the wireline franchise.

allows them to be more lenient with each other than cellular carriers that only meet once.

Estimations

I estimated three models. The first one has as a dependent variable the logarithm of the price of each carrier (LINDEXMIN = LOG(INDEXMIN)), that does not require signing a contract. Here the assumption is that, because prices are statistically different between carriers, there should be some variable, for example, product differentiation that will capture this effect. Therefore I have to include the prices of both carriers.

The second model has as a dependent variable the logarithm of the price of each carrier (LINDEXCON = LOG(INDEXCON)) that requires signing a short-term contract. I control for contract length with the variable "CON."

The third model has as a dependent variable the logarithm of the price of each carrier (LINDEX = LOG(INDEX)) that requires or does not require the signing of a contract. I control for contract length with the variable "CONTRACT."

I estimated the model including all the variables, but because there is high linear correlation among several of the explanatory variables, I tested for multicollinearity. Based on the tests several variables have been excluded in the final model.

To test for heteroscedasticity, White's (1980) test to detect it is used. This test is based on the fact that if heteroscedasticity is present, the least squares covariance matrix of the estimators is inconsistent. A statistic is built that follows an asymptotic χ^2 distribution with degrees of freedom equal to $k(k+1)/2-1$. Table 2.4 summarizes the results for the three final models estimated. The null hypothesis that there is homoscedasticity can be rejected with 90% confidence for the case of LINDEXMIN and with 99% confidence for LINDEXCON and LINDEX. Therefore corrections for heteroscedasticity were made.[39]

Results

General Results Table 2.5 summarizes the results, where only the significant variables and regulation have been maintained.

In the three models the signs of the statistically significant variables are as expected:

- The denser cell-sites are (POPCELL), the more expensive it is to provide the service and the higher its price. Because this variable also expresses capacity, the larger the number of people per cell, the more expensive it will be to serve the area, up to the capacity limit, above which price rationing will be used. A cellular carrier in an area with 5% more people than the average (about 5,000 people above the population mean) will charge a price

39. Once heteroscedasticity was corrected, the parameter estimates did not significantly change. The *t statistics* and the overall consistency of the model did not change significantly either. These results can be obtained from the author.

TABLE 2.4
Heteroscedasticity Analysis

Dependent Variable	*Degrees of Freedom*	χ^2	*Prob* > χ^2
LINDEXMIN	26	35.14	0 .1086
LINDEXCON	34	55.56	0.0112
LINDEX	34	54.62	0.0139

TABLE 2.5
Results

	LINDEXMIN		*LINDEXCON*		*LINDEX*	
Variable	*Estimated Coefficient*	*t statistic*	*Estimated Coefficient*	*t statistic*	*Estimated Coefficient*	*t statistic*
CONSTANT	4.18	62.3	4.02	67.28	4.03	68.56
POPCELL	2.58 e - 7	1.42	3.13 e - 7	1.97	3.21 e - 7	2.03
NONRDS3	2.40 e - 5	4.51	3.11 e - 5	6.64	3.12e - 5	6.74
MONTHOPE	-0.00236	-3.65	-3.20 e - 3	-5.64	-3.00 e - 3	-5.21
BLOCK	-0.03	-1.7	-0.04	-2.7	-0.05	-2.84
CONTRACT			-0.02	-1.42	-0.03	-2.71
COMMUTE	7.93 e - 3	4.62	0.01	6.93	0.01	6.87
INCOMEPC	1.21 e - 5	3.63	1.40 e - 5	4.7	1.36 e - 5	4.65
REGULATION	-0.03	-1.39	-0.01	-0.63	-0.01	-0.73
No. of observations	391		400		398	
R^2	0.179		0.329		0.3412	

per month 0.16% higher (about $0.15 above the estimated mean price of $95.41).[40]

- The trunks variable (NONRDS3) has a positive effect. The more expensive the connections to the cell-sites are, the higher the cost to provide cellular service and the higher the price has to be. A cellular carrier in an area in which the price of a DS3 connection is 5% more expensive than the average price for all the areas (about $89 above the average DS3 connection price) will charge a price per month that is 0.28% higher (about $0.26 above $95.41).
- The number of months of operation (MONTHOPE) has a negative effect, meaning that there might be some learning by operating. The more months a cellular carrier has been operating, the lower its price is. A cellular carrier that has been operating in an area 5% longer than the average (about 3

40. All the comparisons in this section are made with the estimated equation for LINDEX, which is the one that has the highest explanatory power.

months more than the average) will charge a price per month that is 0.99% lower (about $1 below $95.41). The hypothesis of collusive behavior that postulates that longer operation lets the firms know more about each other is therefore rejected.

- The nonwireline variable (BLOCK) indicates that nonwireline carriers charge lower prices. Wireline carriers might charge higher prices because they have higher costs or due to "brand recognition," and customers pay a premium to get services from them. This might happen because they are affiliated with the local landline telephone company and their brand name is familiar to the customer because it in most cases provides the local telephone service in the area. A cellular carrier that is nonwireline charges 4.9% less than the wireline carrier (about $5 below the price of the wireline carrier that is $95.41).
- The contract variable (CONTRACT) has a negative effect. Cellular carriers try to "lock in" customers by offering lower prices if customers sign a contract, and the longer the contract, the lower the price. The price of cellular service under a contract is 2.96% less than a noncontract service (about $3 below the price of the noncontract service, that is $95.41).
- The commute variable (COMMUTE) has a positive sign. The greater the average time spent commuting in an MSA, the more willing customers are to pay a higher price. A cellular carrier in an area in which the commute time is 5% longer than the average of all areas (about 2 more minutes) will charge a price per month that is 2.11% higher (about $2 above $95.41).
- The income per capita variable (INCOMEPC) also has a positive sign, meaning that the higher the income in the area, the higher the price that the consumers can pay. A cellular carrier in an area with income per capita 5% higher than the average (about $816 above the average) will charge a price per month 1.12% higher (about $1 above $95.41).

I dropped several of the variables because they were not statistically significant.[41] Because of multicollinearity between several of them, I did not include them in the final estimations I present. Although regulation (REGULATION) was not statistically significant in either model, I kept it in the estimated models in order to show its irrelevance. The price of cellular service in a state that is regulated is 1% lower than the price of a carrier in a nonregulated state (about $0.96 below the price of the nonregulated carrier price that is $95.41).

Capacity Constraints. In addition to testing capacity constraints with the variable that indicates how many people per cell there are (POPCELL), I estimated models for the three indices (LINDEXMIN, LINDEXCON, LINDEX) considering capacity constraints. This has been defined as an indicator variable (FULL) that takes the value of 1 if the cellular carrier is facing capacity constraints and

41. For example, CELL, the number or cell-sites per carrier, was not statistically significant. If we consider that this variable is a "proxy" for product differentiation, then we can conclude that product differentiation in the cellular market is not significant.

TABLE 2.6
Estimation Considering Capacity Constraints

	LINDEXMIN		*LINDEXCON*		*LINDEX*	
Variable	*Estimated Coefficient*	*t statistic*	*Estimated Coefficient*	*t statistic*	*Estimated Coefficient*	*t statistic*
CONSTANT	4.24	13.14	4	12.63	4.06	12.92
FULL	-0.02 e - 2	-0.17	-0.03 e - 2	-0.27	-0.03 e - 2	-0.33
NONRDS3	2.24 e - 5	1.99	2.88 e - 5	2.58	2.63e - 5	2.38
MONTHOPE	-1.78 e - 3	-0.92	-2.69 e - 3	-1.38	-2.28e - 3	-1.2
BLOCK	-0.08	-1.2	-0.06	-0.84	-0.07	-1.01
CONTRACT			6.9 e - 3	0.12	-0.04	-1.05
COMMUTE	9.62 e - 3	2.17	0.01	2.55	0.01	2.73
INCOMEPC	4.54 e - 6	0.28	1.19 e - 5	0.72	8.0 e - 6	0.5
No. of observations	40		40		40	
R^2	0.32		0.4435		0.4618	

zero otherwise. A carrier is considered to be facing capacity constraints if the number of customers it is serving is greater than 300,000.[42] I only had the number of customers available for the 20 largest MSAs, therefore this hypothesis was tested for a small sample.

TABLE 2.7
Estimation Considering Multimarket Interaction

	LINDEXMIN		*LINDEXCON*		*LINDEX*	
Variable	*Estimated Coefficient*	*t statistic*	*Estimated Coefficient*	*t statistic*	*Estimated Coefficient*	*t statistic*
CONSTANT	4.32	53.2	4.2	50.89	4.17	56.55
POPCELL	2.08 e - 7	1.15	2.58 e - 7	1.64	2.49 e - 7	1.59
NONRDS1	-1.35 e - 4	-3.11	1.01 e - 4	-2.47	0.000132	-3.71
NONRDS3	1.53 e - 5	2.98	2.69 e - 5	5.89	2.71e - 5	6.04
MONTHOPE	-2.51 e - 3	-3.11	-3.07 e - 3	-5.19	-2.74 e - 3	-4.65
BLOCK	-0.04	-1.97	-0.04	-2.7	-0.05	-2.78
CONTRACT			-0.03	-1.97	-0.03	-3.15
MULTIM	2.75 e - 3	1.14	4.57 e - 4	0.21	2.17 e - 5	0.01
COMMUTE	8.07 e - 3	4.69	0.01	7.03	0.01	6.96
INCOMEPC	1.34 e - 5	3.95	1.69 e - 5	5.34	1.64 e - 5	5.32
No. of observations	392		399		399	
R^2	0.1985		0.3479		0.3607	

Table 2.6 shows these results. Contrary to what we found with the variable (POPCELL), the results show that capacity constraints are not statistically significant. In the three cases the t statistics for the capacity constraint indicator variable are very small. This contradictory result is probably due to the fact that I only used data for the 20 largest MSAs.

Multimarket Interaction. The multimarket interaction variable (MULTIM) was included in the set of exogenous variables to test whether cellular carriers that compete against each other in other markets price higher or lower than the ones who meet in only one market. The variable MULTIM reflects the number of multimarket interactions and therefore a positive sign will prove that cellular carriers that have multimarket contact price higher than the ones that do not. The estimations give a positive sign to this variable, but it is not statistically significant. Table 2.7 shows the results for the three price indices (LINDEXMIN, LINDEXCON, LINDEX). This might be the case because, either (a) cellular carriers that meet with each other do not behave strategically, or (b) although several cellular carriers meet with each other in other markets, there are too few instances to be statistically significant.

CONCLUSIONS

The conclusions from the theoretical and empirical analysis can be grouped in to two sets: theoretical and empirical. Whereas the theoretical conclusions are for any duopoly in which some characteristics are analyzed (product differentiation and regulation), the empirical conclusions are specific to the cellular telecommunications services in the United States.

Theoretical Findings

The theoretical analysis provides a framework to analyze the pricing strategies in the cellular telecommunications industry. The analysis based on a pricing model of a duopoly included product differentiation, time lags, capacity constraints, and regulation. The analysis was centered on what is the effect of these variables on the sustainability of cooperation between the duopolists.

The following conclusions can be drawn:

- *Product differentiation.* The condition for sustainability of cooperation in prices of duopolists depends nonmonotonically on the degree of product differentiation. When the degree of homogeneity rises from completely independent goods ($z = 0$) to less heterogeneous goods ($z = .7321$), the conditions for the cooperation equilibrium to arise become more stringent. Therefore, product homogeneity makes cooperation less sustainable. However, when products get more homogeneous ($z > .7321$) up to the point of

42. This is based on the estimates of the current capacity of the cellular systems considering the available technology (see Donaldson, Lufkin, & Jenrette, 1990).

perfect substitutability ($z = 1$), the condition for the cooperation equilibrium to be sustainable turns less stringent. Therefore the belief that more homogeneity leads to more sustainability of tacit collusion is true only for high product homogeneity. In the case of cellular telecommunications, product differentiation is not significant. This puts cellular service in the upper end of product homogeneity (z is close to 1). Thus, the conditions for the cooperation equilibrium to arise are less stringent.

- *Regulation.* Two types of regulatory regimes were analyzed: (a) Firms have to file price changes or new pricing plans with a regulatory agency. This is either to lower or raise a price, or to offer a special "high user" plan. (b) The regulatory agency sets a price cap for a given period of time. The condition for sustainability of cooperation in prices of duopolists varies nonmonotonically with respect to the level of product differentiation and the degree of regulation for both kinds of regulation. The exact relationship for the case of cellular telecommunications is tested empirically. The empirical analysis is done only for the first kind of regulation because this one is the one that is currently applied. Price cap regulation was theoretically explored because it has been recently introduced to regulate landline telephone communications.

The results and conditions, however, give no specific answers about what the signs and significance of the variables are because most of the theoretical results give nonmonotonic relationships. Thus, it is the empirical evidence that gives to the final conclusions.

Empirical Findings

The following conclusions can be drawn from the empirical analysis:

- *Capacity constraints.* Capacity constraints raise prices. This result was obtained when capacity constraints were approximated with population density per cell-site. Nonetheless, when a test was done for a small sample using an indicator variable, capacity constraints were not statistically significant. The contradiction in these results is probably due to the difference in sample sizes. The results with the larger sample, which show that capacity constraints raise prices, are probably more significant.
- *Regulation.* Regulation is not statistically significant. This can mean two things: (a) prices are competitive so regulation does not play any role; or (b) even though pricing might be cooperative, regulation neither brings them more in line with competition nor turns them more cooperative. However, one has to be cautious in concluding that regulation does not play a role, because the analysis only considered the basket of an average consumer. Further estimations for, let us say, low-usage consumers and high-usage consumers can give more evidence on the effect of regulation on cellular pricing. It might also be the case that there is no regulation where it is not needed, so that the sample might be biased. Given that regulation itself might depend on how pricing is done in this market or might also depend

on other unobserved market parameters, which were not considered in the analysis, the results might be biased.

- *Multimarket interaction.* Multimarket interaction is not statistically significant. This might be the case because: (a) cellular carriers that meet with each other do not behave strategically; (b) although several cellular carriers meet with each other in other markets, there are too few instances to be statistically significant.

ACKNOWLEDGMENTS

The views here presented are those of the author and should in no way be attributed to the World Bank or any of its affiliated organizations. This chapter is based on the author's PhD dissertation done under the supervision of Professor Ingo Vogelsang at Boston University. I would like to acknowledge his comments and encouragement as well as those of my second reader, Glenn A. Woroch. I would also like to acknowledge David Wheeler's careful and valuable comments. Of course, I am solely responsible for any errors.

APPENDIX A: REGULATION

TABLE A.1.
State Regulation of Cellular Telecommunications Services

		Wholesale			*Retail*		
State Name	*Overall*	*CPCN*	*Tariff*	*USOA*	*CPCN*	*Tariff*	*USOA*
Alabama	None	No	No	No	No	No	No
Alaska	Full	Yes	Yes	Yes	Yes	Yes	Yes
Arizona	Partial	Yes	Yes	N/A	No	No	No
Arkansas	Partial	Yes	No	No	Yes	No	No
California	Full	Yes	Yes	Yes	Yes	Yes	Yes
Colorado	None	No	No	No	No	No	No
Connecticut	Partial	Yes	Yes	No	No	No	No
Delaware	None	No	No	No	No	No	No
Dist of Columbia	None	No	No	No	No	No	No
Florida	None	No	No	No	No	No	No
Georgia	None	No	No	No	No	No	No
Hawaii	Full	Yes	Yes	Yes	Yes	Yes	Yes
Idaho	None	No	No	No	No	No	No
Illinois	Full	Yes	Yes	Yes	Yes	Yes	Yes
Indiana	None	No	No	No	No	No	No
Iowa	None	No	No	No	No	No	No
Kansas	None	No	No	No	No	No	No
Kentucky	Partial	Yes	Yes	Yes	No	No	Yes
Louisiana	Full	Yes	Yes	N/A	Yes	Yes	N/A
Maine	None	No	No	No	No	No	No
Maryland	None	No	No	No	No	No	No
Massachusetts	Full	Yes	Yes	Yes	Yes	Yes	N/A
Michigan	None	No	No	No	No	No	No
Minnesota	None	No	No	No	No	No	No
Mississippi	Full	Yes	Yes	N/A	Yes	Yes	N/A
Missouri	None	No	No	No	No	No	No
Montana	None	No	No	No	No	No	No
Nebraska	None	No	No	No	No	No	No
Nevada	Full	Yes	Yes	Yes	Yes	Yes	Yes
New Hampshire	None	No	No	No	No	No	No
New Jersey	None	No	No	No	No	No	No
New Mexico	Partial	Yes	Yes	N/A	No	No	N/A

Table continues on next page

New York	Partial	Yes	Yes	N/A	No	No	N/A
North Carolina	Full	Yes	Yes	No	Yes	Yes	No
North Dakota	None	No	No	No	No	No	No
Ohio	Partial	Yes	Yes	Yes	No	No	No
Oklahoma	None	No	No	No	No	No	No
Oregon	None	No	No	No	No	No	No
Pennsylvania	None	No	No	No	No	No	No
Rhode Island	None	No	No	No	No	No	No
South Carolina	Partial	Yes	Yes	Yes	No	No	No
South Dakota	None	No	No	No	No	No	No
Tennessee	Partial	Yes	Yes	N/A	No	No	No
Texas	None	No	No	No	No	No	No
Utah	Partial	No	No	N/A	No	No	N/A
Vermont	None	No	No	No	No	No	No
Virginia	Partial	Yes	Yes	N/A	No	No	N/A
Washington	None	No	No	No	No	No	No
West Virginia	Partial	Yes	Yes	No	Yes	Yes	No
Wisconsin	Partial	Yes	Yes	N/A	No	No	N/A
Wyoming	Full	Yes	Yes	N/A	Yes	Yes	No
Puerto Rico	N/A	N/A	N/A	N/A	N/A	N/A	N/A

Note: Overall: Public Utility Commission (PUC) fully, partially or does not regulate cellular communications.
CPCN: PUC requires Certificate of Public Convenience and Necessity.
Tariff: PUC requires filing of tariffs.
USOA: PUC requires Uniform System of Accounting.
N/A: Not available.

Source: NARUC (1992).

APPENDIX B: LIST OF VARIABLES AND DATA SOURCES

INDEXMIN: Price index of each cellular carrier in the Metropolitan Statistical Area (MSA) computed for a monthly usage of 200 minutes. Corresponds to the minimum price when no term contract is required. Source: *Cellular Pricing and Marketing Newsletter* (*CPMN*) published by Information Enterprises Inc.

INDEXCON: Price index of each cellular carrier in the MSA computed for a monthly usage of 200 minutes. Corresponds to the minimum price when a contract of 1 year or less is required. Source: *CPMN*.

INDEX: Price index of each cellular carrier in the MSA computed for a monthly usage of 200 minutes. Source: *CPMN*.

CELL: Number of cell-sites of each cellular carrier in the MSA. Source: *Cellular Business*.

DENSITY: MSA population density = MSA population / MSA area. MSA population is the 1989 population Source: Regional Economic Information System (REIS) of the Bureau of Economic Analysis (BEA). MSA area is the square mile area. Source: City and Metropolitan Area Data Book by the Bureau of the Census.

DENCELL: Density of cells = CELL / MSA area.

POPCELL: MSA population per cell-site = MSA population /CELL.

MONDS1: Monthly price of DS1 trunks in MSA. Source: *Special Access Study* by Bellcore.

MONDS3: Monthly price of DS3 trunks in MSA. Source: *Special Access Study* by Bellcore.

NONRDS1: Nonrecurring charge for a DS1 service in MSA. Source: *Special Access Study* by Bellcore.

NONRDS3: Nonrecurring charge for a DS3 service in MSA. Source: *Special Access Study* by Bellcore.

MONTHOPE: Number of months each cellular carrier has been operating in MSA. Source: *Cellular Business*.

WAGES: Annual wages of telecommunications workers in MSA in 1989. Source: REIS of the BEA.

BLOCK: One if carrier is wireline, zero otherwise. Source: *Cellular Business*.

AREA: Area of the MSA. Source: REIS of the BEA.

COMMUTE: Minutes of daily commute in MSA. Source: *Places Rated Almanac* by Rand McNally.

POP89: Population of MSA in 1989. Source: REIS of the BEA.

PERIN89: Total personal income in MSA in 1989. Source: REIS of the BEA.

INCOMEPC: MSA annual income per capita in 1989. Source: REIS by BEA.

CON: length of contract for variable INDEXCON for each cellular carrier in MSA. Source: *CPMN*.

CONTRACT: Length of contract for variable INDEX for each cellular carrier in MSA. Source: *CPMN*.

REGULATION: One if cellular carriers have to file any tariff change. Source: NARUC.

MULTIM: Number of times cellular carriers in each MSA meet with each other in other markets. Source: *Cellular Business*.

FULL: One if the cellular carrier has more than 300,000 customers, zero otherwise. Data on number of customers for the 20 largest MSAs was published in *Radio Communications Reports* (RCR).

REFERENCES

Calhoun, G. (1988). *Digital cellular radio*. New York, Artech House.

Cellular Telecommunications Industry Association (CTIA). (1994). *CTIA's 1994 wireless sourcebook: Three books in one, announces year-end 93 data survey results* [Press release].

Chamberlin, E. (1929). Duopoly: Value where sellers are few. *Quarterly Journal of Economics, 43,* 63–100.

Deneckere, R. (1983). Duopoly supergames with product differentiation. *Economics Letters, 11*, 37–42.

Donaldson, Lufkin, & Jenrette (1990). The cellular communications industry. [Special report.]

Federal Communications Commission. (1975). *FCC Docket No. 18262*. Washington, DC.

Federal Communications Commission. (1981). *Report and order, FCC Docket No. 79-318*. Washington, DC.

Friedman, J. (1990). *Game theory with applications to economics*. Oxford: Oxford University Press.

Jacquemin, A. & Slade, M. (1989). Cartels, cooperation, and horizontal merger. In R. Schmalensee & R.D. Willig (Eds.), *Handbook of industrial organization (Vol. I* pp.) Amsterdam: North-Holland 415–473.

Joskow, P. & Rose, N. (1989). The effects of economic regulation. In R. Schmalcnscc and R.D. Willig (Eds.) *Handbook of industrial organization, (Vol. I* pp.). Amsterdam: North-Holland 1449–1506.

Kaestner, R. & Kahn, B. (1990). The effects of regulation and competition on the price of AT&T intrastate telephone service. *Journal of Regulatory Economics*, *2*, 263–367.

Lee, C.Y. (1988). *Digital cellular radio*. New York: Praeger.

Mitchell, B. M. & Vogelsang, I. (1991). *Telecommunications pricing, theory and practice*. Cambridge, UK: Cambridge University Press.

NARUC. (1992). *NARUC Report on the status of competition in intrastate telecommunications*. Washington, DC.

Porter, R. H. (1983a). Optimal cartel trigger price strategies. *Journal of Economic Theory*, *29*, 313–338.

Porter, R. H. (1983b). A study of cartel stability: The Joint Executive Committee, 1880–1886. *Bell Journal of Economics*, *14*, 301–314.

Rand McNally. (1989). *Places rated almanac.* New York.

Ross, T. W. (1992). Cartel stability and product differentiation. *International Journal of Industrial Organization, 10*, 1–13.

Ruiz, L. K. (1994). Pricing Strategies and Regulatory Effects in the U.S. Cellular Telecommunications Duopolies. Unpublished doctoral dissertation, Boston University.

Salant, D. J. and Woroch, G.A. (1992). Trigger price regulation. *Rand Journal of Economics, 23*, 29–51.

Stigler, G. J. (1966). *The theory of price.* New York: Macmillan.

Tirole, J. (1988). *The theory of industrial organization.* Cambridge, MA: MIT Press.

United States General Accounting Office. (1992). *Testimony Before the Subcommittee on Communications, Committee on Commerce, Science, and Transportation, United States Senate.*

White, H. (1980). A heteroscedasticity-consistent covariance matrix estimator and a direct test for heteroscedasticity. *Econometrica, 48*, 817–838.

3

Critical Mass and Network Evolution in Telecommunications

Nicholas Economides
Charles Himmelberg
Stezy School of Business, New York University

Many goods, most notably in the telecommunications industry, exhibit *network externalities*. That is, the value of the good to the consumer depends on the number of consumers purchasing the same (or a similar) good. Many other consumption goods, such as computer software, display network externalities, although usually in less obvious ways. All such goods have in common the feature that an increase in expected sales of network services creates positive expected benefits for every consumer of the good.

In this chapter, we show that consumption goods with network externalities are often characterized by the existence of a *critical mass point*. That is, an equilibrium market for the good does not exist unless the installed base is greater than a minimum level. We show that this is a general feature of goods that exhibit network externalities and will be observed in a variety of market structures. Nevertheless, the presence of positive network externalities and positive and significant critical mass have significant impact on the analysis of conduct, structure, and performance of network industries. We analyze single-period as well as multiperiod (dynamic) effects of network externalities both for perishable and durable goods.[1]

We illustrate these ideas using the U.S. market for facsimile machines as an example. This example illustrates the potentially large effects of network externalities on the growth behavior of markets and outlines a strategy for estimating the effect of externalities using such models. Specifically, we focus on the growth of the market following the introduction of the (industry accepted) D3 standard in the late 1970s. During the early 1980s, as fax machines conforming to this standard were just beginning to ship, our data indicate that fax machines were selling at an average price of more than $2,000. This high average price evidently exceeded the

1. Many of the results of this chapter are based on Economides and Himmelberg (1994).

reservation price of most consumers, because by 1983 the estimated installed base was still less than 1 million machines. But by 1983, the average selling price had begun to decline rapidly. Between 1982 and 1985, average prices fell by a factor of four to about $500, and by 1987 had fallen still further to about $250.

This dramatic price drop was driven by dramatic reductions in the price of microelectronic components used in the production of facsimile machines, and it sparked an explosion in demand. Prior to this, most of the demand had come from "early adopters," which had been sufficient to keep the market growing during the first half of the 1980s at a rate of about 20% per year. However, in 1986, the demand for fax machines began to accelerate dramatically; in 1987 it exploded to more than double the previous year, and in 1988 demand more than doubled again. Demand for fax machines remained robust throughout the rest of the decade, and by 1991 the installed base had grown to more than 10 million machines.

Although the dramatic drop in prices clearly played a major role in the growth of the fax market, we argue that the particularly explosive growth during the mid to late 1980s was fueled by both realized and anticipated increases in the size of the installed base. In the following model we derive the aggregate demand function from a discrete choice model of demand at the level of the individual consumer. This "bottom-up" approach allows us to infer the effect of network externalities on consumer purchasing decisions. The model assumes that the *net* value of a fax machine to an individual consumer depends on that consumer's income level, the size of the installed base, the price paid for the machine, and a random element that captures idiosyncratic tastes. We use this model to predict the rate at which the installed base should have grown in the absence of network externalities, given the path of prices and the distribution of consumer income. We then compare the theoretical rate of market growth predicted by the model with the empirical rate of market growth in the data. On the basis of the discrepancy between these two, we back out the implied parameters of consumer utility function that measure the value of the installed base.

In the next section, we introduce a conceptual framework for thinking about network externalities, and we focus on the nature of communication networks in particular. In the following section we introduce the intuition for critical mass points as well as the intuition for the formal conditions under which network goods would display such phenomena. We show that the critical mass of a network is (surprisingly) independent of market structure, and that market equilibria are (unsurprisingly) inefficient relative to the allocation that would maximize total welfare. In the remaining sections, we show that our static framework generalizes to a dynamic setting, and for the benefit of our empirical exercise, we further modify the framework to accommodate the durability of fax machines.

SOURCES OF NETWORK EXTERNALITIES

The key reason for the existence of network externalities is the complementarity between the components of a network. Depending on the network, the externality may be direct or indirect. Networks where services AB and BA are distinct are

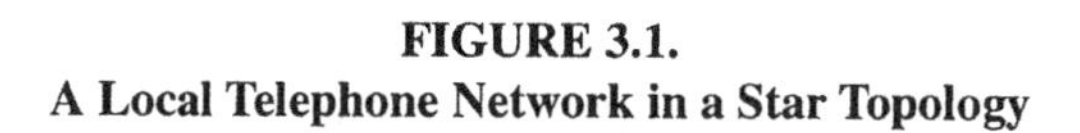

FIGURE 3.1.
A Local Telephone Network in a Star Topology

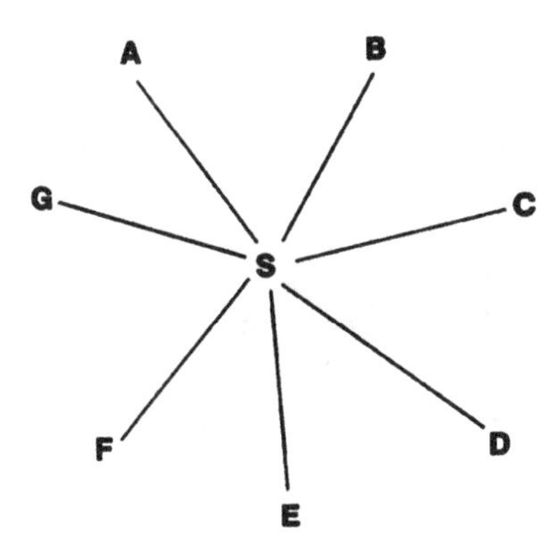

named *two-way networks* in Economides and White (1993). Two-way networks include railroad, road, and many telecommunications networks, such as the facsimile network that we analyze in later sections. In typical two-way networks, customers are identified with components and the externality is direct. Consider, for example, the local telephone network of Fig. 3.1. In the n component network of Fig. 3.1, there are $n(n-1)$ potential goods. An additional customer provides *direct externalities* to all other customers in the network by adding $2n$ potential new goods through the provision of a complementary link (say GS) to the existing links.[2] For example, in a simple fax network with, say, 100 nodes, there are 9900 distinct "goods" available (99 per node), and the addition of the 101 node creates an additional 202 goods (distinct, one-way fax transmissions).

When one of AB or BA is unfeasible, or does not make economic sense, or when there is no sense of direction in the network so that AB and BA are identical, then the network is called a *one-way network*. In a typical one-way network, there are two types of components, composite goods are formed only by combining a component of each type, and customers are often not identified with components but instead demand composite goods. For example, automatic teller machine (ATM) services, broadcasting television (over-the-air and cable), electricity networks, retail dealer networks, and paging are one-way networks.

In typical one-way networks, the externality is only indirect. When there are m varieties of Component A and n varieties of Component B as in Fig. 3.2 (and all A-type goods are compatible with all B-type), there are mn potential composite goods. An extra customer yields indirect externalities to other customers, by increasing the demand for components of types A and B and thereby (because of the

2. This property of two-way networks was pointed out in telecommunications networks by Rohlfs (1974) in a very early paper on network externalities. See also Oren and Smith (1981) and Hayashi (1992).

FIGURE 3.2.
A Pair of Vertically Related Markets

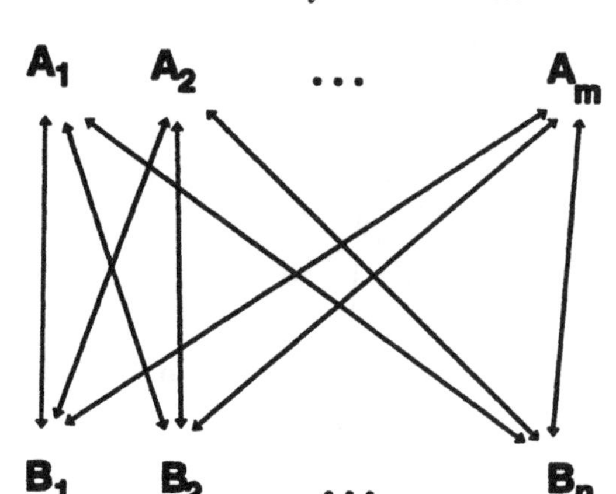

presence of economies of scale) potentially increasing the number of varieties of each component that are available in the market.[3]

CRITICAL MASS

For normal goods that do not exhibit network externalities, demand slopes downward; as price decreases, more of the good is demanded. Conversely, higher levels of consumption are associated with lower prices. This fundamental relationship may fail in goods with network externalities. For these goods, the willingness to pay for the last unit increases as the number expected to be sold increases. If expected sales equal actual sales, the willingness to pay for the last unit may increase with the number of units sold. Thus, for goods with network externalities, the (fulfilled expectations) demand-price schedule may not slope downward everywhere. In such markets, as costs decrease we may observe discontinuous expansions in sales rather than the smooth expansion along a downward sloping demand curve. In particular, we may observe a discontinuous start of the network: As costs decrease, the network starts with a significant market coverage (say 10% of the market) rather than starting with 0.1% coverage.

Critical mass is defined as the minimal nonzero equilibrium size (market coverage) n^0 of a network good or service (for any price). We argue that, for many network goods, the critical mass is of significant size, and therefore for these goods small market coverage will never be observed—either their market does not exist or it has significant coverage.

3. In many industry structures, the addition of new varieties is concurrent with an intensification of competition; in these cases, consumers have the added benefit of price decreases as the number of varieties increases.

The concept of critical mass formalizes the "chicken and the egg" paradox that logically arises in such markets, namely that many consumers are not interested in purchasing the good because the installed base is too small, and the installed base is too small because an insufficiently small number of consumers have purchased the good. Thus, consumers' expectations of no network good provision may be fulfilled. However, for a range of costs, expectations of positive level(s) of sales of the network good are also fulfilled. Often, there are multiple fulfilled expectations equilibria. Consumers and producers can coordinate to reach any one of them. We assume that they will reach the equilibrium of the largest network size. Thus, when more than one network size is supported by the same price, we select as the equilibrium the highest network size supported by that price; this network size Pareto dominates the other network sizes supported by the same price.[4]

Perfect Competition

The essential features of our model are derived from assumptions about the utility that consumers receive from owning the network good. For simplicity, assume that utility that consumers receive from owning the network good is proportional to their income, y, and further assume that their willingness to pay for the good is given by the function $u(y, n) = y^{\gamma}(k + dn^{\alpha})$, where n is the market coverage of the network (installed base), $n \in [0, 1]$, α takes values between zero and one, and d is a parameter with a value greater than zero. Without loss of generality, we assume $\gamma = 1$. The constant term k is the innate value of the good to the consumer when the size of the installed base is zero. This functional form assumption implies that utility is increasing with the size of the installed base. This is a rough approximation intended to capture the intuition that individuals with higher incomes would tend to make more use of fax machines and therefore place a higher value on both the machine itself and the size of the installed base.[5]

From these assumptions about individual utility, we derive the aggregate demand by counting the number of people willing to purchase the good given the market price p and the size of the installed base n^e, that is, the number of people with incomes sufficiently high so that $u(y, n) \geq p$, where p is the price of the good. Thus, aggregate demand takes the general form $n = f(n^e, p)$. For a given level of the installed base, n^e, this demand function is downward sloping in price. As the expected size of the network n^e increases, this demand curve shifts up.

We can invert this demand curve to yield the price that the marginal consumer is willing to pay given the installed base and the number of people who demand

4. Our analysis shows that the equilibrium is also stable.

5. This specification of the utility function can be readily generalized to accommodate an arbitrarily large number of consumer characteristics associated the utility derived from faxes. We chose income because data on the distribution of incomes are readily available and because income is probably the best single measure of willingness to pay. Using a univariate also simplifies the exposition substantially.

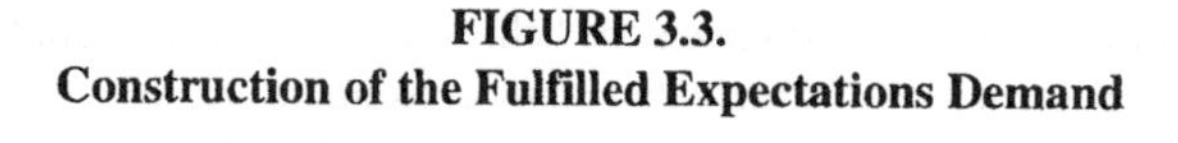

FIGURE 3.3.
Construction of the Fulfilled Expectations Demand

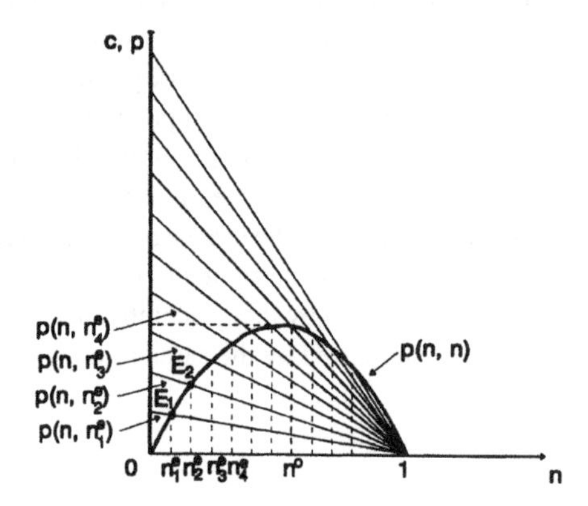

the good, that is, $p = p(n, n^e)$.[6] Thus for a given level of the installed base, n^e, price is a function of n as indicated by the downward sloping curves plotted in Fig. 3.3.

Up to this point in the discussion we have allowed the size of the installed base to differ from the size of the level of network. In equilibrium, this would clearly be impossible, because the level of network demand obviously determines the size of the installed base of the market clears. Therefore, we now impose the equilibrium condition that $n = n^e$, and we refer to the resulting demand curve as the *fulfilled expectations demand, p(n, n)*. Figure 3.3 shows the construction of a typical fulfilled expectations demand. The curves $p(n, n_1^e)$ and $p(n, n_2^e)$ show the willingness to pay, given different sizes of the installed base that consumers expect to emerge in equilibrium, where $n_2^e > n_1^e$. The point labeled E_1 on the first curve represents the point at which n equals n_1^e, and analogously, E_2 on the second curve represents the point at which n equals n_2^e. The locus of all such points traces out the fulfilled expectations demand curve. Observe that the fulfilled expectations demand $p(n, n)$ is not monotonic. Also note that, in addition to the prices indicated by the inverted U-shaped curve in Fig. 3.3, the fulfilled expectations demand curve $p(n, n)$ also includes the entire vertical axis at zero, which is drawn thicker on purpose. This is because at any marginal cost $c > k$ a network of zero size is a fulfilled expectations equilibrium, and Fig. 3.3 is drawn for the special case when $k = 0$.

For $n > 0$, the fulfilled expectations demand is single-peaked (quasi-concave) for a fairly general set of conditions. We can show that $\lim_{n \to 1} p(n, n) = 0$, so that $p(n,$

6. The willingness to pay function can be derived as follows. We assume that total demand is given by the total number of consumers for whom $u(y, n) > p$, that is, the number of consumers with incomes satisfying $y > y^*$, where $y^* = p/(k + dn^\alpha)$. Therefore demand is given by $n^d = 1 - G(p/(k + dn^\alpha))$, where G is the distribution of income. The willingness to pay function is derived by solving for p, so that $p(n, n^e) = (k + dn^\alpha)G^{-1}(1 - n)$, and the fulfilled expectations willingness to pay function is $p(n, n) = h(n) G^{-1}(1 - n)$.

n) is decreasing for large n. This guarantees that the market does not explode toward infinite output. Given single-peakedness, the fulfilled expectations demand can either decrease everywhere, or have an increasing part for small n, as in Fig. 3.3. If $p(n, n)$ decreases for all n (the case of weak network externalities), it exhibits no qualitative difference from an ordinary demand curve. The interesting case arises when $p(n, n)$ increases for small n (the case of strong network externalities).

Figure 3.4 shows the fulfilled expectations demand for strong and weak network externalities. Focusing on the case of strong network externalities pictured on the left side, consider perfect competition with constant marginal cost c. In equilibrium, price equals marginal cost, that is,

$$p(n, n) = c.$$

Let c^0 denote the peak of $p(n, n)$. For $c > c^0$, the only equilibrium is of zero size. For $c^0 > c > k$, there are two other equilibria, besides the zero one, at the intersections of the horizontal at c with $p(n, n)$. The lower of the two is unstable, and the higher one is stable. For $c < k$, there is only one equilibrium, and it is positive and stable. When more than one network size is supported by the same price, we select as the equilibrium the highest network size supported by that price. Thus, the equilibrium we select is always Pareto dominant and stable. In the case of strong network externalities, for $c > c^0$, the equilibrium is of zero size; at $c = c^0$, the network starts at size $n^0 > 0$; and, for $c < c^0$, the network follows the outer part of $p(n, n)$ with size $n > n^0$. Clearly, the market on the left panel of Fig. 3.4 exhibits a positive and significant critical mass of size n^0. Thus, nonzero networks of smaller size

FIGURE 3.4.
The Fulfilled Expectations Demand with Strong and Weak Network Externalities

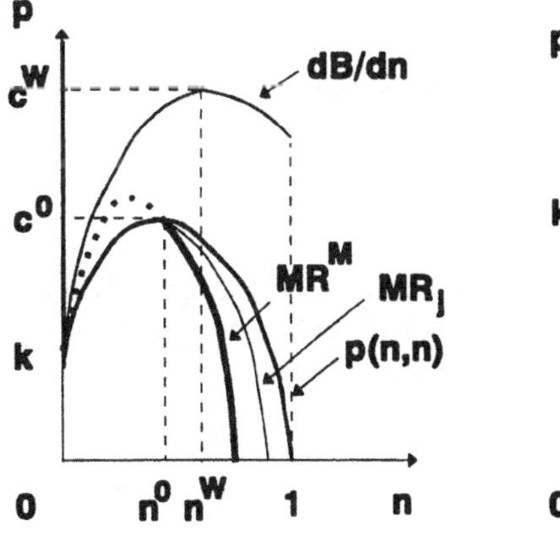

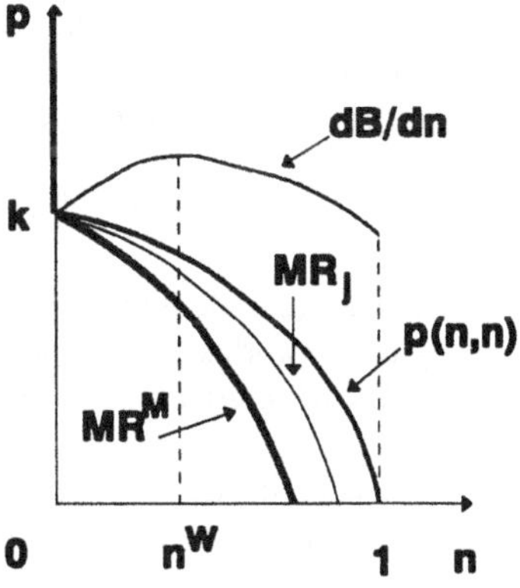

than the critical mass, $n < n^0$ and $n \neq 0$, will not be observed at any prevailing marginal cost or price.

Under what conditions do network services exhibit critical mass? The crucial requirement is that the fulfilled expectations demand is increasing for small n. Economides and Himmelberg (1994) showed that the fulfilled expectations demand is increasing for small n if either one of the three following conditions hold:

1. The utility of every consumer in a network of zero size is zero.
2. There are immediate and large external benefits to network expansion for very small networks.
3. There is a significant density of high willingness-to-pay consumers who are just indifferent on joining a network of approximately zero size.

The first condition is straightforward and applies directly to all two-way networks, where network goods have no value if there are no other participants. The typical example is a telephone or fax network. These networks exhibit critical mass, that is, they start with a significant market coverage. The second condition describes goods that may have some intrinsic value in zero-size networks, but their value increases dramatically as sales expand. A good example of this may be a specialized computer program that relies on support mainly from other users.[7] The addition of even a few users can increase significantly technical support and the value of the product. Another example is a specialized newsgroup on Internet. The third condition describes goods that may have some intrinsic value in zero-size networks and their value does not increase dramatically as sales increase, but they have very widespread appeal. A good example of this may a computer software with large sales, but low externality from each sale. Each extra copy sold of a word processing program, such WordPerfect, creates a small externality; however, once its sales become very large, secretaries get trained in WordPerfect, thus creating a significant externality.

Welfare Maximization

A social welfare maximizing planner can fully internalize the externality. Thus, we expect that the planner will provide a larger network size. The planner maximizes the fulfilled expectations net benefit of a network of size n,

$$W(n, n) = B(n, n) - C(n) = \int_0^n (p(q, n) - c)dq$$

where $B(n, n) = \int_0^n p(q, n)dq$ is the gross benefit of the network. The optimal choice is defined by the network size that makes the marginal net benefit equal to zero,

$$dW/dn = dB(n, n)/dn - c = p(n, n) + \int_0^n p_2\, dq - c = 0$$

where $p_2 > 0$ denotes the derivative of the gross benefit function B with respect to its second argument, that is, with respect to expected sales. Thus, the marginal

7. Some freeware and shareware computer programs fall into this category. Support is provided essentially by other users through discussions on bulletin board services (BBSs).

gross benefit is higher than the fulfilled expectations willingness to pay, $dB(n, n)/dn = p(n, n) + \int_0^n p_2\, dq > p(n, n)$ (see Fig. 3.4). It follows that a planner will start the network for marginal costs, $c \in [c^w, c^0]$, for which there would be no network under perfect competition, and will always support a larger network than perfect competition.

The wedge between price and gross marginal benefit, ignored by perfect competition, but taken into account by the planner, clearly implies that perfect competition is inefficient. Finally, as seen in the right panel of Fig. 3.4, the welfare maximizing solution may exhibit a positive and significant critical mass, even when there is no positive critical mass under perfect competition.

Monopoly

A monopolist may or may not be influenced by expectations of consumers. If the monopolist does not influence expectations, it is clear that it will produce less than perfect competition and have greater inefficiency. On the other hand, there is hope that a monopolist who can influence expectations will support a larger network than perfect competition, thereby resulting in higher total surplus. However, we show that a monopolist who is unable to price discriminate will always support a network that is smaller than perfect competition and results in lower total surplus. The reason for this is simple. The monopolist has two opposite incentives in setting network size. It would like to increase the network size to increase the surplus it can appropriate. On the other hand, the monopolist wants to reduce quantity below the competitive level so that it can increase price. We show that the second incentive is dominant.

The monopolist's profits,

$$\Pi^M(n, n) = R^M(n) - C(n) = n(p(n, n) - c),$$

are maximized when

$$d\Pi^M/dn = MR^M - MC = p(n, n) + ndp/dn - c = 0.$$

The marginal revenue curve is shown in Fig. 3.4 as a bold line for $n > n^0$, and as a dotted line for smaller n.

The monopolist will operate only when price is above marginal cost. This implies $dp/dn < 0$, and therefore $MR^M = p(n, n) + ndp/dn < p$. Thus, the monopolist will only operate on the downward sloping part of the fulfilled expectations demand. As for any downward sloping demand, in this portion, marginal revenue is below price, and the quantity produced by the monopolist falls below that of perfect competition. Therefore the monopolist starts the network service at the same cost as perfect competition and has the same critical mass. For all smaller marginal costs, the monopolist produces less than perfect competition and charges a lower price. Despite his influence on expectations, a monopolist supports a network that is smaller and more inefficient than perfect competition from a social welfare point of view. Therefore the existence of network externalities cannot be claimed as a reason in favor of a monopoly market structure.[8]

Oligopoly

Oligopolists may produce network goods that are all compatible to each other, or some firms may produce goods that are incompatible with some subset of goods of other producers. In this chapter we consider only the compatible goods oligopoly. This is not because of lack of theoretical interest in the incompatible goods case (see Economides, in press), but we want to focus our attention of the application to the fax market where there is no competing technical standard. Suppose that all firms produce compatible and identical goods. Consider Cournot oligopoly among them. Assume that each firm is able to influence the expectation of consumers only about its own quantity of production. Then it is easy to see that this oligopoly will result in an outcome that lies between perfect competition and monopoly.

Firm j's profits in a k-firm oligopoly,

$$\Pi_j = n_j[p(\Sigma_{i=1}^{k} n_i, n_j+\Sigma_{i\neq j}^{k} n_i^{e}) - c],$$

are maximized when

$$d\Pi_j/n_j = MR_j - MC = p(\Sigma_{i=1}^{k} n_i, n_j+\Sigma_{i\neq j}^{k} n_i^{e}) + n_j dp/dn_j - c = 0.$$

It is easy to show that the resulting fulfilled expectations equilibrium is symmetric and $n_i^{e} = n_i = n/k$, for all i, where $n = \Sigma_{i=1}^{k} n_i$, so that the equilibrium is characterized by

$$p(n, n) + (n/k)(dp/dn) - c = 0.$$

Clearly the marginal revenue for Firm j lies between monopoly and perfect competition, as seen in Fig. 3.4. Thus, the oligopoly equilibrium network size lies between the perfectly competitive size and the size chosen by a monopolist who influences expectations. From a social welfare point of view, the equilibrium of oligopolists that influences expectations is more inefficient than the perfectly competitive outcome but more efficient than the choice of a monopolist who influences expectations.

Dynamics and Durable Goods

To describe more accurately the fax market, we develop next a model of durable goods competition with network externalities. The problem is more complex, because now firms and consumers need to predict accurately the whole path of future

8. Of course, the results of this subsection were established only for linear prices, and may not hold in the presence of two-part tariffs and general nonlinear pricing schemes. For an excellent survey of nonlinear pricing see Wilson (1993).

network sizes and prices. Nevertheless, we are able to show a one-to-one correspondence between a dynamic durable goods problem under perfect competition to a single-period problem.

Let the instantaneous utility of owning the network good for consumer of income y and network of size n be given by $u(y, n)$. Assume for simplicity that once a good is purchased, it yields an infinite stream of future utility. Given an expected future time path of network size $n^e(t)$, the present value of a machine purchase at time t for a consumer of income y is given by

$$V(y, t, n^e(t)) = \int_t^\infty e^{-\rho s}\, u(y, n^e(s))ds,$$

where ρ is the discount rate. If the durable good is purchased at time t at price $p(t)$, the present value of its cost is

$$q(t) = e^{-\rho t}\, p(t).$$

The consumer of income y buys at time t^* that maximizes $V(y, t, n^e(t))$ - q(t), that is, he solves

$$V'(y, t^*, n^e(t^*)) - q'(t^*) = 0.$$

This expression simplifies to[9]

$$u(y, n^e(t^*)) = \rho p(t^*) - p'(t^*) \equiv \lambda(t).$$

The shadow price $\lambda(t)$ plays exactly the same role in the durable goods problem as price p plays in the single period problem. In the durable goods case, $\lambda(t)$ represents the opportunity cost of buying the good at t rather than $t + dt$. The first term $\rho p(t)$ measures the cost of waiting one period, assuming that the price remains the same. The second term reduces the cost of buying today by any price increase in the time increment dt. Thus, $\lambda(t)$ represents the opportunity cost of buying today rather than tomorrow. Using this reinterpretation, we can apply and extend results from the nondurable analysis to the durable good case.[10]

In the dynamic setting, a fulfilled expectations equilibrium (rational expectations equilibrium) is a pair of paths of prices and sales $\{p(t), n(t)\}$ such that expectations are fulfilled and supply equals demand at every period, that is, it fulfills:

demand: $n_D(t) = 1 - G(p^e(t)/h(n^e(t)),$

---

9. $V' - q' = -e^{-\rho t}u(y, n^e(t)) + \rho e^{-\rho t}p(t) - e^{-\rho t}p'(t) = 0$, so that $u(y, n^e(t)) = \rho p(t^*) - p'(t^*)$.

10. In particular, given instantaneous utility $u(y, n^e) = yh(n^e)$, the marginal consumer at time t is $y^* = \lambda(t)/h(n^e)$, and therefore the demand at time t is $n(t) = 1 - G(\lambda(t)/h(n^e(t))$. At a fulfilled expectations equilibrium, $n^e(t) = n(t)$, so that $n(t) = 1 - G(\lambda(t)/h(n(t))$ or equivalently $\lambda(t) = h(n(t)G^{-1}(1 - n(t))$.

supply: $p(t) = c(t, n_S'(t))$,

fulfilled expectations of sales: $n(t) = n^e(t)$,

fulfilled expectations of prices: $p(t) = p^e(t)$,

market clearing: $n_D(t) = n_S(t) = n(t)$,

where $c(t, n_S'(t))$ is the marginal cost at time t, which may depend on the size of output $n_S'(t)$ at t. In the next section, we apply this dynamic analysis to the fax market in the United States.

THE GROWTH OF THE U.S. MARKET FOR FACSIMILE MACHINES

As already described in the introduction, the market for facsimile machines in the United States exploded during the mid to late 1980s, with growth rates of the number of units shipped exceeding 150% in 1987. We argue that this tremendous surge in demand was not driven as much by outside shifts in consumer demand and price reductions as much as it was driven by the "feedback" effect induced by both past increases and anticipated future increases in the size of the installed base. The anecdotal evidence is consistent with this interpretation because the most dramatic fall in prices occurred well before 1987. However, this was the year in which the rate at which prices were falling began to taper off. This is an important clue because in the consumer's solution to the dynamic, durable goods problem, the desire to postpone a purchase is proportional to $\lambda_t = \rho p(t) - p'(t)$. This implies that as long as prices are still falling (i.e., as long as $p'(t) < 0$), aggregate demand is weak. This is exactly what the data seem to show.

We now formalize this intuition with a simple calibration exercise. In addition to our data on the average prices and quantities of facsimile machines sold in the United States between 1979 and 1992, empirical estimation of our model requires data on the distribution of consumer characteristics. Ideally, these characteristics would be identified by collecting marketing data on consumers that purchase fax machines and then using these data to estimate a discrete choice model. In practice, however, access to such data is difficult, so we pursue an alternative strategy that is feasible with aggregate data. Most fax purchases are made by firms and not consumers, and many firms purchase more than one machine. We argue that it is nonetheless reasonable and convenient to model the unit demand for fax machines as a function of consumer characteristics. This is because a firm's demand for fax machines ultimately is derived from "employee demand." For example, a firm with a high fraction of highly skilled white collar workers will have a higher demand for fax machines than a firm with a high fraction of production line workers. For simplicity and feasibility, we assume that the employee characteristics related to fax demand can be summarized by employees' income.

In order to characterize the distribution of consumer types as a function of consumer income, we use data on the distribution of income from the Current Population Survey (CPS) for survey years 1976, 1981, 1986, and 1991. Because income is approximately lognormal, we transform the data using natural log to obtain normally distributed log income. We then calculate the mean and variance for each of those 4 years and interpolate the distribution of log income for each year between 1979 and 1992.[11] This gives us a time-varying estimate of the distribution function described in the previous section, that is, $G(y; t)$. We use the notation μ_t and σ_t to denote the mean and standard deviation of log income, and the notation $\Phi(x)$ to denote the standard normal distribution. This notation allows us to represent our empirical estimate of G by

$$G(y; t) = \Phi((ln(y) - \mu_t)/\sigma_t).$$

Next we normalize the size of the fax network by assuming that maximum potential network size is 20 million fax machines. This number implies a maximum ratio of about one fax machine for every five workers in the United States (we estimate the number of fax machines in 1992 to be about 7.6 million). In the results reported in the following, we experimented with both larger and smaller values of the maximum network size and this did not affect the calibration results reported here. To construct the "stock" of fax machines, that is, the installed base, we assume that fax machines depreciate at a rate of 13.3% per year, and use a perpetual inventory method to accumulate unit sales. Here, too, we experimented with various depreciation rates and this did not significantly affect the calibration results reported here.[12] Finally, we deflated our price series for fax machines using the GDP deflator reported in Table 1.1 of the Current Survey of Business.

With this empirical estimate of the distribution of consumer income $G(y; t)$ and our data on normalized network size, n_t, real prices, p_t, we calibrate the model by choosing values of the remaining unknown parameters to fit the model. We simplify the model somewhat by assuming that fax machines are pure network goods, so that they give no utility in a network of size zero. That is, we assume $k = 0$. This is not strictly true, because fax machines can also double as telephones, but given the widespread availability of telephones in most places where fax machines are

11. To be specific, this procedure also requires that we account for top coding in the CPS. For each of the 4 years, we truncate above at $100,000, except for 1976, which is truncated above at $80,000. We also truncate below at $2000 in order to avoid data problems with outliers. We then use formulas for the truncated mean and variance of a normal distribution to calculate means of log income (nominal) of 9.33, 9.38, 9.70, and 9.73 for the 4 years, respectively. The standard deviations of income in these 4 years are 1.18, 1.27, 1.31, and 1.33, respectively. The interpolation is done using a cubic polynomial. Finally we converted these numbers to real 1987 dollars using the GDP deflator reported in Table 1.1 of the Current Survey of Business. We are extremely grateful to our colleague Rick Flyer for providing us with the estimates from the CPS.

12. We chose the depreciation rate of 13.3% by applying the average service life of telephones (7.5 years), which is estimated using life expectancy tables for consumer possessions used by insurance adjusters in responding to claims for fire and theft damage. We are grateful to Peter Klenow for providing us with this estimate.

used, it seems reasonable to assume that the fax machines are valued only for their ability to make fax transmissions.

We relax the simplifying assumption made in the previous section that $\gamma=1$. Thus, our general Cobb–Douglas utility specification is

$$u(y_t, n^e{}_t) = Ay_t^{\gamma}n_{t-1}{}^{\alpha}.$$

Note that we have made the assumption $n_t^e = n_{t-1}$, that is, our empirical specification assumes that the expected size of the network this year is a linear function of the network size at the beginning of the year.[13]

Recalling our use of the notation $\lambda_t = \rho p_t - p_t'$, we construct a data series for λ_t by assuming $\rho = 0.2$. Our results in the previous section show that the value of the marginal consumer is calculated by setting utility equal to λ_t and solving for y_t. Taking natural logs of the resulting expression yields

$$ln\, y_t = \gamma^{-1}(ln\, \lambda_t - \alpha ln\, n_{t-1} - ln\, A)$$

Using our empirical estimate of the distribution of consumer income, the equilibrium network size is given by

$$n_t = 1 - \Phi((ln\, y_t - \mu_t)/\sigma_t).$$

Inverting Φ and solving this expression for $ln(y_t)$ yields

$$\sigma_t\Phi^{-1}(1 - n_t) + \mu_t = ln\, y_t.$$

Because the inverse of the cumulative standard normal is easily calculated, and because σ_t, n_t, and μ_t are all variables in our data, the term on the right side of the above expression is a variable that we construct. We define this variable using the notation $g_t = \sigma_t\Phi^{-1}(1 - n_t) + \mu_t$. Finally, substituting our expression for the marginal consumer yields our estimating equation

$$g_t = \beta_0 + \beta_1\, ln\lambda_t + \beta_2 ln\, n_{t-1} + e_t$$

where $\beta_0 = -\gamma^{-1}ln\, A$, $\beta_1 = \gamma^{-1}$, $\beta_2 = -\gamma^{-1}\alpha$, and e_t is an error term that represents approximation errors in the functional form assumptions as well as errors in the measurement of g_t.

We estimate this specification of the model using OLS. We point out that this is essentially a demand equation in which a nonlinear transformation of the quantity variable appears on the left side of the equation and a price term (λ_t) and a demand shifter (n_{t-1}) appear on the right side of the equation. This interpretation exposes a potential econometric problem with the use of OLS to estimate the model. The prob-

13. Imposing a coefficient of one is arbitrary and reflects the fact that the constant term A absorbs this scaling factor in any case.

lem is that the error term e_t could also contain unexplained variation in g_t due to surprises in the realization of prices (i.e., realizations of λ_t). That is, the price term is endogenous. For this reason, we also report results estimated with generalized method of moments (GMM) using lag values of $ln(\lambda_t)$ and $ln(n_t)$ as instruments.

We also experimented with various ad hoc modifications of the specification to assess the robustness of our calibration estimates. Table 3.1 reports the estimates of the model for several variations of the basic specification already described. For Model 1, the estimates reveal a large positive coefficient on the price term, as predicted, and a large negative coefficient on the network term, also as predicted by the model. Both coefficients (as well as the coefficient estimates reported for Models 2–6) are estimated with tight standard errors, although we hasten to emphasize that these standard error estimates can be misleading given the very small size of our sample. We include them merely to indicate that the parameter estimates do a very satisfactory job of matching the data. We note that goodness of the calibration fit is also revealed by the high R^2 value of 0.902.

The coefficient estimates for Models 2 through 6 largely confirm the results for Model 1, and reveal them to be fairly robust to alternative assumptions. The estimates for Model 2 reveal that the estimates (particularly the coefficient on the price term) are robust to the use of instrumental variables. In Model 3 we included the lagged value of the price term (rather than the current) as an alternative means of controlling for the endogeneity of price. In particular, because one component of this term is the expected price change, one could argue that the lagged value is a good proxy for the expected value. As we expected, this slightly increased the coefficient on the price term, but the magnitude of the increase is not dramatic, nor did it change by much when we estimated the model using instruments (Model 4).

Finally, Model 5 includes a time trend to control for the possibility of omitted variables such as trends in the distribution of consumer characteristics that are not well proxied by income, or changes in the quality of fax machines over time. The inclusion of a time trend somewhat reduces the coefficient on price, but again the magnitude is small, the standard error of the trend coefficient is relatively large, and the magnitude of the adjusted R^2 actually falls slightly. Hence, we conclude that there is little evidence that the price term and network term are contaminated by omitted variable bias, or that they do not do a good job of matching the model to the data.

The virtue of our "structural" approach to the specification of the empirical model is that we can interpret the coefficients on the price and network variables in terms of the preference and technology parameters of the model. The estimates in Table 3.1 allow us to identify values of the model parameters α and γ. For example, the model implies that the coefficient on $ln(\lambda_t)$ is the inverse of γ. Hence, our estimates imply values of γ that range from 6.5 to 10. These estimates are of interest, because they allow us to infer a value for α. In particular, the coefficient on $ln(n_t)$ is minus the inverse of γ times α. Hence, our estimates imply values of α that range from 3.6 to 6.2.

TABLE 3.1.
Calibration Estimates

Variable	*Model 1* *OLS*	*Model 2* *GMM*	*Model 3* *OLS*	*Model 4* *GMM*	*Model 5* *OLS*	*Model 6* *GMM*
Constant	-0.501	-0.527	-0.632	-0.633	-1.401	-1.103
	(0.050)	(0.031)	(0.051)	(0.034)	(1.386)	(0.964)
$\ln \lambda_t$	0.118	0.100	...	...	...	...
	(0.023)	(0.017)				
$\ln \lambda_{t-1}$	...	...	0.142	0.153	0.129	0.144
			(0.020)	(0.015)	(0.032)	(0.023)
$\ln n_{t-1}$	-0.582	-0.619	-0.574	-0.555	-0.616	-0.581
	(0.034)	(0.022)	(0.026)	(0.016)	(0.081)	(0.055)
Year	...	...	...	...	0.008	0.005
					(0.015)	(0.011)
# Obs	15	14	14	14	14	14
Adj. R^2	0.902	0.914	0.951	0.951	0.947	0.944

Note. Asymptotic standard errors appear in parentheses.The instruments used for the GMM estimates in Models 2, 4, and 6 are two lags each of $\ln \lambda_t$ and $\ln n_t$ plus the mean of log income and a time trend. Hansen's test of the over identifying restrictions generated p values of 0.363, 0.540, and 0.324, respectively.

CONCLUSION

In this chapter, we discussed the equilibrium size of networks under alternative market structures for both nondurable and for durable goods. In the presence of network externalities, we showed that, for high marginal costs, the size of network is zero; as costs fall, the network size abruptly increases to a positive and significant size (the critical mass) and thereafter it increases gradually as costs continue to fall.

We generalized these results to a dynamic multiperiod setting and to durable goods. In this framework, the abrupt increase of the network from zero to critical mass of the single-period model is replaced by a continuous but steep increase in network size. We applied our model to the U.S. fax market. Calibration of our model for this market suggests that its growth was strongly influenced by network externalities.

REFERENCES

Antonelli, C. (1992). *The economics of information networks.* Amsterdam: North-Holland.

Arthur, W. B. (1988). Self-reinforcing mechanisms in economics. In P. W. Anderson, K. J. Arrow, and D. Pines (Eds.), *The economy as an evolving complex system* (pp.). Cambridge, MA: Addison-Wesley.

Arthur, W. B. (1990). Positive feedbacks in the economy. *Scientific American.* pp. 92–99.

Cabral, L. (1990). On the adoption of innovations with 'network' externalities. *Mathematical Social Sciences*, 19, 229–308.

Cabral, L. and Leita, A. (1989). Network consumption externalities: The case of Portuguese telex service. In C. Antonelli (ed.), *The economics of information networks*, September (pp. 129–40). Amsterdam: North-Holland.

Economides, N. (in press). *The economics of networks.* International Journal of Industrial Organization

Economides, N. (in press). Network externalities, complementarities, and invitations to enter. *European Journal of Political Economy.*

Economides, N. and Himmelberg, C. (1994). Critical mass and network size. Unpublished manuscript, New York University.

Farrell, J. and Saloner, G. (1985). Standardization, compatibility, and innovation. *Rand Journal of Economics*, *16*, 70–83.

Hayashi, K. (1992). From network externalities to interconnection. In C. Antonelli (Ed.), *The economics of information networks.* Amsterdam: North-Holland.

Katz, M. and Shapiro, C. (1985). Network externalities, competition and compatibility. *American Economic Review, 75*(3), 424–440.

Rohlfs, J. (1974). A theory of interdependent demand for a communications service. *Bell Journal of Economics*, *5*(1), 16–37.

Wilson, R. (1993). *Nonlinear pricing.* Oxford: Oxford University Press.

II

LOCAL COMPETITION AND INTERCONNECTION

4

Competition in Local Telecommunications: Implications of Unbundling for Antitrust Policy

Robert G. Harris
University of California, Berkeley

Gregory L. Rosston
Federal Communications Commission

David J. Teece
University of California, Berkeley

In the past two decades, competition has increased substantially in telecommunications equipment and interexchange services, through a combination of technological advances and changes in public policies. Innovations in microwave communications, for example, combined with the allocation of radio frequency spectrum (the "above 890" decision by the FCC), enabled Microwave Communications, Inc.—now MCI—to enter into interexchange services. More recently, technological innovations in fiber optics have stimulated additional entry into interexchange services. That process has been long and drawn out, in part because a sea change in public policy was required, from a belief that "the system is the solution" to the view that competition is the best method of providing quality services at lower prices. The waves of change—"gales of creative destruction," Schumpeter would say—are now hitting the beaches of local telecommunications, first through *targeted* competition from competitive access providers (CAPs), then through *ubiquitous* competition from cable systems and mobile communications.

Technological change is increasing the range of services that can be economically provided by each mode of communications, thereby increasing the potential for intermodal competition in communications. During this decade, intermodal competition will greatly intensify in communications, just as it has in transportation (e.g., railroads, motor carriers, waterways, pipelines, and air freight). Intermodal competition will emerge from

- Gas and electric utilities deploying optical fiber and wireless technologies to exploit their extensive rights of way, which reach virtually every home and office.
- Cable systems operators deploying new digital technologies to significantly increase the capacity of cable systems, and enable two-way communications over those systems.
- Cellular personal communications services (PCS) carriers deploying digital technologies that will dramatically increase capacity, reduce costs and prices so that cellular service competes directly with wireline.
- Satellite-based communications services, including VSAT, DBS (direct broadcast satellite) and LEOs (low earth-orbiting satellites), expanding rapidly with digital technology.

Combinations of communications modes through strategic alliances, cross-ownership, and intermodal mergers will further facilitate competitive entry and intermodal competition. In addition to the growing size and increasing resources of competitors, most competitors have undertaken a variety of acquisitions, mergers, joint ventures, and strategic alliances to further strengthen their competitive capabilities.

The revolution in telecommunications, though, is not just a story of escalating competition, it is also a story of complementarity and cooperation. Even as intramodal and intermodal competition increase, most communications will span networks, the interconnection and interoperability of which are absolutely essential. In this chapter, we address the emergence and acceleration of competition and complementarity in local access and exchange services, the unbundling and interconnection of competitive networks, and the implications of these developments for antitrust policy in telecommunications. As reviewed in the next section, the deployment of fiber optics in local telecommunications networks by CAPs has begun to break through the market dominance of LECs, by targeting areas with high concentrations of intense users of local telecommunications services. The CAPs are growing at markedly faster rates now than did MCI at a comparable stage in its development, owing in part to the radical character of the technological innovations in the use of fiber optics in the local loop, as well as regulatory changes that have lowered entry barriers.

Yet, the "gales of creative destruction" have just begun to blow in local telecommunications. In the next few years, through the widespread deployment of fiber optics and interactive communications capabilities by cable system operators, the potential for wireline competition in local telecommunications services will explode. As discussed later, this process is well underway, and is likely to accelerate markedly within the next few years. As it does, local exchange companies (LECs) will face ubiquitous competition for local access and exchange services. Conversely, LECs will be upgrading their networks to provide broadband services to end users, which means that customers will have at least two wireline options for both video and telephone services.

In a later section, we discuss the impacts of current and pending changes in wireless technologies for competition in local telecommunications services. With the advent of digital wireless systems and a multiple increase in the allocation of spectrum for mobile communications, true head-to-head competition between wireline and wireless services will develop over the next 5 to 10 years. Whereas the prices of cellular service remains well above local wireline service, radical innovations and heightened competition in wireless communications will surely drive prices way below current levels, with substantial increases in capacity and substantial improvements in service quality.

It is evident from these technological and competitive developments that communications will be provided over an increasingly complex array of networks offering competing and complementary services. In many cases, one network will both provide inputs to, and compete with, another network. Cable companies and LECs, for example, will both sell transport services to PCS providers, as well as compete with them in the provision of local telecommunications services to end users. In another section, we describe and discuss the Ameritech proposal for unbundling and interconnecting their network in exchange for relief from the Modification of Final Judgment (MFJ) interLATA (Local Access and Transport Area) restriction.[1] It is our view that the adoption of appropriate unbundling and interconnection rules will facilitate the advance of competition and promote economic efficiency in local telecommunications services. Finally, we consider the implications of these developments and the Ameritech plan for antitrust policy in telecommunications, specifically, the MFJ interLATA restriction.

TARGETED COMPETITION FROM COMPETITIVE ACCESS PROVIDERS

The demand for local telecommunications services is highly concentrated: A small percentage of customers, lines, and geographic areas account for a very large share of the revenues in most service categories because the intensity of access and usage varies dramatically across customers and space. In addition, the density of customers varies dramatically across space; that is, the most intensive customers tend to be highly concentrated geographically. Because demand has also become very highly concentrated, entrants with geographically limited networks can reach a very substantial share of access revenues. Business customers located in just 1% of the total land area served by LECs in 10 large states constitute 30% of total LEC revenues; 75% of total revenues are located in just 8% of the land area. It should also be noted that, because user demands are so highly con-

1. Professor Teece has testified in support of the Ameritech plan to the U.S. Department of Justice and the Federal Communications Commission. Dr. Rosston assisted Dr. Teece in the preparation of his testimony as a Senior Economist at the Law & Economics Consulting Group. Professor Harris has testified in support of the plan to the Illinois Commerce Commission.

centrated in telecommunications services, one of the most important forms of competition is "self-supply" or "contract carriage" by large, intensive users.[2]

A substantial portion of LEC revenue is derived from business customers.[3] LECs derived approximately 41% of their local revenues from commercial customers in 1993. Business customer growth is expected to be 80% greater than residential customer growth over the next 5 years.[4] Thus, business customers are an important part of the telephony market, and will become even more so in the future. Because basic residential service is not as profitable to LECs as other services, it is likely that business customers represent the principal source of profits to LECs.

Because revenues are highly concentrated in local telecommunications services, these markets are easily segmentable and targetable. A new entrant does not need to serve all geographic or customer segments to compete effectively in one or a few segments. Instead, the rational entrant will target its initial entry at the small share of the customers who account for a large share of revenues. Moreover, although LECs have been allowed to deaverage their prices to a small degree, there are still customers with very different costs of service who pay the same prices. Hence, profitability is even more highly concentrated than revenues, because the highest volume customers and those in the most densely populated areas are also, typically, the lowest cost customers. Whereas an LEC has an obligation to serve all customers, entrants and competitors can and do target their investments, facilities, operations, and marketing efforts at those segments with the highest expected returns.

By successfully targeting the most profitable geographic areas and customers, CAPs are growing at extraordinary rates. CAPs are currently operating networks in 222 cities and have announced plans to enter 41 more. CAPs have begun to install switches and thus can provide switched as well as special access services and have formed alliances with cable companies and interexchange carriers (IXCs) to help extend the reach of their networks. MFS has switching capabilities in New York City and has authorization to provide switched services in Chicago and Baltimore;[5] Teleport also provides switched services in New York City, Boston, Chicago, and San Francisco.[6]

2. Although there is nothing inherently wrong with the rapid growth in private networks, there is reason to believe that at least in some cases, they are stimulated by regulations that require uneconomic pricing and/or inhibit the offering of new services by the LEC. In those cases, self-supply through private networks is contrary to economic efficiency and other public policy objectives.

3. Network access revenues account for nearly 25% of LEC revenues and other services (e.g., directory advertising and equipment sales) account for an additional 22% of revenues. Long-distance service, such as intraLATA toll calling, comprises an additional 12% of revenues. Thus, nearly 59% of LEC revenues come from sources other than local service.

4. INSIGHT Research Corporation.

5. See "MFS Intelenet Launches Full Service Phone Company Providing Both Local and Long Distance Services," MFS Communications Company News Release, October 5, 1993.

6. "Teleport Communications Prepares for Local Service Offensive," *Local Competition Report*, October 4, 1993.

Once a CAP has built its core fiber ring in a metropolitan area, the incremental cost of serving additional customers is quite low, relative to the potential gain in revenue. Having established strong footholds in downtown urban areas, one should expect continuing rapid growth by CAPs, as they sign up more customers and expand their networks over larger geographic areas. Because CAPs target their entry selectively to high-volume, high-density business customers (or smaller customers located in the same or adjacent buildings), they can exploit LECs' price averaging requirements. Because CAPs choose not to serve high-cost areas, they have a distinct cost advantage over LECs. CAPs can exploit these advantages of asymmetric regulation as they expand into switched access and exchange services as well. CAPs have expanded beyond central business districts in major metropolitan areas: Linkatel from Los Angeles to Anaheim and Santa Monica; Intelcom Group from Denver to Boulder and Colorado Springs; Tampa Electric Company from Tampa to Sarasota. The ability to serve customers in a concentrated geographic area allows the CAPs to maintain relatively low start-up costs. CAPs can establish a fiber ring in a downtown area for a relatively small investment, as low as $1 million in certain cases.[7]

The targeting strategy has enabled CAPs to grow rapidly. CAP networks, as measured by route miles, multiplied by 24 times between 1987 and 1992. CAP investment in local loop networks is now well in excess of $1 billion.[8] CAP revenues increased by 43% between 1992 and 1993. The annual route growth rate for MFS and Teleport, two of the largest CAPs, equalled 65.9% and 94.3%, respectively, between 1987 and 1992. In the past 3 years, MFS has grown at the phenomenal rate of 919%, from $10 million to $140 million in revenues, indicating that the company is competitive and that the environment for competition is hospitable to the new entrant. For that reason, some sources expect CAP revenue to more than triple between 1993 and 1996.[9] Perhaps the strongest evidence of the rapid growth prospects of CAPs is their extraordinary market valuations. MFS ranked second in *BusinessWeek*'s market value ranking of firms with under $150 million in sales with a 1993 value of $1.9 billion on sales of $141 million.[10]

In a study commissioned by the Regional Bell Operating Companies (RBOCs), Quality Strategies analyzed LEC high-capacity service (special access and intra-LATA point-to-point services for DS0, DS1, DS3, etc.) in 10 metropolitan areas in which CAPs were operating. Based on 4,500 customers, they found that CAPs have captured approximately 30% of high-capacity transport services. A recent study conducted for Pacific Telesis found that CAPs have captured 36% and 32% of revenue for high-capacity transport services from point (customer or POP) to point in downtown Los Angeles and San Francisco, respectively. In response to

7. Peter Huber, *The Geodesic Network: Report on Competition in the Telephone Industry*, 1992, p. 2.69.

8. *A CAP Market Update*, The Yankee Group, July 1993, p. ii.

9. *Ibid.*

10. "The Business Week 1000," *Business Week*, March 28, 1994, p. 69.

these CAP inroads, LECs have substantially reduced their special access rates; since 1991, for example, Illinois Bell's price for DS1 service has fallen by 39%.

CAPs have been sufficiently successful to attract the IXCs into local telecommunications services. MCI, for example, has announced plans to spend $20 billion developing "network MCI," a national network providing local and long-distance telephony services. Included in these plans is "MCI Metro"—a $2 billion plan to build local networks in 20 major cities. Through its purchase of Western Union conduits, MCI already has rights of way to build networks in these cities.

We do not mean to suggest, by concentrating this discussion on CAPs and IXCs, that there are not other significant competitors in local telecommunications. Gas and electric utilities, for example, have rights of way to almost every residential and business customer within their service area, and have begun to install and utilize broadband networks to provide local telecommunications services to themselves and others. For example:

- Entergy Corp., whose subsidiaries serve 1.9 million customers in Arkansas, Mississippi, Texas, and Louisiana, is testing an energy management technology called "PowerView." In addition to its "intelligent utility" capability, the technology enables the utility to support cable TV service and telephone services.[11]
- The Electric Plant Board of Glasgow, Kentucky has installed a coax-based broadband network from which spare capacity is being used for cable television, wide-area public data networking, local telephony, and long-distance access.
- A subsidiary of Citizens Utilities Company, ELI, has filed with the Washington State Utilities and Transportation Commission to offer intrastate interexchange and exchange switched services. ELI currently owns and operates networks in Seattle and Portland and is constructing networks in Phoenix, Salt Lake, and Sacramento.

EMERGING COMPETITION FROM CABLE SYSTEMS

Cable companies have an existing wire-based network that passes 90% of all homes and businesses in the United States.[12] Increasingly, the backbone distribution network of cable companies is fiber-based and thus capable of handling two-way communications. In fact, cable operators' use of fiber optics has increased 600% since 1988.[13] Thus, cable companies either possess or are installing the

11. "Utilities Emerging Role in Local Telecom Markets," *Telco Competition Report Special Report*, February 17, 1994, p. 15.

12. Paul Kagan Associates, *Cable TV Financial Databook*, 1992. About 60% of all homes actually subscribe to cable TV.

13. Cable companies in 1993 planned to install approximately 465,000 miles of new optical fiber cabling in their networks, for a cumulative installed total to data of about 28 yards of fiber per subscriber. Equivalently, telephone companies planned to install about 1.8 million miles of additional fiber in 1993, for a cumulative total of roughly 111 yards per subscriber (*Lightwave*, August 1993).

physical plant required to provide telephony services. In addition, the fiber optic cable used in their backbone loops for the provision of video services generally has unused capacity, which greatly reduces the cost of offering telephony services. Cable networks are already used for the backhaul of voice and data transmissions for cellular providers and CAPs. For example, AirTouch-Detroit has replaced some RBOC-provided local loop circuits with leased cable TV fiber to connect to IXCs' facilities and uses fiber in combination with microwave for its network.[14] In Kansas City, FiberNet, a cable venture begun in 1988, provides data and voice services to interexchange carriers, several airline reservation subsidiaries, and financial brokerage houses and other large firms.[15]

Moreover, cable companies are beginning to provide telephony services to end users directly over their cable networks. Time Warner is upgrading its facilities to offer telephony services in Rochester. Cablevision (in conjunction with AT&T) won a competitive bid over NYNEX to provide local telephone and cable services to Long Island University's C. W. Post campuses. Cablevision continues to build a fiber optic based network on Long Island and in New York City with the capability of offering video on demand, interactive games, and an alternative phone service to subscribers.[16] In addition, Cablevision has constructed on Long Island the fiber backbone of a high-speed communications network linking Stony Brook University and Brookhaven National Laboratory, termed FISHNet, using an ATM technology that allows voice, video, and data images to be processed together.[17]

Cable companies have also formed alliances with other telecommunications companies. MCI recently announced a joint trial with Jones Intercable to test phone service over the Jones cable network in Alexandria, Virginia. In June 1993, Teleport Communications Group (TCG) announced that it had signed letters of intent to establish joint ventures with 11 major cable operators to build new fiber networks and expand existing TCG networks (using some cable capacity for both projects).[18] In February 1993, Southwestern Bell purchased Hauser Cable in Montgomery County, Maryland, and has announced plans to offer telephone service to compete directly with Bell Atlantic. In May 1993, US West bought a 25% stake in Time Warner for $2.5 billion and BellSouth acquired 22.5% of Prime Management, which operates Prime Cable.[19] Bell Canada has purchased Jones Intercable. These "intermodal" alliances provide cable companies with significant financial backing and the technological know-how concerning the provision of two-way telephony and will thereby accelerate entry by cable companies into telecommunications.

A recent study has estimated that the costs of upgrading existing cable plant to provide telephony services (assuming the cable company has already upgraded its

14. Peter W. Huber, *The Enduring Myth of the Local Bottleneck*, March 14, 1994, p. 39.

15. Fred Dawson, "In Teleportís Shadow," *Cablevision*, September 21, 1992, p. 31.

16. Joshua Quittner, "Cableís Vision," *Newsday*, February 25, 1993, pp. 3, 18.

17. See "Cablevision Seeks to Catch Big Fish in its High-Speed Long Island Net," *Communications Engineering and Design*, April 1994, p. 8, and "Information Superhighway Adds Lane," *Currents*, April 1994, p. 1.

18. 1993 Connecticut Research, VII-80.

19. Huber (1994), p. 26.

backbone transmission plant to fiber optics) would be about $207 per subscriber. If both telephone and distributed video services were provided, the cost per subscriber would only increase to $297 due to significant economies of scope in the provision of telephony and distributed video services. The analysis further demonstrates that upgrades to existing plant represent a large cost advantage to deployment of new networks and that there may be economies of scope between distributed video services and PCS. The author concludes "this outcome increases the value of the incumbent cable television network."[20] Similar conclusions have been reached by a leading investment analyst:

> The reason for the enormity of the financial implications of this technological change is that . . . *the cost of adding telephony to a cable system is far less than the cost of the existing telephone plant* (italics in original). The cost of upgrading a cable system by adding fiber trunks is less than $150 per subscriber. The cost is so low that the reduction in maintenance expenditures alone is adequate to more than pay for the upgrade, so effectively the cost is zero. The cost of adding telephony to an upgraded cable system is less than $400 per subscriber, and it is only incurred for the subscribers who purchase the new service.[21]

The cable industry is well aware of the enormous opportunities ahead. In July 1994, Cable Television Laboratories issued a "Request for Proposals for a Telecommunications Delivery System over a Hybrid Fiber/Coax (HFC) Architecture." The purpose of the RFP is to "expedite the design, test, production and phased implementation of practical, cost-effective approaches to telecommunications services over the evolving cable infrastructure."[22] It also announces that TCI, Comcast, Continental Cablevision, Cox Cable, Time Warner and Viacom—the majority owners of CableLabs, "intend to individually purchase equipment under this RFP."

The transformation of cable systems to interactive, broadband networks capable of offering a wide range of telecommunications and video services means that, within a decade or so, the millions of miles of LECs' existing copper-twisted pair cables will be thoroughly obsolete. As noted by Philip Sirlin:

> The telecommunications industry is about to undergo a technology-driven earthquake of enormous magnitude . . . The financial epicenter of this metamorphosis will be in the . . . local loop [because] copper twisted pair is a very high cost, low functionality, archaic technology . . . The new

---

20. See David P. Reed, *The Prospects for Competition in the Subscriber Loop: The Fiber-to-Neighborhood Approach*, presented at Twenty-First Annual Telecommunications Research Policy Conference, September 1993.

21. Philip J. Sirlin, *The Digital Battlefield: Bellopoly—The End of the Game*, Investment Report by Wertheim Schroder & Co., March 22, 1994, p. 13.

22. Cable Television Laboratories, "Request for Proposals for a Telecommunications Delivery System over a Hybrid Fiber/Coax (HFC) Architecture," Boulder, Colorado, p. 6.

> technologies—high capacity fiber circuits to large businesses, wireless (new cellular, SMR, and PCS) systems and telephony and video on fiber/coaxial cable systems—have lower costs and higher functionality than the existing copper twisted pair local loop . . . New entrants who can deploy the new technologies and gain market share will be very successful.[23]

As suggested by Sirlin, there is a strong affinity between CAPs and cable companies, both in terms of ownership structure and network deployment. Several of the leading CAPs are owned by cable companies; others have formed alliances with them, including the following:

- Cox, TCI, Continental, Comcast, Time Warner Cable, and Teleport: Cox and TCI acquired TCG, the largest CAP, and sold minority stakes to the two other multiple system operators (MSOs) in 1993. Cox owns a 25.05% stake, followed by TCI with 24.95%, and Time Warner, Comcast, and Continental with 16.67% each.
- TCI, ATC, and TeleCable: The MSOs have participated in a joint venture known as FiberNet since 1989 in and around Kansas City, MO. TCI, American Television and Communications (ATC), and TeleCable jointly own the all-fiber network, covering close to 200 route miles on both sides of the Missouri River. The network now serves upwards of eight interexchange carriers, several airline reservation subsidiaries, financial brokerage houses, and other large firms requiring diverse paths to carry their traffic.[24]
- Monmouth Cablevision, Adelphia Cable, and Comcast Cable Communications: The three cable operators in Central New Jersey began setting up an inexpensive fiber interconnect in 1993 through a joint venture that will open new business opportunities for them such as alternative access to long-distance services. Each company expects its cost for the interconnect to be less than $50,000.[25]
- Continental Cable and Hyperion: The MSO and the telecommunications subsidiary of Adelphia agreed to set up a metropolitan area network through a joint venture in Jacksonville, FL. The network will utilize Continental's existing fiber backbone and will require some construction of a series of fiber rings and fiber hookups to the premises of potential users. Between 30 and 40 large business users have been identified as likely connection points for the operation.[26]
- Continental Cable and Teleport: The MSO and CAP began building loops around greater Boston and in the Wilshire corridor of Los Angeles through a joint venture since 1992. TCG has been able to extend its business beyond the city limits via fiber routes available over Continental's suburban sys-

23. Sirlin, op. cit., p. 5.
24. "In Teleportís Shadow," *Cablevision*, September 21, 1992.
25. "RBOCs? Who Needs RBOCs?" *Cablevision*, December 6, 1993.
26. "In Teleport's Shadow," *Cablevision*, September 21, 1992, p. 31.

tems, allowing the MSO to enter the business without devoting a tremendous amount of startup effort.[27]

- Comcast and Eastern Telelogic: Comcast agreed to acquire a 51% stake in the CAP in July 1992 and subsequently expanded the CAP's operations in Philadelphia.[28]

The rate at which these deals are being struck indicates both the strong complementarity between CAP and cable networks and the mutuality of their business interests in competing with LECs. Not surprisingly, CAPs are, in many cities, laying fiber to connect the head-ends of cable systems and, therefore, connecting cable customers to the CAPs switches. These extended networks will be able to provide wireline services to end users, and transport services to interconnect wireless cell-sites and switching centers, both in competition with LECs.

EMERGING COMPETITION FROM WIRELESS

Wireless telecommunications providers already have infrastructure in place that allows them to provide telephone service to more than 90% of the population. With the introduction of satellite service, coverage will be ubiquitous and coverage will be available from a number of different providers. The two major questions about whether wireless can provide a real competitive alternative to wireline telephone service focus on price and capacity: Will costs be low enough so that competitive prices will entice enough customers to demand wireless service in place of wired service and will there be sufficient capacity to meet demand at those prices to force competitive pricing for the wireline alternatives?

Although wireless communications has been used for a relatively long time, there are two significant changes that will impact the nature of wireless communications.[29] Capacity increases of more than an order of magnitude resulting from continuing improvements in technology and new allocations of spectrum will yield enough capacity so that wireless can be more of a threat to wireline telephone service. In addition, regulatory flexibility will allow wireless service providers to use this capacity to serve an increasing variety of different consumer needs. The confluence of these technical and regulatory changes will have significant impacts on competition within and between wireless and wireline communications. This section examines the technological changes that are transforming wireless communications into a rapidly growing, high-quality service; the regulatory changes that have accompanied and complemented the changing technology to prompt the growth and competitiveness of wireless services; and the effects of these two forces on competition for local telecommunications services.

Wireless communications already plays a significant role in telecommunications. The success of wireless is much greater than was anticipated 10 years ago

---

27. *Ibid.*

28. *Ibid.*

29. Historically, wireless communications preceded wireline: The original telegraph system in France was a wireless system. See Noam (1993).

when the first cellular systems were licensed. At that point in time, analysts projected that mobile telephony would only serve a limited market and that there would be about 100,000 cellular subscribers by the year 2000.[30] The number of cellular subscribers has grown almost eightfold over the past 5 years;[31] there are currently more than 19 million cellular subscribers. In the last 6 months, more than 100,000 net new subscribers were added each week.[32] Cellular revenue in 1993 was almost six times as great as its level 5 years ago, totaling over $6 billion. Cellular traffic is increasing significantly as a result of the new subscribers, even though the average minutes of use per subscriber is decreasing. To meet the demand, cellular service providers have significantly increased system capacity through the installation of additional cells and the implementation of digital signaling. Existing analog cellular systems have added or "split" cells to substantially increase capacity. System operators employ small radius cells where there is significant demand so they can reuse frequencies more in these areas. Because each cell has historically cost approximately $1 million for the capital investment, cell splitting is only undertaken in areas where the traffic justifies the added expense. In areas where demand is relatively low, cells cover wide areas. Adding cells to current systems can increase capacity until the minimum size cell is reached. Because of the propagation characteristics of spectrum, there appears to be a limit to decreasing cell size and the additional costs of increasing the number of cells is causing systems to switch to higher capacity digital systems. To determine the ability of a cellular system to provide a significant competitive alternative to landline service, it is possible to determine the fraction of minutes of use accounted for currently by cellular systems and then determine the increase in capacity available from additional spectrum and digital signaling to estimate a minimum capacity available over cellular. This estimate is likely to provide a significant lower bound because there are many areas of the county where cell splitting has not even begun to approach the limit using current technology. In fact, some cellular systems have such low demand that they have not even used the additional 5 MHz of spectrum they were allocated in 1986.

It is possible to determine the current share of minutes of use from cellular systems by using data available on the number of "lines" and the average minutes of use for both cellular and wireline service. The ratio of cellular service to the total population is over 7% compared to about 56% for landline service. The average minutes of use for a cellular phone is about 70 minutes per month compared to more than 1,600 for the average landline phone.[33] Using these figures, it is

30. Huber (1992), op. cit., p. 4.22.

31. Cellular telephony's astounding growth is expected to continue. Link Resources Corp. estimates that the annual growth rate for cellular voice services through the year 1998 will be 20.2% and that the corresponding rate for cellular data services will be approximately 33.0%. *Wall Street Journal*, February 11, 1994, p. R5.

32. Cellular Telecommunications Industry Association, July 1994.

33. FCC Statistics of Common Carriers, July 1994, and Cellular Telecommunications Industry Association data. Calculation assumes $35 per month average service charge and $0.35 per minute average usage charge.

straightforward to calculate that cellular minutes of use account for approximately 0.5% of total minutes of telephone use. The cumulative capital investment in the cellular industry is now almost $14 billion, implying that the cellular infrastructure is becoming increasingly well positioned to compete against wired telephony.

The "new" wireless services can be both complementary to and competitive with existing communications services. The ability to provide high-quality voice and data services with wireless technology derives directly from advances in electronics that provide the backbone for the networks. Early wireless communication usually emanated from a high-power tower to multiple mobile units. The quality and privacy were very low. Advances in signal processing and microelectronics led directly to the development of higher quality cellularlike systems by making it possible for a mobile switching office to track thousands of mobile units, manage the allocation of spectrum and hand-offs, and perform normal telephone end office switching functions. In addition, the advances in circuitry and miniaturization combined with economies of scale in manufacturing have led to the decline in mobile telephone prices from $2,500 in 1984 at the initiation of cellular service to around $200 now.[34] The advances have increased the quality of service, and the capacity of service and have reduced the cost of the consumer and network equipment needed for service.

A number of cellular systems are currently converting to digital technology because they need to further increase their capacity. There appear to be two different "camps" within the cellular community—those who support Time Division Multiple Access (TDMA) and those who support Code Division Multiple Access (CDMA). Some systems are already in the process of converting to digital signaling and using TDMA because it is commercially available. Those who support CDMA must wait for the technology to become commercially available, but they feel that ultimately it will offer higher capacity and quality.[35] Both of the technologies result from advances in signal processing that derive directly from microelectronic advances. Also, digital phones will be able to be smaller, lighter, and use less power than their analog counterparts.

Although the ultimate capacity increase from the digital conversions is unknown, the initial increase from TDMA is expected to triple current capacity. PCS systems are expected to employ digital technology from the outset and smaller cell sizes than cellular so they will also have significant capacity available. A number of Specialized Mobile Radio (SMR) companies are converting their high-powered analog systems to cellular configurations with digital technology, which they estimate will increase their capacity by a factor of 15. Finally, mobile satellite providers expect to be able to offer a variety of different services on a nationwide or even global basis. Over time, digital technology is expected to significantly more than triple the capacity of current analog systems. This magnitude of in-

34. "Cellular Market and Profit Opportunities through Year 2000." Herschel Schostek Associates, Ltd, February 1992, Silver Spring, MD.

35. Qualcomm presentation at Telecommunications Policy Research Conference, Solomons, MD, October 1994.

crease will allow cellularlike systems to significantly increase sales, some of which are likely to displace local wireline calls.

There is 50 MHz of spectrum allocated to cellular service. There is also 14 MHz of SMR spectrum in the 800 MHz band that is currently being aggregated by operators so they can form more efficient and higher quality cellular systems. In addition, the FCC has allocated 120 MHz of additional spectrum for broadband personal communications services. The earlier calculation showed that current analog cellular accounts for about 0.5% of minutes of use. The additional 134 MHz of spectrum available for wireless telephony adds to the current 50 MHz of cellular spectrum to more than triple capacity. The introduction of digital technology will triple that increase again so that the effective increase will be approximately a factor of 10. It should be noted that this simple calculation does not take into account the fact that current analog cellular is not near its theoretical capacity, nor does it take into account the ability to implement significantly smaller cells using the 1.8 GHz spectrum associated with PCS. As a result, this capacity increase should be viewed as a significant lower bound. In any event, it shows that wireless systems would easily have the capacity to provide for more than 5% of the minutes of use across the country.

The previous exercise did not account for the fact discussed earlier that traffic is not uniform across the country, but rather is concentrated in specific areas. In 1985, Hatfield undertook a study showing that analog cellular with 20 MHz of spectrum could provide a competitive alternative to landline local loops for a significant portion of the San Francisco Bay area.[36] His study vividly illustrated the point that local loop costs are directly related to loop length, whereas the costs of wireless "loops" are relatively insensitive to loop length and much more sensitive to traffic. As a result, in less densely populated areas where loop lengths tend to be longer, the competitive threat from wireless loops will be stronger. In addition, in these less densely populated areas, the traffic demands on spectrum tend to be lower so that the opportunity costs for spectrum devoted to wireline competition will also be lower.

The FCC recently reclassified a variety of wireless services into a category called Commercial Mobile Radio Services (CMRS). CMRS encompasses a variety of radio services that historically faced different regulatory frameworks and service restrictions. By including all of these and adopting a forward-looking framework for establishing that all of these services should be considered "substantially similar" the FCC has determined that it expects that all of these services will compete with each other either now or in the future. The major components of CMRS include PCS, cellular, SMR, paging, and possibly mobile satellite services.

---

36. "A Comparison of the Costs of Providing Ordinary Telephone Service Using Conventional Wireline and Cellular Radio Technology," Hatfield, D. Ax, G. and Dunmore, K. Boulder, CO, October 1985.

The FCC specifically adopted a flexible definition for PCS.[37] In addition, it has issued a Notice of Proposed Rulemaking to increase flexibility for cellular and SMR providers.[38] These rules and changes will allow service providers greater ability to use their wireless capacity to respond to marketplace demands—whether they be for high quality mobile voice, dispatch, paging, or service more similar to residential fixed service.

Because of the flexible definition of PCS, a large number of companies are pursuing different visions of what PCS will be. To date, 187 companies have obtained experimental PCS licenses. About 10% of those companies are cable companies and cable companies are among the most active in a broad range of cities. For instance, Comcast is conducting trials in five cities, Hauser Communications is testing in five cities, Prime II is testing in six cities, Time Warner is testing in five cities, United Artists Cable is testing in five cities, Viacom is testing in five cities, Cable USA is testing in four cities, and Cablevision is testing in four cities. In addition, CableLabs, a research and development consortium of North American cable companies, is investigating using preexisting broadband cable infrastructure for PCS. Continental Cablevision, Cablevision of Boston, and Time Warner Cable became the first cable TV companies to interconnect their systems to demonstrate how PCS could be offered over cable TV systems in Boston in late 1993. The demonstration allowed the MSOs to overcome their problems with interconnecting with differing systems and offering telecommunications subscribers extended and seamless service by using a combination of technology and cooperation. The companies had to do very little to their basic cable infrastructure to offer wireless services and to bypass the local telephone company.[39]

New wireless PCS competitors are likely to begin service within the next 1 to 2 years. The FCC scheduled broadband auctions to begin on December 5, 1994. The new PCS providers will have to submit high bids at the auctions, go through the FCC licensing process, acquire and construct cell-sites, and, in some cases, negotiate for and effectuate the relocation of incumbent microwave users. In addition, because of the propagation characteristics of the 1.8 GHz PCS spectrum compared with the 800 MHz cellular spectrum, PCS providers will be required to use more cells than their cellular and Enhanced Specialized Mobile Radio (ESMR) counterparts to cover the same area. In addition, early PCS operators may enter into roaming agreements with neighboring cellular providers because PCS systems are not likely to become operational across the country at the same time. Based on these events, it will be from 6 months to 2 years before there is significant service from PCS providers.

These providers will have to compete with existing cellular and ESMR providers who are continuing to build their customer bases. Current cellular providers

37. The Commission adopted the following definition for broadband PCS: "Radio communications that encompass mobile and ancillary fixed communication services that provide services to individuals and businesses and can be integrated with a variety of competing networks." FCC Second Report and Order, GEN Docket No. 90-314, para. 24.

38. Notice of Proposed Rulemaking, GN Docket No. 94-90.

39. "CATV networks join to offer PCS," *Telephony*, November 22, 1993, p. 8.

have constructed systems that provide coverage for more than 90% of the U.S. population so that mobile wireless service is available nearly universally. The current belief is that service providers compete on coverage area in addition to price. For example, McCaw heavily advertises its "City of Florida" where customers can travel anywhere in the state and pay home rates and also call anywhere in the state without paying toll charges. PCS systems will start with smaller coverage areas, even though two of the major systems will be licensed on a Major Trading Area (MTA) basis. In fact, as discussed already, the early PCS systems are likely to be forced to negotiate roaming agreements with incumbent cellular providers in order to give their customers equivalent geographic coverage. To take advantage of the expanded coverage area, their customers will be forced to buy dual-mode handsets so they can operate on the different systems. Customers on existing cellular systems will only require a single-mode handset. When PCS is fully deployed, its handsets are expected to be somewhat smaller and lighter and to have a longer battery life because they will transmit with lower power. However, to acquire customers in the interim, PCS providers will either offer a more limited coverage area with a "better" phone or offer equivalent service with a somewhat less desirable phone. As a result, to acquire customers, PCS providers may have to compete with lower prices to compensate for their initial disadvantages.

In addition to targeting customers who do not require roaming, the cost characteristics of PCS favor a less mobile customer base. PCS will be transmitting with lower power and less propagation, which will require a larger number of cells to cover an area than a current cellular system. As a result, the number of hand-offs required for each call will be higher. The less "mobile" the user, the lower the burden will be on the switch and the lower the costs will be. As a result, the target for PCS is likely to be somewhat less mobile than the current cellular customer.

Regulation has made two contributions to the radical change taking place in wireless communications—the allocation of a significant amount of additional spectrum and the flexibility to allow service providers to respond to market forces in determining their product offerings. The increase in spectrum allocation and spectral efficiency have changed the ability of mobile radio service providers to compete with each other and with local wireline service. The technical characteristics are now such that the demand for mobile service may not be sufficiently high to demand all of the wireless spectrum. In addition, the change in capacity may make the costs characteristic of wireless service more applicable to wireline service. To encourage the efficient use of spectrum, the FCC has allowed providers to configure their product offerings to target the highest value use. Such policies are important to the development of competition in areas where it is warranted. By allowing flexible use of spectrum, wireless service providers will be able to target those areas of wireline service where their cost advantages allow them to charge lower prices than current providers. Motorola, for example, has already announced a PCS product designed to compete directly with wireline: "Tele-Density" is a low-power, high-quality, low-mobility handset combined with small, low-power base stations.[40] Alternatively, mobile service providers may charge somewhat higher prices than conventional landline services, but entice consumers away from wireline through increased functionality and more enhanced services.

The combined effects of these changes, technical, regulatory and competitive, lead experts to predict a very substantial role for wireless communications. Mercer Management recently conducted a national survey, analyzed several market and cost possibilities, and interviewed telecommunications industry experts:

> Nearly half of the industry experts that Mercer interviewed projected wireless service would become a "viable substitute" for traditional wireline service within 10 years . . . Half of those experts interviewed predicted more than 15 percent of the public would be using a wireless handset in five years, compared with the current 7 percent. They expect that figure to rise to more than 30 percent in 10 years."[41]

In order to reach anywhere near these predicted market penetration levels, wireless services must compete with—and take share from—traditional wireline providers of local access and exchange services.

UNBUNDLING AND INTERCONNECTING LOCAL TELECOMMUNICATIONS NETWORKS

In the prior sections, we have shown that competition is developing rapidly in local telecommunications services. In the next few years, there will be a proliferation of communications technologies and networks. To a substantial degree, these networks will compete directly with each other. They will also serve as complements, in the sense that the communications needs of any given customer will be served by a combination of networks and service providers. Note that interconnection and interoperability will be essential, even if every single end user is served by a completely integrated service provider, so that customers served by one integrated provider can interact with customers served by other providers. To accommodate this "network of networks," public policies must promote both competition and cooperation. In this section, we describe and assess one company's proposal for doing so.

The Ameritech plan, as presented to the U.S. Department of Justice, Federal Communications Commission in 1993 and to the Illinois Commerce Commission in 1994,[42] is an effort by one company to address both key policy concerns and business realities.[43] The plan is, in essence, a quid pro quo: Ameritech would take steps to unbundle its local telecommunications services, offer a standard set of in-

40. Motorola press release, September 23, 1994.

41. *The New York Times*, February 9, 1994.

42. To be implemented throughout the Ameritech region, state regulatory approval in Indiana, Michigan, Ohio, and Wisconsin would also be required.

43. Rochester Telephone has a somewhat different proposal for unbundling its local network.

Figure 1: Current Status of the Local Exchange

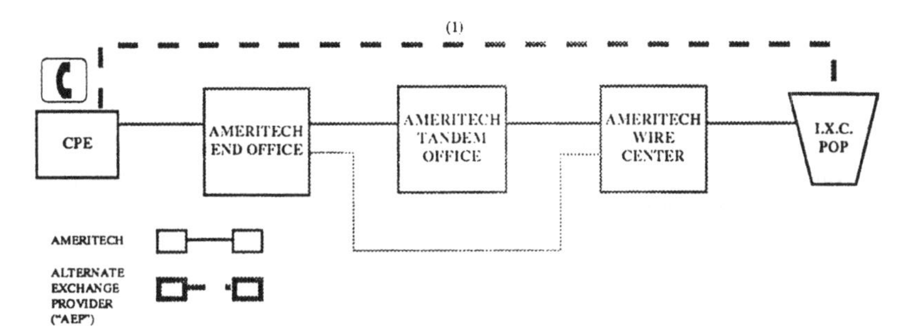

terconnection prices and arrangements, make the necessary network investments to promote further competition, and generally support removal of regulatory restrictions that may inhibit entry. In return, Ameritech seeks removal of the MFJ restriction so that it could offer interLATA services to customers in its region. Under the plan, Ameritech will open its network as follows:

- *Loops:* Offers local loops on an unbundled basis at tariff rates approved by state regulatory agencies; proposes rates above long-run incremental costs, not to exceed fully distributed costs; access to local loops at the main distribution frame or the digital cross-connect frame.
- *Switching:* Offers interconnection to its local switching with loops provided by others, enabling all providers to seamlessly connect to a "network of networks."

Figure 2: Local Exchange with Unbundling

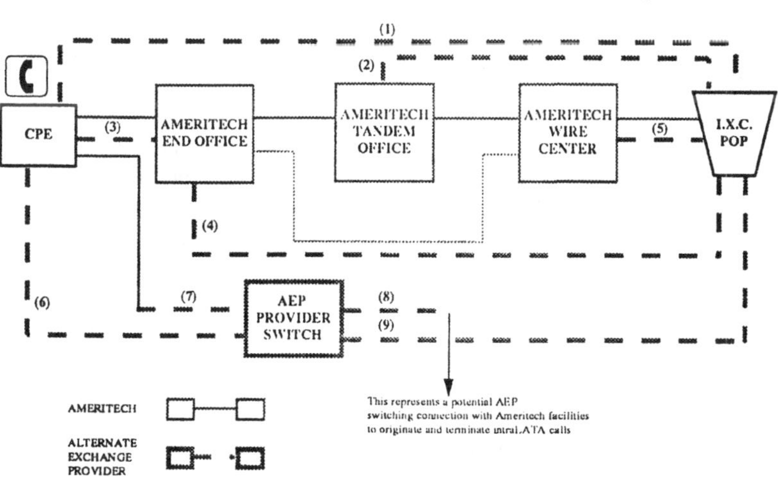

- *Signaling:* Unbundles SS7 call set-up capabilities and permits competitors to access the SS7 signaling network without subscribing to Ameritech's transport or switching service.
- *White Pages listings, 911 service, deaf-relay services:* Offers these network support services to competitors on an optional basis, at wholesale prices.
- *Cooperative engineering:* Offers cooperative engineering, operation, maintenance, and administrative practices.
- *Rights of way:* Will continue to make conduit and pole attachment space available on a non-discriminatory basis to authorized interconnectors where sufficient space permits.
- *Mutual compensation:* Offers mutual compensation arrangements for termination of traffic by state certified alternate exchange providers at reciprocal rates.
- *Numbering plans:* Makes available complete NXX codes to other qualified providers through a third-party administrator.
- *Local telephone number portability:* Provides portability to the fullest extent permitted by current technology and a commitment to support development of more robust options through industry forums.
- *Usage subscription:* Allows others to use Ameritech's loops and local dial tone provision while they carry all outbound traffic on their networks (in essence, intraLATA presubscription); a new entrant can offer alternative service without requiring the customer to change telephone numbers.

Unbundling and switch integration would facilitate entry and interconnection. A simplified graphical representation of the current status of the local exchange is shown in Fig. 4.1. Typically alternate exchange providers such as MFS connect directly to customers with their fiber loop and transport the traffic directly to the IXC point of presence (IXC POP). This is depicted by the bold dashed line labeled (1) that connects customer premise equipment (CPE) directly to the IXC POP. Figure 4.2 shows the potential effects of unbundling, with nine different places where alternative exchange providers can connect to and/or make use of portions of the local exchange network. The lines labeled (2), (4), and (5) show potential connections by competitive access providers to take advantage of traffic aggregation by the LEC and provide transport of the traffic from the end office (4) , tandem office (2), or wire center (5) to the IXC POP. Because of unbundling, the CAP can have its customers' traffic routed over the local exchange network until the point where it has facilities. Line (7) represents a provider who uses a portion of the local exchange provider's local loop and then transports the traffic to its own switch. Under the Ameritech plan, a provider might use the local loop, gather its traffic at the frame of the end office, and transfer the traffic to its own switch. Lines (3) and (6) represent alternative local loops. For example, a cable company may be able to provide a local loop, but wish to use some of the local exchange provider's switching capabilities, leading to a scenario represented by Line (3). A cellular company or cable company connecting to its own switch would be represented by Line (6), which directly connects the customer to the alternative exchange provider's own switch. Lines (8) and (9) represent connections from the

Figure 3: Possible IXC Entry with Unbundling

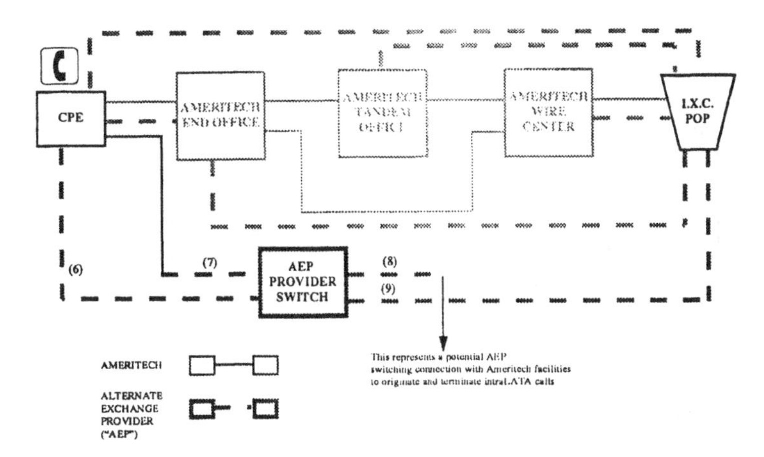

alternative exchange provider's switch to either Ameritech's network or to an interexchange provider.

Ameritech has stated in its filings that it will price its noncompetitive facilities between incremental and fully distributed cost,[44] although the actual measurement of these costs is likely to be a source of contention.[45] For the implementation of the Ameritech plan to support economic efficiency, prices for these noncompetitive services must be set so that efficient producers that use noncompetitive services as inputs are encouraged to provide their services and inefficient producers are similarly discouraged from providing theirs. A "monopoly" provider of an input to a competitive product market can generally use the input itself to provide downstream products, sell the input to other firms who provide the downstream products, or both.[46] To ensure that the firm prices the input so that the most efficient firms provide the downstream product, price should be set equal to the average incremental cost plus the opportunity cost of that input.[47] In this way, prices will not be set below cost, nor will they be set so as to discriminate against competitors. In the presence of scale and scope economies, prices need to be greater

44. Supplemental Materials to Ameritech's Petition for Declaratory Ruling and Related Waivers to Establish a New Regulatory Model for the Ameritech Region, DA 93-481, Attachment 2 of 4, "Ameritech Customers First Expanded Network Interconnection Proposal," pp. 22–24. filed April 16, 1993; Ameritech's Reply to Comments on its Petition for a Declaratory Ruling and Related Waivers to Establish a New Regulatory Model for the Ameritech Region, DA93-481, pp. 36–39, filed July 12, 1993.

45. Kelley, Daniel, & Mercer, Robert, "A General Approach to Local Exchange Carrier Pricing and Interconnection Issues," Hatfield Associates, Inc., Working Paper, September 19, 1992, pp. 20–24.

46. Much of this discussion is based on Baumol, William J., & Sidak, J. Gregory, *Toward Competition in Local Telephony.* AEI Studies in Telecommunications Deregulation, Washington, 1994.

47. At this level, the sale will make the appropriate contribution to joint costs without distorting downstream provision of alternative products.

Figure 4: Possible CAP Entry with Unbundling

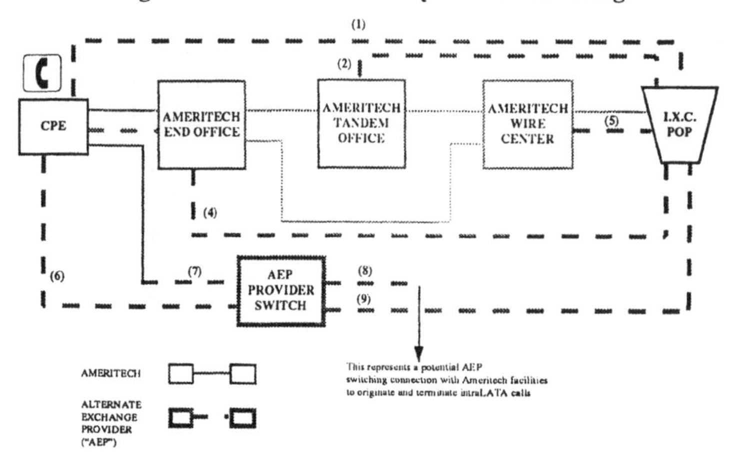

than incremental cost for at least some services in order to pay for the joint and common costs.

In conjunction with providing equal access, the plan has the feature of offering to competitors the scale and scope economies of the Ameritech network. Pricing is based on the long-run incremental costs of operating the Ameritech network. This is especially germane for smaller firms looking to enter specific niche services, as they will be able to compete with Ameritech and other large providers on a more equal footing. Perhaps most importantly, they will avoid the capital outlays and incumbent risks that typically accompany forays into new lines of business. As revenue streams are realized from initial ventures, the cash flow can be used to secure additional customers and offer new services.

Based on current market conditions, there are several classes of competitors positioned to take advantage of the unbundling, switch integration, and usage subscription offered under the plan. These competitors, depending on their specific capabilities, are poised to compete for either the entire market or for distinct subsets of customers. Because each potential competitor has different competitive advantages, the range of customers benefiting from new entry and expanded competition nearly spans the gamut of Ameritech's customer base. In addition, the ability to enter with minimal investment and to act as a reseller gives an entrant complete market presence with little risk.[48]

48. See Porter, Michael E., "Competition in the Long Distance Telecommunications Market," p. 9, Appendix A to "Motion for Reclassification of American Telephone and Telegraph as a Nondominant Carrier." In the Matter of Policy and Rules Concerning Rates for Competitive Common Carrier Services and Facilities Authorization Therefor, CC Docket No. 79-252. He discusses the entry of WilTel and others into interLATA service by employing a niche strategy in combination with resale to expand service to the entire marketplace.

Figure 5: Possible Cable Entry with Unbundling

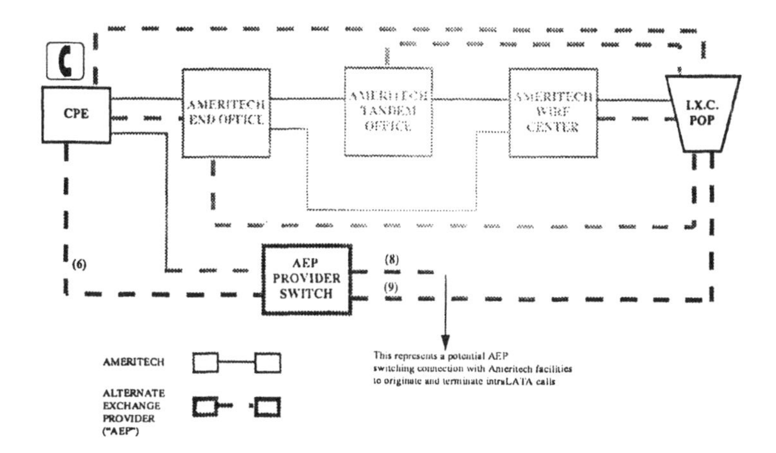

The most likely source of immediate and influential entry into local service will be the IXCs, especially the large, nationwide carriers like AT&T, MCI, and Sprint.[49] All three companies have the ability to self-supply transport, and, once the necessary construction and right-of-way expenses are incurred, the incremental cost to add traffic is quite small.[50] Specifically, once the IXCs have successfully developed the transport segment of their network, they will be able to sign up additional subscribers at little added cost because they can rent loops from Ameritech and transport the traffic to their own switches.[51] Figure 4.3 illustrates the network configuration of possible IXC entry under the plan.

IXCs enjoy their highest margins in the small and mid-size business segment.[52] Consequently, IXCs are likely to pursue these customers first for their provision of end-to-end service.[53] AT&T, as well as other large IXCs, could compete by in-

49. Indeed, Sprint already provides local wireline service. In 1991, the company had local service revenue of $2.3 billion for the nation, $478 million in the Ameritech region alone. Table 29, *FCC Preliminary Statistics of Communications Common Carriers*, 1991. AT&T, despite its protestations to the contrary, has also entered the local service business with its acquisition of McCaw Cellular.

50. MCI has purchased a significant amount of right of way from Western Union. *Telecommunications Alert*, May 11, 1992. MCI has also recently filed for state certification as a competitive access provider in Indiana.

51. According to an MCI expert economist, Kenneth Baseman, "the marginal activation costs and marginal operating costs for new circuits activated on facilities already in place are generally quite low and do not differ significantly depending on whether the IXC is collocated or the IXC's POP is several miles away." Affidavit of Baseman to FCC in CC Docket No. 91-141.

52. "Long Distance—A Healthy Industry Ready To Conquer New Territory," *Bernstein Research*, May 1993, p. 10.

53. IntraLATA margins are also quite high for this customer class. The average revenue per line, at $60 to $80 (which can be computed from Ameritech's access revenues by customer class), is far above the overall per line average of $45 to $50.

Figure 6: Possible Wireless Entry with Unbundling

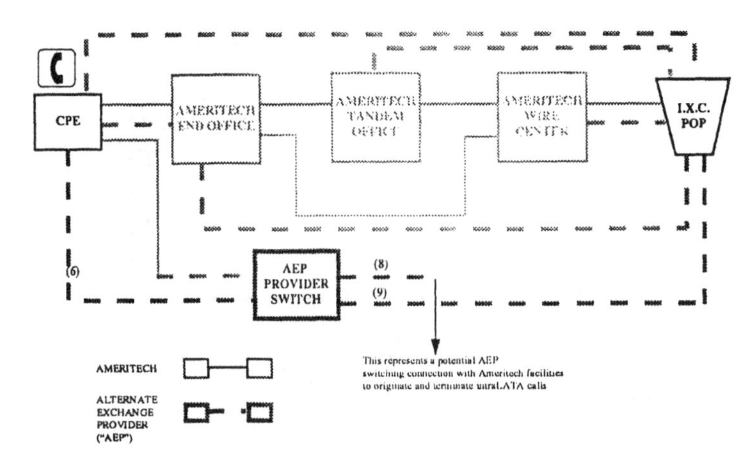

stalling switches (or using excess capacity on its existing switches) to supply dial tone and usage services and routing the traffic to one of their many existing POPs. This could be economical even in an area with a small amount of traffic because the large IXCs could either share capacity on a nearby existing long-distance switch or economically use a somewhat distant switch to provide a local dial tone until traffic justifies a truly local switch. Adding switch capacity is relatively simple with modern modular switches such as the 5ESS. Because the IXCs have fiber facilities in place with excess capacity, the cost of transport to take advantage of a distant "local" switch would be minimal. Moreover, the IXCs have customer-specific demand information that would enable them to target high-margin neighborhoods; they also have strong brandnames.[54] Existing IXC POPs are located within 242 of the 1,183 wire centers in the Ameritech region, thereby providing the IXCs with direct access to 44% of Ameritech's revenue base. The threat of such widespread competition would diminish Ameritech's ability to preserve and subsequently leverage market power. Much of Ameritech's customer base would be exposed to significantly greater competition.

CAPs have entered many major cities by deploying fiber loops through dense downtown areas. These loops give them access to a large number of customers with a relatively high demand for telephone service. With the switch integration portion of the Ameritech plan, CAPs with switches can easily become the local phone service provider to those businesses passed by their network. In addition, the ability to rent loops means that they can provide service, with little incremental investment, to any business or residence that is served by the end offices they pass with their loops. CAPs can also expand their geographic coverage sequentially

54. As a share of revenues, AT&T, MCI, and Sprint spend roughly three times as much on advertising as the average RBOC—15% versus 5%.

and determine the optimal path for their new fiber loops by leasing capacity in the short term while determining where to install plant expansions. Finally, the CAPs will be able to compete to serve multilocation businesses even when they do not have a physical presence near each of the satellite offices. Figure 4.4 depicts this possible CAP entry strategy.

CAPs will be able to increase their target customer base significantly upon implementation of the plan. CAPs currently have networks in 111 of Ameritech's 1,183 wire centers, giving them immediate access to 29% of Ameritech's revenue base. Under the plan, CAPs may deploy fiber in other areas, giving them even more potential customers. CAPs can use the plan to determine demand for their services and perform true market research by purchasing pieces of Ameritech's network before determining where to construct their own facilities. They can greatly reduce the risk of new construction by acquiring an active customer base prior to completion of their facilities. The unbundling and switch integration will make the CAPs' presence significant and immediate.

As shown in Fig. 4.5, a cable operator will be able to begin providing telecommunications services, even if it serves only a portion of a metropolitan area. It can employ its own cable to operate in currently served areas, providing competitive local loops attached to its own, Ameritech (or a third party) switch. It can purchase loops (and/or switching) from Ameritech to serve portions of the region where cable systems have not been upgraded to provide telephone service. If other cable systems are able to provide voice service, cable companies will then have a choice of purchasing facilities from Ameritech or from other cable companies. Under the plan, cable companies will be able to compete for local telephone service without having to ensure that other cable operators will also upgrade their facilities. This will allow cable companies to avoid the uncertainty of the payoff from investment in upgrading their systems. Cable companies are situated to immediately address a substantial portion of Ameritech's revenue base. The top four MSOs[55] have cable networks in 748 of Ameritech's 1,183 wire centers, giving them direct access to 77% of Ameritech's revenues. Their broad coverage of the Ameritech region reduces the incremental capital outlay associated with entering local telephone services and makes them a viable future competitive alternative.

Throughout the Ameritech region, "nonwireline" cellular carriers provide nearly ubiquitous service. Although their "loops" may not currently provide a complete competitive alternative to Ameritech's loops, they are positioned to take advantage of the plan, as shown in Fig. 4.6. They have sophisticated switches and, in some cases, fully functional networks and office support in place that will allow them to lease loops from Ameritech and provide local exchange services with little incremental investment. Their ubiquitous presence allows them to access almost all of Ameritech's revenue base immediately through leased loops, and use their wireless capacity to substitute for these loops in the future. Cellular carriers also possess a select list of customers with a high demand for telecommunications ser-

55. *Multiple System Operator.* The top four MSOs are Time Warner, TCI, Continental, and Comcast.

vices. Cellular carriers also possess a select list of customers with a high demand for telecommunications services.

In the future, the combination of leased wireline access and wireless access may give the cellular carriers a unique advantage in marketing to customers. If they succeed in their drive to receive PCS licenses, wireless carriers would provide customers with three options for "loops."[56] Under one example, the cellular provider can position a cell-site directly adjacent to a wireless private branch exchange (PBX) serving a large corporate complex. The cellular carrier could handle local mobile traffic and serve as the local carrier for all interLATA traffic originating and terminating at the PBX. Although the coverage for the cellular portion of the traffic would be lower than for wireline traffic, the volume of traffic, combined with the absence of interconnect charges for the cellular carrier, would make up some of the gap.[57] With the imminent conversion to digital signaling for cellular, there are a number of cellular operators that will have significant excess capacity. They can market this capacity for use as simple local service. In fact, products are being developed to allow cellular operators to sell service to wireline customers that is transparent to the user. Tellular's "magic box," for example, would enable a cellular customer to easily interconnect a cellular hand unit to its inside wiring (merely by inserting the unit into its charger-base station), thereby routing all long-distance calls from any phone in the house over the cellular carrier, rather than the customer's wireline carrier.[58]

As these examples suggest, the Ameritech plan facilitates many different entry possibilities. Some new entrants may be better suited for niche plays; others may choose more comprehensive strategies. Both can coexist in the marketplace. Although Ameritech ought to be able to competitively respond to new entrants, entrants can target high-profit customers by supplying a small dedicated system catered to the specific customer's needs. This would make them more difficult for Ameritech to dislodge. Such niche plays are likely to be very successful because the plan enables the niche player to take advantage of Ameritech's scale economies.

IMPLICATIONS FOR ANTITRUST POLICY AND INTERLATA RELIEF

Compared to other economically advanced nations, the telecommunications policymaking in the United States is highly fragmented, with substantial involvement by the various states and, especially since the AT&T divestiture, a division of federal policy responsibility between the FCC, the U.S. District Court, and the De-

56. Goldman Sachs, analyzing the recent AT&T/McCaw deal, wrote that the "relationship opens up a major opportunity for McCaw to provide bypass services for AT&T" Goldman Sachs Investment Research, *The McCaw/AT&T Alliance*, November 24, 1992, p. 1.

57. See Goldman Sachs Investment Research, *The McCaw/AT&T Alliance*, November 24, 1992, p. 14, for an example of such a strategy.

58. See the discussion of Tellular Inc.'s "magic box." *Wall Street Journal*, October 4, 1993, p. B1.

partment of Justice.[59] In that environment, Ameritech has proposed a multijurisdictional proposal intended to further the transition to competition in local telecommunications. One of the chief difficulties facing Ameritech is that no single decision maker has the authority to approve all, or even, most aspects of the plan, even though there is a high degree of interdependence among the main elements of the plan. The business, economic, and policy logic underlying the plan is dependent on each of the integral policy components being implemented by the appropriate jurisdictions, the District Court, the FCC, and the Illinois Commerce Commission, respectively. If any one of the integral components of the plan is not adopted by one jurisdiction, for whatever reasons, adoption of the other parts of the plan could have negative economic consequences for Ameritech.

From Ameritech's perspective, the plan is intended to break the "logjam" over the interLATA line of business restriction. As competition in local telecommunications grows and intraLATA service becomes increasingly competitive, Ameritech sees itself boxed into a corner—namely, the LATA boundaries.[60] Its competitors are taking steps to offer bundled or integrated packages of communications services designed to meet specific customer needs. In return, Ameritech is seeking authority to offer interLATA services. It does so in recognition of the increasing tendency for customers, especially business customers, to purchase network services to meet integrated voice, data, and video needs on a global basis. Through acquisitions and alliances, AT&T and MCI will be able to provide end-to-end services to consumers, whether it be information transmitted by PCS, cellular telephony, cable, or fiber optics. These cooperative relationships have special competitive significance when they involve, as they often do, complementary assets and resources. The AT&T-based alliance spans telecommunications equipment and wireline and wireless services. It has enormous financial, human, and technical resources and one of the best known name brands in the United States. The BT–MCI-based alliance includes global and domestic interexchange services, cable and wireless interests, and strong relationships with many competitive access providers. Most recently, France Telecom and Deutsche Telekom have announced their intent to buy substantial equity interests in Sprint.

Of course, Ameritech is also developing global corporate relationships spanning wireline and wireless services (e.g., equity interests in New Zealand Telecom; Ameritech Mobile, a cellular carrier). There is a fundamental difference, however, between Ameritech on the one hand, and the AT&T, MCI, and Sprint agglomerations on the other. The interLATA restriction geographically consigns Ameritech and other RBOCs to local transport areas. In the evolving world of corporate communications and personal mobility, that is an enormous competi-

59. For an extended discussion of jurisdictional conflicts in U.S. telecommunications and a comparison to the "synoptic" policy regimes of leading competitor nations, see Harris, Robert G., "Telecommunications Services as a Strategic Industry: Implications for United States Policy," in *Competition and the Regulation of Utilities*, Michael A. Crew, editor, Boston: Kluwer Academic Publishers, 1990; and Harris, Robert G., "Telecommunications Policy in Japan: Lessons for the U.S.," *California Management Review, 31* (3, Spring), 1989.

60. For idiosyncratic reasons, LATAs in Illinois are called MSAs.

tive liability. The management of Ameritech realized that to succeed in the global telecommunications marketplace, it must also offer integrated network services to its customers. Yet the MFJ and the interLATA restrictions, prevent it from doing so.

Because the premise of that restriction is that the RBOCs can and would use their monopoly power in local telecommunications to compete unfairly in interLATA services, the Ameritech plan is designed to reduce or eliminate its alleged "bottleneck monopoly" over local access and exchange services. Other RBOCs are taking the position that, even without unbundling, growing competition in local markets, regulatory and antitrust safeguards, and the ability of IXCs to detect discrimination are sufficient conditions for removing the interLATA restriction. Conversely, the IXCs argue that the restriction should be removed only after the RBOC can demonstrate that there is full and effective competition for access and exchange services in its region. Ameritech is taking a middle position, which is that the "market opening" elements of their plan—unbundling the network and numbering, plus usage presubscription with the LATAs—are sufficient to reduce or eliminate the threat that it could compete unfairly in interLATA by leveraging its position in intraLATA services. The main rationales for their position are that:

1. There are several different options for an IXC seeking to obtain local switching services under the plan. The carrier can provide the switching itself, or obtain it from another party. The carrier can access the customer directly by cross connecting to its (or a third party's) switch or the carrier can access the customer through Ameritech's usage subscription plan, whereby the customer loop is connected through the Ameritech switch and all traffic is directed to the carrier (or third party). The carrier can also resell Ameritech's exchange service.
2. Even after the implementation of the plan, Ameritech might appear to retain some transitory market power regarding the local loop until cable companies, CAPs, PCS, and other wireless providers complete their drives to create competitive local "loops." However, sources of new competition act as a restraint on current pricing because of the importance of customer goodwill and the desire on the part of the provider to enhance its long-run profitability. Furthermore, even without the new competition, loops will be subject to pricing constraints such as price caps to prevent anticompetitive abuses.
3. The entry-facilitating nature of unbundling will generate a larger pool of competitors upon its implementation. Thus, if a hypothetical monopolist had control of the local exchange, and were to attempt to raise prices for local exchange services under the plan, the price increase could be quickly defeated in a number of ways, including entry from outside the local exchange business. Hence, the relevant market must include the capacity of firms outside the local exchange that can readily enter.[61] In evaluating competition,

61. This supply-side component to market definition is recognized by the Department of Justice and the Federal Trade Commission in the *Merger Guidelines*, April 2, 1992, p. 47.

one must at a minimum include a portion of the capacity that would be added by those classes of competitors presently outside the local exchange that could easily enter in response to a significant and nontransitory price increase in local exchange services.

4. Ameritech entry into interLATA services would be procompetitive, given the evidence of rising prices and increasing price discrimination by IXCs. MacAvoy, for example, has found that the price–cost margins of AT&T, MCI, and Sprint all have increased in four long-distance service groups (MTS, inbound/outbound WATS, and VNS) and are high in absolute value relative to those in other industries.[62] These high margins would explain the political dynamics of interLATA relief: both why the RBOCs so badly want to enter interLATA services and why the IXCs want to keep them out. So too does the growth and profitability differentials between local and long-distance companies: Whereas RBOC revenues are growing just 2.3% a year, and profits only .9%, the long-distance revenues of AT&T, MCI, and Sprint are growing at an annual rate of 4.3%, and their profits are rising 10% per year.[63]

The Ameritech plan is currently under review by the Department of Justice (DOJ). It is expected to make its decision on the merits of the plan sometime soon. Typically, there are negotiations between DOJ and petitioners that could result in modifications to the plan. If the DOJ decides in the affirmative, it would then recommend the plan to Judge Green, who might accept or reject the recommendation. In either case, his decision would likely be appealed by the "losing" party. At the same time, consolidated proceedings on the Ameritech plan, an AT&T petition and proposed staff rules for interconnection and intraLATA presubscription are underway at the Illinois Commerce Commission. There, the IXCs have argued that the unbundling, dialing parity and number portability and other elements of the Ameritech plan should be adopted, whether or not the company is granted interLATA relief.

The long delay in considering—much less implementing—the Ameritech plan is quite unfortunate, considering that it offers an opportunity to conduct a great experiment in local telecommunications. There is much debate about what the effects of unbundling the local network will actually be, with disparate predictions of outcomes. Yet the best that can be offered are predictions based on economic theory and empirical inferences, because the plan proposed by Ameritech has never been done before—here or elsewhere. With no experience in unbundling, the best way to test the proposition is to try it and see what happens:

62. Price cost margins across 284 industries averaged 27%, or about one third of the 1993 margins for MTS, inbound WATS, and outbound WATS, and about 52% of the margin for VNS; MacAvoy, Paul W., affidavit to United States District Court for the District of Columbia Civil Action No. 82-0192, June 6, 1994, filed in support of motion to remove interLATA restriction submitted by four RBOCs.

63. Measured over the years 1990 through 1994 (estimated). Kupfer, Andrew, "The Future of the Phone Companies," *Fortune*, October 3, 1994, pp. 95–105.

- Does entry into local access and exchange actually occur? Does the unbundling of the local network facilitate entry or not?
- Do the proposed rules promote interconnection and interoperability? Is there progress in creating the network of networks?
- Do prices and service offerings in interLATA services indicate healthy competition? Are there benefits from Ameritech's entry?

In assessing the pros and cons of such an "experiment," one should remember that any relief granted from the MFJ could be temporary, conditional on Ameritech's performance under the waiver. One would expect the IXCs—to say nothing of Judge Green—to be vigilant in their monitoring of Ameritech's compliance with the regulatory safeguards included in the plan. The threat of removing the waiver would presumably be strong incentive for Ameritech to avoid conduct that would justify such a result. Conversely, evidence from the experiment could help resolve the issue of interLATA restrictions once and for all.

5

The Regulation of Pricing of Interconnection Services

Bridger Mitchell
Charles River Associates

Werner Neu
Wissenschaftliches Institut für Kommunikations Dienste (WIK)

Karl-Heinz Neumann
Wissenschaftliches Institut für Kommunikations Dienste (WIK)

Ingo Vogelsang
Boston University

This chapter should be read with the decision of the European Commission to achieve a fully liberalized telecommunications market in the European Union by the end of the decade in mind. Interconnection is an essential precondition for reaching the goal of a viable competitive environment. To this end, the European Commission intends to provide a proper regulatory framework for the interconnection of networks and services. It has asked a study team, including the authors of this chapter, to provide the analytical basis for such a framework. This chapter consists of excerpts from the final report containing the results of this study (see WIK/EAC, 1994).

The chapter is thus a policy paper that aims to draw up a blueprint for action and develop the basis for a regulatory framework within which a regulator can deal with the pricing issues arising from the interconnection problem in telecommunications. As such it draws on the results of economic analysis and seeks to make the many insights from this analysis operational for actual application.

The chapter consists of two sections. The first reviews the economics of the pricing (and cost) of interconnection services and presents the reasons that compelled us to select the approach that we propose. The second section, in turn, derives from the results of the first guidelines for an actual regulatory policy for interconnection.

PRICING ISSUES OF INTERCONNECTION

Theoretical Pricing Models and Their Implications

A number of pricing rules have been proposed in the literature, based on theoretical models. These models make various assumptions, on which the relevance and range of their results depends. Because of the wide range of possible outcomes and because of the complicated interaction of variables that regulators find difficult to observe, the theoretical models are likely to provide insight but little practical applicability for the regulator.

The generic case for interconnection used in most theoretical models has an incumbent telecommunications operator (TO) that produces an intermediate service called *interconnection* or *access* and one or two final outputs (local and trunk services). The incumbent is either a monopolistic (bottleneck) supplier of the intermediate input or can be bypassed at some (higher) cost. An interconnector therefore either needs interconnection as an essential input or has some choice. Interconnectors sell a single final output (trunk services) that can be a perfect or imperfect substitute (or a complement) to the TO's final output(s).[1] Prices that need to be determined in the models include the two or three final output prices and the price for interconnection (and possibly for bypass). The models generally do not treat the case of unconstrained profit maximization by the TO. Rather, the TO is usually assumed either to maximize social surplus or to be regulated by a surplus-maximizing regulator, whereas interconnectors maximize profits. We now discuss some of the resulting outcomes.

Unconstrained Socially Optimal Interconnection Charges

Assume that the regulator sets all of the TO's prices optimally but can influence the interconnector's behavior only via the TO's pricing. The pricing outcomes under social surplus maximization without constraints then depend on the type of competition in the "trunk" market (Armstrong & Doyle, 1993).

If competition in the trunk market is of the Bertrand type and if all types of trunk services are perfect substitutes for each other, then interconnection charges should equal the marginal costs of interconnection services. Under Bertrand competition the outcome in the trunk market is fully efficient, so interconnection charges should be fully efficient as well. The reason is that under Bertrand competition firms take price as their strategic variable. The regulator will induce the TO to charge a marginal cost price in the long-distance market as long as the TO is at least as efficient as the interconnector. Otherwise, the TO will charge slightly more

1. With the exception of Willig (1979), the models generally do not treat network consumption externalities in the context of interconnection.

than the interconnector's marginal costs to assure that the interconnector serves the long-distance market at an efficient price. However, this result is of limited significance because trunk services are unlikely to be perfect substitutes for each other (and the same holds for other retail services in telecommunications).

If trunk services supplied by the TO and by interconnectors are imperfect substitutes and if interconnectors compete with each other in Cournot fashion then interconnection charges should be set below the marginal costs of interconnection services. Under Cournot competition (with output or capacity as the strategic variable) each firm has some market power and profit-maximizing firms will price above marginal costs. The reason for setting the regulated interconnection charge below marginal costs then is that the trunk services supplied by the interconnector have an inefficiently high market price that can be (partially and imperfectly) corrected through lower interconnection charges. Thus, the tendency of firms with market power to price above marginal costs is counteracted by reducing their marginal costs. This counterintuitive result is efficient, although not distributionally satisfying.

We conclude:

1. Interconnection pricing at or below marginal costs of providing interconnection services can be optimal.
2. Marginal cost pricing of interconnection is strictly optimal only for very specific cases.
3. Interconnection charges below marginal costs, which have to be financed out of taxes or higher prices elsewhere, are likely to be distributionally unacceptable.

Ramsey Pricing of Interconnection Charges

Pricing at or below marginal costs can lead to losses for the incumbent. Thus, a balanced budget constraint guaranteeing an efficient incumbent TO normal profits may need to be imposed on the regulator's objective function. If the incumbent TO does not itself provide retail telecommunications services (vertical separation), interconnection charges should equal average costs of interconnection services, as long as the balanced budget constraint is binding.[2]

If the incumbent does produce retail services and faces a balanced budget constraint then both the retail services and interconnection contribute to balance the budget. Under constrained social surplus maximization, contributions to the TO's profits have to be traded off against contributions to consumer surplus and to the interconnector(s)' profits. This leads to possibly complicated Ramsey pricing formulae. A regulator already finds ordinary Ramsey prices hard to deal with because (a) marginal costs and (super-) elasticities have to be measured and (b) consumers with inelastic demands oppose high prices associated with their purchases. In addition to the demand relationship between the final products, the Ramsey pricing

2. If the incumbent TO sells several interconnection services, the average costs of interconnection are no longer well defined. In this case mark-ups over marginal costs may optimally differ for the various interconnection services. This may become a problem for unbundling of interconnection services.

formulae with interconnection charges depend (a) on the type and intensity of competition between the TO and the interconnector(s), (b) on relative sizes of the firms, (c) on differences in costs of supplying the final output(s), (d) on the strength of the budget constraint, and (e) on the cost of interconnection. Most of these variables are extremely hard to determine even by insiders to the firms, let alone by a regulator.[3] In order to assess the importance of these terms, we briefly go through them.

In the simplest case of a dominant TO and a competitive fringe of other operators in the trunk market, the optimal interconnection charge a equals $a = MC_{IX} + (p/\eta)(\lambda/(1+\lambda))$, where MC_{IX} is the marginal cost of providing interconnection, λ is the Lagrange multiplier of the TO's budget constraint, p is the price charged by the interconnector in the trunk market, and η is the super elasticity faced by the interconnector. Reflecting the inverse elasticity rule, the less elastic the interconnector's trunk market segment, the higher the interconnection charge should be.

A more elaborate case that is typical for the Ramsey approach to interconnection charges is provided by Masmoudi & Prothais (1994, building on Laffont & Tirole, 1993a). They assume the incumbent is a vertically integrated TO that sells interconnection to an interconnector who competes with the TO in mobile services in a differentiated Cournot duopoly. In setting the interconnection charge and the incumbent's other prices the regulator acts as a Stackelberg leader, meaning that the regulator acts optimally, knowing that the market participants solve their own optimization problems (and the market participants take the regulated prices as given). The resulting formula for the interconnection charge is the sum of four components:

- MC_{IX}
- A Ramsey term as already mentioned, but including market shares and competitive interaction in the trunk market. The more competitive the trunk market, the smaller should be the mark-up of the interconnection charge over marginal costs of interconnection. Thus, a more competitive trunk market is treated like a market with a higher elasticity.
- A differential efficiency term. This term contains (a) the difference in efficiency between the incumbent and the TO, (b) competitive interaction, and (c) λ. The main effect is that the more efficient the interconnector relative to the incumbent, the more the interconnector should produce and therefore the smaller the interconnection charge should be.[4] On the other hand, the larger the concern for the incumbent's profit, the lower the concern for the interconnector's efficiency.

---

3. Under an informational constraint and for surplus maximization with welfare weights, Laffont and Tirole (1993a) derived informationally adjusted Ramsey prices. These prices, in addition, contain an incentive adjustment that may already be taken care of by other regulatory tools, such as the use of price caps for the TO's retail services.

4. If the final goods market is not regulated it usually increases social welfare to have a competitor enter and provide competition for the final stage even if that competitor is less efficient than the incumbent monopolist.

- An interconnection charge revenue effect. This effect is nonnegative and contains the reciprocal of the elasticity of the interconnector's output with respect to the interconnection charge. Thus, the less elastic this output is, the higher the interconnection charge.

Because the different parts of the pricing formula have different signs, the cumulative outcome cannot be assessed in general terms but only for specific parameter values. The authors therefore resort to simulation in order to evaluate the sensitivity of the outcome to parameter values. Their judgment suggests that the interconnection charge can exceed MC_{IX} by a substantial margin, but this is not a general result.

We conclude for Ramsey pricing models of interconnection:

1. Interconnection charges should be lower, the more competitive the trunk market, the more efficient the interconnector relative to the incumbent, and the less endangered are the incumbent TO's profits.

2. Although the principle of Ramsey pricing models is fairly straightforward, its explicit application to interconnection pricing leads to complicated and opaque results that do not lend themselves to straightforward prescriptions for regulatory action.

3. The Ramsey pricing models can, however, be used to assess whether markups for purposes other than universal service obligations (USOs) are advisable or not. Criteria suggested by these models include the competitiveness of the trunk market, the relative efficiency of interconnectors in supplying trunk services, and the financial condition of the incumbent.

Efficient Component Pricing for Interconnection

For a TO monopoly provider of interconnection services that it sells to a competitor of the TO's trunk services, Baumol (1983) and Willig (1979) developed the efficient component pricing rule (ECPR). Compared to the complicated Ramsey pricing rules for interconnection charges, developed by Laffont–Tirole and others, the ECPR has the attraction of looking simple and, according to its proponents, of being widely applicable and optimal. According to this rule, the interconnector should pay to the incumbent monopolist the "opportunity costs" of interconnection. These opportunity costs include the incremental costs of providing the interconnection plus the profit contributions that the monopolist forgoes by selling interconnection rather than retail services. Because the contributions foregone depend on the TO's retail price, efficient component pricing is based on retail prices.

The ECPR was described extensively by Baumol and Sidak (1994), who claimed four properties:

- It sends the right signal to potential entrants who will only enter if their services are more efficient than those provided by the incumbent.
- It is revenue neutral for the incumbent.
- It does not interfere with (desired?) cross-subsidization.

- It eliminates incentives for incumbents to keep rivals out. In particular, there is no incentive for a price squeeze because competition by interconnectors cannot hurt the incumbent TO.

The ECPR is highly controversial. It has been applied explicitly by U.S. regulators to railroad regulation and, less explicitly, by U.K. regulations to interconnection with British Telecom. It is under review as a competition policy standard for interconnection charges in New Zealand. Among economists, it has been backed by Kahn, Taylor, and von Weizsäcker. It has been vehemently opposed by Ergas, Ralph Tye, and W. Tye, and mildly opposed by Laffont and Tirole. Laffont & Tirole (1993a) argued that the ECPR is irrelevant because, under its assumptions, there would be no sales by the interconnector if the interconnector were less efficient than the incumbent and there would be no sales in the final goods market by the incumbent if the interconnector were more efficient. In the latter case, what would be the meaning of the incumbent's "opportunity costs" of interconnection as defined for the ECPR?

The claimed optimality of the ECPR is based on explicit or implicit assumptions, including the following:

- The retail trunk market is homogeneous.
- Interconnectors are price takers in the trunk market.
- Interconnectors have no fixed costs (only the TO does).
- The incumbent's prices in the local and trunk markets are Ramsey optimal.[5]

How restrictive are these assumptions? In a number of different cases analyzed by Armstrong & Doyle (1993), the ECPR survives as a principle, although the actual pricing outcomes may vary substantially by case. For example, if trunk services offered by the TO and by interconnectors are imperfect substitutes, an adjustment to the ECPR needs to be made in order to capture the effects of market power of the interconnectors and in order to capture the fact that the TO will not lose its trunk demand one-for-one when providing interconnection services. Whereas Armstrong and Doyle provided cases that still can be interpreted as the ECPR, Ergas & Ralph (1994) provided cases in which the ECPR is dominated by lump-sum interconnection charges. The reason appears to be that, in the Ergas & Ralph cases, interconnection charges are used to raise a specified amount of money for USOs. In contrast, Baumol and Willig had the costs of USOs embedded in efficient final goods prices.

The analysis by Ergas and Ralph also indicates a fundamental problem with the approach behind the ECPR. The ECPR assumes that the only socially beneficial role of interconnectors is to provide trunk services at costs lower than the TO. It otherwise assumes that regulation is perfect. In contrast, most economists would

5. The problem with this assumption is that it recognizes that the problem is really one of simultaneous optimization of pricing in all markets served by the incumbent. Because interconnection charges feed back into trunk charges, one cannot take the latter as given and then optimize with respect to trunk charges. Laffont and Tirole (1993a) claimed that there must also be a general assumption of constant returns to scale. This is not explicitly made, for example, in Baumol and Sidak (1994). Under this assumption the ECPR would coincide with Ramsey prices and marginal cost pricing.

tend to see the social function of competition by interconnectors in its ability to supplement or supplant inherently imperfect regulation. In particular, one function of competition is to determine prices. It is reasonable, under this view, to assume that prices in the markets served by the incumbent TO are inefficient and that competition by interconnectors may help correct this problem. However, the ECPR provides no tool to improve final output pricing. On the contrary, it invites price rigidity and collusion in the final output market.

In summary, we conclude:

1. The ECPR has been developed for an almost perfectly regulated monopoly situation that needs only some fine tuning. It does not deal with a situation where competition is to be introduced into a market dominated by an inefficient incumbent with heretofore insufficiently regulated monopoly rights.

2. The ECPR evolves from one term, the incumbent's opportunity cost of providing interconnection. This cost is potentially complex and hard to measure. The ECPR therefore provides no more guidance for the assessment of access charges than the Ramsey models.

Implications of the Theoretical Pricing Models for Interconnection

We believe that a fair summary of the theoretical models is as follows:

- The pricing prescriptions depend crucially on the assumptions made. The ranges for possible pricing outcomes in the Ramsey (i.e., Laffont–Tirole) and the ECPR (Baumol–Willig) frameworks are large and can extend above and below all reasonable cost measures.
- The results themselves are derived under a number of restrictive assumptions. In particular, the ECPR assumes price-taking behavior, an absence of sunk costs on the part of the interconnectors, and efficient pricing on the part of the TO. It is not a priori clear what the effects of lifting the assumptions are.
- The models, in particular ECPR, have little to say if prices in other markets served by the TO are not optimal.
- The more general Ramsey pricing models for interconnection provided by Laffont and Tirole are extremely hard to interpret and even more difficult to implement.

In summary, the guidance from the theoretical models is limited.[6] The models do agree on some outcomes for special cases. We may therefore be able to narrow the outcomes by observing certain market features and desiderata in the form of constraints not imposed in these models. These observations include the following:

- The main basis for allowing and promoting network entry is that economies of scale and scope are not so pronounced that they will make network competition infeasible or inefficient.

6. We have not explicitly discussed revenue sharing between interconnecting parties. Such an interconnection arrangement reduces conflicts and therefore moves them toward joint profit maximization. This may be advantageous if the parties face sufficient competition by others. If not, it is likely to lead to collusion.

- A second basis for allowing network competition is that entrants are likely to be more efficient than the incumbent. This would compensate for entrants' smaller size, should economies of scale continue to play some role.
- The results desired from network competition are lower resource costs to the economy and lower prices to customers. Changes in the incumbent TO's retail price structure as a consequence of network competition are inevitable.
- Even at interconnection charges as low as MCIX or average incremental costs of interconnection (AICIX) the incumbent is unlikely to lose its market quickly. Usually, there are sunk costs (that entrants have to expend), switching costs by customers, name recognition, brand loyalty, and other advantages the TO possesses that prevent consumers from switching to entrants even at substantially lower prices. For example, in the UK Mercury only gained about a 10% market share in its first 10 years.
- Pricing distortions (unbalanced tariffs) in other markets served by the TO are desirable only in exceptional cases and therefore deserve special scrutiny. Because the individual outputs should generally contribute different amounts to common and overhead costs of the TO, the existence of a valid "local access loss" cannot simply be established by using fully distributed costing rules.
- Making it difficult to add a contribution element to interconnection charges puts pressure on national regulators and TOs to rebalance retail rates and to justify payments for USOs to be incorporated in interconnection charges.
- Interconnection charges under a Ramsey rule or ECPR are extremely hard to calculate. This problem is compounded if interconnection services are unbundled (as we believe they should be). Such explicit calculations should therefore be the exception rather than the rule.
- Interconnectors offer services that are differentiated from the incumbent TO's services and from each other.[7] Differentiation can occur physically by delivery (wire or wireless) or through pricing policies (at a high price wireless services may be complementary to fixed-link services, whereas at a low price wireless services may be substitutes to fixed-link services). Using Ramsey models or the ECPR would lead to individualized interconnection charges, depending on the competitive relationship of interconnectors to the TO and other factors. Such severe price discrimination in the market for interconnection is likely to be inequitable, invite arbitrage, become arbitrary, and interfere with sound competition policy.
- A mature telecommunications market is likely to be characterized by price and service competition in retail markets. However, bottlenecks in reaching individual customers are likely to persist and with them the necessity to intervene in the market for interconnection through regulation or competition policy. Because of the diversity of uses for interconnection services, the

7. Price differences between incumbent TOs and entrants indicate that retail telecommunications services are virtually always differentiated.

principles for interconnection charges should be independent of the specific service created with interconnection.

From this we generally infer that interconnection charges should be cost based, with the basic standard being either MC_{IX} or AIC_{IX}. Incremental costs are the costs incurred by adding an output increment, where the increment can range from infinitesimal to a whole set of outputs. Thus, incremental costs have wider relevance for interconnection than marginal costs. For a purist, the incremental costs of interconnection would be all the costs incurred by the incumbent TO in addition to the costs of providing its other outputs. The average incremental costs would then be the incremental costs divided by the quantity in which the interconnection service is measured (e.g., in minutes, busy hour erlang, or call attempts). Our approach is less pure in that we take as AIC_{IX} the sum of average costs of the incumbent TO's facilities used for interconnection weighted by the interconnector's share in (busy hour) use of these facilities. The cost standard for pricing decisions that must generally be satisfied in any enterprise is the one of long-run costs. For an enterprise in a competitive market, this is the relevant standard, because without prices that cover at least all costs in the long run, an entrepreneur will sooner or later have to exit the market. Long-run costs should therefore also be the standard in the telecommunications industry.

The issue of contribution to overhead and common costs must be addressed as it affects the viability of the incumbent. Whereas the entrant's viability should, in general, not be increased by forcing the incumbent to provide interconnection below costs, the incumbent's viability may legitimately have to be safeguarded through interconnection charges above costs. Such a mark-up would be in line with the Ramsey approach already described and would have to depend on the demand relationships, the state of competition, and the seriousness of financial shortfalls. The burden of proof for justifying these mark-ups must be on the incumbent.

There is also an issue of the quality of interconnection services that needs to be considered in this context. If the incumbent's return from providing interconnection is substantially below that of providing final goods and if the two types of services are substitutes, then the incumbent may resort to hidden quality deterioration of interconnection. Such behavior may require costly regulatory action and monitoring that may be worse than higher interconnection charges.

Concluding from these observations:

1. We call for cost-based interconnection charges (based on MC_{IX} or AIC_{IX}).

2. We believe that cost-based charges should form the base-line case but that mark-ups above MC_{IX} or AIC_{IX} may be justified depending on the incumbent's legitimate revenue requirements.

3. The burden of proof for determining these mark-ups must be on the incumbent.

Cost-Oriented Pricing

In order to base interconnection charges on the costs of interconnection, one has to know what the costs are and how they are related to interconnection as a service. The incumbent incurs five types of costs for interconnection:

1. Costs of conditioning the system of the incumbent TO for competition.

2. Costs of establishing the physical interconnection between the two networks, as well as one-time costs of compatibility testing and making numbering and routing arrangements.

3. Costs of providing sufficient capacity for switching, transmission, and related network components to accomodate traffic from the interconnecting network at the TO's peak period.

4. Variable costs of call set-up, billing, and so on.

5. Overhead costs for accounting, management, legal expenses, and those overhead costs associated with the four other cost types.

How do these types of costs relate to interconnection services? Examples of the first type are costs of providing additional network security and of possibly far-reaching standardization measures made necessary by the very fact of interconnecting other systems, and the cost of installing equal access facilities. These costs are related to interconnection in general and not to a particular interconnection arrangement. The second type of costs, in contrast, occurs each time a new interconnection arrangement is made. The third type of costs is due to the use of facilities that are usually shared with other interconnectors and final users. The first three types of costs are essentially one-time capital costs. The fourth type of costs arises because of the shared use of resources. However, it varies directly with usage and it contains little or no capital costs. The fifth type again is shared and has only a small capital component.

Different types of interconnection services will incur these five types of costs in different amounts (actually, the first type of cost is not incurred by any particular interconnection service). So, the cost structure will differ between call conveyance, emergency services, directory assistance and so on, and there will be aggregation problems for the sum of services purchased by an interconnector.

Basing interconnection charges on costs means that the interconnection services have to be related to the five types of costs. Only the fourth and part of the fifth type of costs can be directly related to the quantity of usage conveyed via interconnection, where the quantity is the number of successful or unsuccessful call attempts. These variable costs are important but they are dwarfed by the first three types, which are all capital costs.

As already indicated, costs of the first type are assignable to the very fact that there is interconnection. They represent identifiable additional costs due to the introduction of competition, which are justified by the beneficial effects that competition brings to consumers. Given that all consumers, and especially the incumbent's customers using services similar to those that the competitors are going to offer, will benefit from these effects of competition, there is a good argument that all customers participate in paying for them. This means that the costs would have to be shared by both the interconnector(s) and the incumbent alike. One possible method of dividing the costs would be in relation to the shares of volume of the relevant businesses, which would initially burden the incumbent with most of them. As competitors gain market shares, the burden would decline as the costs would then be divided more and more evenly among all competitors.

The second type of costs is directly related to the particular interconnection agreement in question and is a one-time capital cost. These costs would have to be borne by the interconnector. Only if it is evident that these costs, which are similar to the first type in that they are additional costs of introducing competition, keep new competitors from entering the market should the regulator consider that they be borne by the incumbent. There is also an issue of cost sharing if the facilities can be used for interconnection in both directions and the incumbent TO can benefit from additional traffic generated by interconnection. The share of costs to be included in the charges for the interconnector must then be determined according to the relative weights of the traffic flows in question. Beyond this, this second type of costs does not usually cause major pricing problems. In principle, the interconnector could provide the facility itself if that were not interfering with property rights of the incumbent TO.[8] Because of such rights, the incumbent TO will normally want to build such facilities and will have the interconnector pay for them through a one-time payment or rental. The type of payment can reflect property rights issues and comparative advantage in financing. In particular, the incumbent TO may want the interconnector to pay a lump sum if the interconnector can readily choose interconnection locations and equipment (virtual collocation). An advantage of rental fees is that they can include follow-up costs of maintenance and can provide incentives for the incumbent to provide high-quality interconnection facilities.

The third type of capital costs is the most problematic, both in its large size and because the facilities are shared with others. Controversies concern the calculation of the costs and the way to charge them. In the following we therefore concentrate on these costs of shared network facilities. In this connection it is important to differentiate between two kinds of network capacity costs, nontraffic sensitive costs (NTS) and traffic sensitive costs (TS). Whereas TS costs vary with (busy hour) interconnected traffic, the NTS costs are unaffected by interconnection. These NTS costs include the lines connecting subscribers to TO local switching offices, parts of the local switches, and so on. They are not influenced by interconnection. However, they are used by interconnected traffic and the value of an interconnection agreement to an interconnector clearly depends on them. The question, therefore, is to what extent the interconnector should share in paying for these costs. Two views are worth considering. The U.S. practice has been to share the cost according to relative use (although the numbers were distorted in actuality). Contrary to U.S. practice, the costs could be collected as a lump sum from the interconnectors. The second view considers the prime beneficiaries to be the subscribers. Payment of NTS by interconnectors then becomes an issue of residuals not paid by subscribers.

We conclude:

1. Of the five types of costs associated with interconnection, three are fairly easy to relate to interconnection. They are the costs of conditioning the system of the incumbent TO for competition, the costs of establishing and maintaining the

8. Such property rights have been a major issue in the United States for physical collocation.

interconnection of the particular interconnector, and the variable costs associated with interconnection services.

2. The more difficult costs to deal with concern costs for network capacity that are shared with other users and overhead and common costs. (In the following sections we concentrate on these two types of costs.)

3. The share of the costs of conditioning the system of the incumbent TO for competition to be paid by the interconnector should reflect to what extent his customers (through his prices) are to pay for the introduction of competition. In deciding this, consideration must be given to the beneficial effect that competition will have on all consumers, including those who are customers of the incumbent.

4. All other types of costs related to interconnection are true input costs for the interconnector and should be borne in full by him.

Capacity-Based Versus Peak-Load Pricing

The fact that the TO's capacity is shared between various users raises the question about what shall be priced? Two categories of interconnection have gained prominence. First, treated in the current section is the question of defining the output "interconnection." What are the units in which interconnection shall be measured? This question leads to capacity versus usage pricing and the relationship of this issue to peak-load pricing. It also leads to an inquiry into the role of cost drivers. Second, not treated in this chapter, is the issue of unbundling interconnection services into smaller categories that can be sold and priced separately.

Pricing would be *capacity based* if a user paid at each point of time in relation to the depreciation charges for that part of the capacity that, either at the time of investment, or at the latest revaluation of the asset because of changing market conditions, was "reserved" for him. We need not go into a discussion of the usefulness and/or competitive implications of this idea when applied to prices for normal users. The enormous transaction costs involved would by themselves forbid such an approach. The consumption patterns of individual users are highly varied and almost impossible to predict (by the individuals and by the supplier). It would, however, not be unrealistic to apply capacity-based pricing to large users, in particular interconnecting network operators. Such users have predictable consumption patterns (derived from the law of large numbers). One probably beneficial aspect would be that in the long run the interconnecting operator would shoulder the true economic risk of the right expectations regarding demand of end users having been made, that is, the operator who is in a better position to evaluate the risk shoulders it. From this it also follows conversely that as long as interconnecting operators or service providers are new and small they should not be required to pay capacity-based charges. In their case the incumbent operator may actually be in a better position to assume the risk, especially if there are many such small demanders and the average risk per demander, due to the effect of correlation, is smaller than the individual risk.

Capacity-based costing would not preclude charging for operating costs on the basis of actual usage with which this kind of cost varies. Nor would it preclude, of course, charges based on actual usage if the latter exceeds the capacity that was reserved for the demander.

Capacity-based costing is also commensurate with pure peak-load pricing. Under peak-load pricing, the capacity paid for is the amount needed at the peak. The charges due for using more services at the peak than had been reserved would have to equal the share of capacity covered by the capacity price plus a charge for the extra costs caused by exceeding that share. Charges during off-peak periods would have to cover only operating costs. Because interconnecting operators' total costs are highly sensitive to interconnection charges and because such operators themselves make their output pricing decisions based on such charges, it is important that the system peak be accurately reflected in peak-load pricing of interconnection.

The capacity costs relevant for interconnection depend on whether capacity expansion is required or not. If no capacity expansion is required we recommend using the TO's historic stand-alone costs of the network as the basis. The proportionate share in capacity use by the interconnector during the busy hour would be a simple and acceptable allocator. This would mimic the result of joint ownership of network capacity in proportion to peak use. If capacity expansion is required, the basis for charging should be the incrementall cost of expanding capacity.

Total network capacity costs can increase nonlinearly in network capacity. In particular, there may be increasing returns to scale. Pricing network capacity at its marginal costs could then lead to insufficient cost coverage. We believe that the best way to avoid this problem is to price capacity (both for peak-load pricing and for capacity-based pricing) in proportion to peak capacity utilization. This corresponds to pricing by AIC of capacity. Information on increasing returns to scale is hard to come by, and two-part tariffs or other nonlinear pricing schedules are likely to burden entrants.

In practice, charges for call conveyance (conveyance charges) are often priced by the minute, and off-peak charges are distinctly positive, in spite of the almost total absence of usage-related network costs. One argument is that these off-peak prices reflect overhead costs. However, any overhead costs that are not taken care of in one-time interconnection charges, in charges for call attempts, and in capacity charges are likely to be very small. What remains, then, is an argument that the time profile and sizes of peaks are uncertain. But given the regularity of system peaks, this argument is certainly not convincing. It is more likely that per-minute charges have simply been transplanted from end-user charges to interconnection charges and that minutes of use are easily measured.[9]

We consider capacity-based interconnection charges to be the optimal approach for interconnection between a sophisticated TO and a sophisticated interconnector. In practice, however, it may be difficult to move directly toward such capacity-based charges. Also, the demands of interconnectors on the charging system may differ, depending on whether the interconnector is a TO, a mobile operator, or a

9. Nonlinear per-minute pricing schedules have also been discussed in the literature. See Mitchell and Vogelsang (1991).

service provider. We therefore conclude by suggesting a flexible and optional approach to the type of charging.

1. If possible, capacity-based charges should be offered, and they could be applied either ex ante or ex post. *Ex ante* application would mean that the interconnector and TO would agree on the busy hour and the busy hour contribution(s) to be paid by the interconnector. This might require both parties to make point estimates, and there could be penalties for exceeding limits and bonuses for staying below. *Ex post* application of capacity-based charging would be very similar to peak-load pricing, only that it would apply to the TO's actual system peak hour rather than to a predefined schedule.

2. As an alternative or option, sophisticated peak-load pricing of interconnection services would be offered. This would be based on the expected system peak. Uncertainty could be taken care of by spreading the capacity charges according to the probability with which the system peak occurs at different hours of the week.

3. As a further option the interconnector could choose to be charged according to the time-of-day schedules offered to retail customers. These schedules are likely to be unsophisticated and driven by concerns other than the system peak (e.g., regular business hours).

Access Charges and USOs

Normally the incumbent operator has no alternative but to accept when the regulator significantly changes its opportunity set by opening the market. When interconnection is ordered, the operator is allowed to charge interconnection charges that in principle would be cost oriented and possibly are derived on the basis of capacity costs. It would probably include common costs and average costs of capacity. As regards opportunity costs, the possible exception is to include an additional contribution element in the interconnection charge to collect revenues not directly derived from the costs of interconnection. Such additional charges are termed *access charges* in the European discussions of interconnection policy.

In our view, the regulator's position on the access charge should depend on evaluation of the extent to which opening of the market has in fact eroded the regulated operator's market power and will in the future prevent it from reaping extra benefits, either in terms of supranormal profits or in unwarranted X-inefficiency. In other words, during the period of transition from monopoly to effective competition, one should leave the incumbent with the obligation to shoulder the burden of the local network cross subsidy, to the extent that the regulator has convinced itself that the incumbent is still enjoying advantages that competitors do not have. If these remaining advantages are not considered sufficient, there may have to be some additional source of revenues to help to cover the costs of USOs.

Note that this approach rejects the notion that with the mere formal opening of the market a kind of parity is reached between incumbent and new competitors. Under that notion, if tariffs are unbalanced the market entrants should automatically be considered contributors to the funds needed to fill any gap and thus incur the same competitive handicap, so that not requiring them to contribute automatically means that they are provided with entry assistance. The contrary notion is that the continued obligation placed on the incumbent to cover the costs of USOs, at least in the short and medium run, is to be seen as an additional means to make entry conditions commensurate with what would exist if the incumbent had not enjoyed its protection for decades in the past. In addition, not including specific access charges in interconnection charges puts pressures on the incumbent TO and on the regulator to rebalance the retail rate structure,[10] to justify additional burdens carried by the incumbent TO in terms of USOs, and to find other ways to pay for USOs. The incumbent TO may actually have comparative advantages in fulfilling USOs, and that can be determined during this initial time period.

How can one know whether there is a local access loss or not? This obviously depends on how such a deficit is defined and how it can be measured. One straightforward definition is that a local access loss exists to the extent that a profit-maximizing TO would want to cease supplying access to customers at the current tariffs. Contrary to widespread practice and political arguments, this definition is directed at customers rather than at individual services. The reasons we prefer this definition are:

- Whether or not access is a separate service, it is jointly consumed with other services by the same TO subscriber. Consumers benefit from low connection and rental charges which may translate into higher usage (via the network externality). Thus, the TO usually finances local access costs jointly from connection charges, rentals and usage fees.
- It is incremental (avoided) cost based and thereby avoids arbitrary cost allocations.
- The TO may even benefit from subscribers that are individually not cost-covering but increase usage by others.

Thus, the customer (group) is the relevant unit of observation for measuring an eventual local access loss. Whether a local access loss actually exists can then be established in a two-step procedure: First, calculate whether the incremental costs of subscriber classes and combinations of subscriber classes are covered by the revenues from those customers. If all subscriber classes (and combinations) cover their incremental costs, there can be no local access loss. Second, if there are any subscriber classes whose incremental costs are not covered, calculate the TO's total return from all network services on network assets, based on interconnection charges without any access charges. If this return is at or above the TO's cost of capital, then the local access loss has been covered elsewhere and does not need to be covered by interconnection charges. In principle, all these calculations should be made for the case after the effect of interconnection has been factored in. However, the effects of interconnection on market shares and sales of the incumbent TO are usually gradual (for the case of no vertical or horizontal divestiture of the TO). Therefore, current figures can be used instead of projections. It

10. In our view, the current Access Deficit Contribution practice in the United Kingdom, where BT loses its right to ADCs if it does not fully utilize its ability to rebalance rates, could invite BT to price consumer access above cost.

only has to be assured that interconnection charges can be changed in case of large changes in the TO's market share.

Although we clearly prefer a customer-specific approach to USOs, many countries have been using a service-specific approach. As the reason for this, we conjecture a mixture of regulatory inertia (fear to change price structures even if the average customer ends up paying the same as before) and accounting convenience. If one accepts that regulators in the short run are unable to move to a customer-specific approach to USOs, the local access loss may also have to be measured for local access as a service. In this case the direct costs of local access should be calculated and compared to the income from subscriber connection charges and line rentals paid by subscribers. Any resulting deficit (and the costs of general overheads) then needs to be shared by all the remaining services, of which interconnection is only one. It becomes a regulatory decision as to how this sharing should occur, but it is not clear a priori that interconnection should bear proportionally more than any other service. This is where the regulator may be guided qualitatively by the Ramsey approach. Recall that the required mark-up under this approach depends on the demand elasticities, the amount that needs to be raised and the state of competition. In our view, however, the regulator may want to use the desired competitiveness of the sector rather than its actual competitiveness to determine the mark-up for interconnection charges.

If a decision has been made to cover USOs or a local access loss through interconnection charges, it still needs to be addressed how this can best be done. In their opposition to the ECPR (mentioned earlier) Ergas and Ralph (1994) provided an example of lump-sum access charges. Such lump-sum charges may not be acceptable due to the high costs they may impose on small entrants. However, under presubscription to competing trunk carriers, there exist more efficient and acceptable charges in the form of charges paid on the basis of numbers of customers interconnected rather than on the basis of calls conveyed. In this way, a long-distance company would pay a fee for each of its customers that need to be accessed via the incumbent TO. Similarly, the incumbent TO would impute the same charge to each of its customers. These access charges could be passed on to the customers in any way deemed optimal by each of the operators (and would be in addition to the connection charge currently collected by the TO from all its customers).

We conclude:

1. Access charges are not our preferred way of financing USOs, including local access losses.

2. A rebalanced tariff structure would make unnecessary a large share of current USOs. It would make economic sense if regulators use the time between starting an interconnection regime and the time access charges become a pressing issue to rebalance tariffs and find other ways of financing USOs.

3. If that cannot be achieved, the standards for calculating the amounts to be raised by access charges should be high and the burden of proof upon the incumbent.

4. If access charges are nevertheless necessary, they should be imposed in the least distortionary manner, preferably not on a per-minute basis.

REGULATORY TREATMENT OF INTERCONNECTION PRICING ISSUES

The Role of the Regulatory Authority With Respect to Interconnection Charges

There is a wide range of possibilities of involvement by the regulatory authority (RA) with respect to charges for interconnection services. This involvement may range, at the one end, from leaving the fixing of charges completely to negotiations between the parties concerned and, at the other, to determining them on the basis of the RA's own evaluation of relevant costs and market conditions. In our empirical analysis covering 16 countries we have found examples of both extremes.[11] The presumption underlying the analysis of this chapter is that the RA has the mandate to exercise its regulatory control over interconnection charges but that the involvement should be differentiated depending on the requirements of the cases in question.

Individual interconnection arrangements and the corresponding charges may prima facie be left to negotiations between the parties. The roles that the RA could assume in the context of such negotiations are:

- Participate in the negotiations as a facilitator.
- Initiate arbitration if negotiations threaten to fail.
- Make ex-post determinations if negotiations do in fact fail.

As a facilitator of negotiations, the RA may be present at the meetings, either as an observer or an adviser to prevent the negotiations from getting stalled or proceeding in a dead-end direction. When using the instrument of arbitration, the RA may assign the role of arbitrator to outside parties. This approach has certain advantages over one in which the RA itself makes a determination.

Irrespective of the scope available for facilitation of negotiations and arbitration by alternative agents, there should be a right to an ex-post determination. So that the prospect of the potential use of this instrument develop its full effect, the RA should indicate clearly before negotiations start what the standard is on the basis of which its determination would be made (see the following sections for what this standard should be). There should also be a policy statement regarding whether charges that have been agreed on without active intervention of the RA will need to be approved, and, if this is the case, what the criteria are for such an approval. If there is no explicit approval procedure, the charges may nevertheless be subject to the RA's scrutiny for anticompetitive conditions.

The adjustment over time as cost conditions of individually negotiated charges change may be done by negotiating anew at specified time intervals or at times when prespecified conditions are fulfilled. There is then no reason to proceed differently than when the charges were negotiated the first time. It may also be agreed between the parties that charges will adjust following a scheme like the price-cap approach. If the RA had to pass a determination, it might impose the price-cap regime if it expects that on the next occasion there may be no prospect of a negoti-

11. See WIK/EAC (1994).

ated result. If there is no reason for the latter expectation, the establishment of new charges may again be left to negotiations between the parties.

Socially-Optimal Versus Cost-Based Interconnection Charges

We have established the need for regulatory intervention whenever a telecommunications network operator has control over bottleneck facilities and interconnection would result in the realization of substantial positive network externalities. The very presence of network externalities prevents a normally functioning market mechanism from achieving solutions that are optimal from a social point of view, for the market mechanism is, by definition, unable to take externalities into account. This problem still needs to be faced by the RA after it has guaranteed the right to interconnection and imposed the duty to offer it whenever bottleneck conditions prevail.

Our economic analysis has demonstrated the difficulties of deriving socially optimal interconnection charges. The results depend on a whole range of different, plausible conditions, and their calculation would also require the availability of information that is normally not at the disposal of the RA. This conclusion holds in particular for the much discussed ECPR and the sophisticated versions of Ramsey pricing, so-called second-best pricing approaches that provide pricing rules that are socially optimal after one has taken into account constraints dictated by reality.

We came to the conclusion that, although socially optimal pricing rules allow much insight, the attempt to implement them outright would in all likelihood be infeasible. In the following we therefore propose an approach that we think would achieve a reasonable approximation to results that follow from these rules.

The RA should require a methodology for determining interconnection charges that is based on costs. Costs should be basically divided into two parts: costs caused by the service in question (or, more realistically, those that can be traced or attributed to the service), and costs that cannot be so traced and are therefore common costs. The first type of costs would determine the lower limit for the charge of a particular interconnection service or facility. The second type of costs would have to be covered through contributions from all services where the corresponding percentage mark-ups on direct costs would, however, for good reasons not necessarily be equal across the different services.

In the economic analysis of the previous section, we identified the standard of long-run average incremental cost (LRAIC) as the one that best meets the regulatory requirement of a standard for the direct cost of interconnection services. This standard should be applied to all categories of services supplied by the TO. All costs not accounted for when long-run incremental costs of all services are added up (i.e., the sum of amounts arrived at by multiplying LRAIC by the volume of the relevant service) should be counted as overhead and common costs. This would in particular also include the difference between the historic costs of a service irreversibly sunk in the past (because investments made in the past are irreversible) and the costs of the service evaluated at current, possibly lower prices of inputs.

Interconnection charges set at LRAIC would fail to provide contributions to the regulated firm's truly common costs and other justified revenue requirements. Therefore mark-ups on this cost standard should be allowed on the basis of feasible Ramsey pricing. This would require that in setting markups above LRAIC for interconnection services, one takes into account the relevant market conditions for all services (in particular also end-user services) and the mark-ups for all these services as well. In other words, mark-ups for interconnection services should not be determined in isolation. Setting the relative mark-ups differently according to perceived market conditions would then amount to a rough and ready application of the Ramsey rule.

The percentage mark-ups on top of LRAIC for interconnection services should vary between zero, as lower limit, and, as the upper limit, the minimum uniform markup, that is, that common mark-up that, when applied to the LRAIC of each service, would lead to revenues that cover all costs, including common costs, and all other revenue requirements. Mark-ups for interconnection services should be constrained within this range because in the market for the typical end-user services for which interconnection services are needed there is generally an above average degree of competition (meaning an above average price elasticity), which in the Ramsey calculation would make for lower mark-ups. Furthermore, lower interconnection prices mean lower input prices for imperfect competitors, which will intensify competition and thereby increase welfare. This also argues for the mark-up for interconnection services to be less than the average mark-up. At the limit, when the degree of competition is very high, a mark-up of near zero may indeed be appropriate.

From this, we conclude:

1. The RA should not aim to impose interconnection charges that claim to correspond exactly to socially optimal prices.
2. The RA should define lower and upper limits within which interconnection charges must be set.
3. The lower limit of an interconnection charge should be that of LRAIC.
4. The upper limit of an interconnection charge should be a charge calculated by adding to LRAIC a markup that, when applied to the LRAIC of each service, would lead to revenues sufficient to cover all revenue requirements (minimum uniform mark-up).

The Mechanisms for Arriving at Interconnection Charges

We divide the discussion in this section into two parts, one dealing with the case in which there are private party negotiations with the aim of agreeing on interconnection charges for an individualized set of services, the other with the case in which charges for a standardized set of interconnection services are to be fixed by way of a proposal and approval process between RA and TO.

On entering negotiations, parties to a prospective interconnection agreement should be instructed by the RA to find charges that are within the bounds defined in the previous section: The lower limit should be the LRAIC and the upper limit should be the LRAIC plus the minimum uniform mark-up. The RA should indicate that, if no agreement was reached and it were asked to make an ex-post determination, it would determine a charge in that range on the basis of its assessment of the demand conditions in the market. This would provide proper incentives for the two sides in the negotiations. Not knowing what the RA would do in case of failure of negotiations, both would have a preference to reach a settlement on their own accord. Of course, if either party speculated that it would have a good chance that its view on charges would be confirmed by the RA, it might opt to let negotiations fail and rely on the RA's decision. This would have to be accepted as a legitimate part of the process.

As regards charges for standardized services, the fixing of these charges would have to wait until agreement on the relevant set of these services had been reached, which would most probably occur only after some time and after consultations involving demanders as well as the TO and the RA. The interconnection charges would then have to be set in a process in which the TO submits a proposal for the charges and the RA examines and approves them. They, too, would have to fall within the bounds defined earlier. Most probably, the process could profit from the prior experience on charges in private party negotiations and possibly ex-post determinations by the RA. Conversely, the standard charges, once established, could serve as benchmarks for individual negotiations in the sense that any interconnector could request to be served under them.

For the sake of supporting the emergence of competition, there may be a case for relatively low mark-ups on LRAIC in the case of charges for standardized interconnection services. It is granted that this approach would at first provide smaller contributions to common costs and the revenue requirement of the incumbent TO than is warranted. The regime that would provide for the adjustment of charges over time as demand and cost conditions change (preferably the price-cap regime) could then be specified in a way that the firm is able to generate sufficient contributions over time. As mentioned, this approach would amount to an explicit policy of facilitating competitive entry. The extent of entry assistance would abate with time, however, as the price adjustment regime would increase the regulated firm's margin between charges and costs.

Our conclusions are:

1. The parties in negotiations for interconnection charges should be given ranges within which the charges are to be fixed.

2. The lower limit should equal LRAIC and the upper limit LRAIC plus the sufficient common percentage mark-up.

3. If negotiations fail, the RA should determine charges that fall within the given range using its assessment of demand conditions on the different markets.

The Provision of USOs

In general, the placing of USOs on TOs and the method of financing them need not be a topic for designing an interconnection regime. The relevant regulatory measures could be discussed completely outside the confines defined by interconnection issues. The principal reason that USOs are important in the interconnection context is the so-called local access loss that arises when connection and recurring charges that end users in total pay do not cover the cost of the local loop. The argument is that the loss that local telephone companies incur thereby should in part be recovered from access charges paid by competing carriers and service providers as part of their interconnection charges.

Our position is that burdens placed on TOs in the name of USOs should be compensated out of a general universal service fund (USF). Here USOs would mean all obligations, including the local access loss. In principle the USF could be financed from any source, however, the realistic approach would be to require all telecommunications network operators and service providers to pay into it on the basis of their volume of activity in the market. The approach would have the important advantage of decoupling the issue of interconnection charges from the issue of financing USOs.

The scheme has a number of additional merits. For one, it broadens the focus by considering who actually supplies USOs. As already indicated, this need not be the incumbent TO alone but may also be one or more of the new competitors. Furthermore, attention is drawn to the total costs of USOs. This is important information for all who have to worry about their justification. Once one knows total costs of all USOs, independent of who has met the obligations, one could determine to what extent the various TOs or service providers are to shoulder the burden of financing them. This should be determined on the basis of the overall ability to pay. Presumably, incumbent TOs, with their advantages accumulated in the past over many decades and their continued ability to charge tariffs in excess of costs, are better placed to contribute to the USF than new market entrants. The effect of a policy taking this into account would be similar to the one intended by the U.K. regulation not to require market entrants to pay access deficit contributions as part of their interconnection charges until a certain market share is reached.

If the approach of a USF is not feasible and the costs of USOs are to be covered from interconnection charges, these costs should then be treated like a common cost. Interconnectors' contributions to them would be obtained through the mark-up above LRAIC, as discussed earlier. The implicit access charge would thus be included in the mark-up.

Independent of whether a USF or access charge approach is used, the costs of USOs—the amount to be recovered—need careful determination. This is particularly valid for the USO considered to be the most important, that is, the local access loss. The following is the outline of a suggested methodology: Determine a forward-looking cost measure for the local loop, differentiated by types of local networks (metropolitan, medium city, small city, rural) and taking particularly into account existing infrastructure such as ducts and other very long-lived facilities. The costs of large parts of local networks are sunk costs that were incurred far in the past and that will not recur for a long time in the future. For this evaluation, special studies would need to be carried out, as the normal cost accounting systems would not be able to provide this kind of differentiated information. Having made the determination on costs, one would need to decide to what extent the sunk costs of the past should be considered having already been covered by past profits. One could presume this if the TO in question had made particularly large profits in the past. Only after having also answered this question one would proceed to calculate an access loss to be covered from current sources.

This determination should be considered as one of the most difficult tasks of the RA, particularly if the access loss is to be covered from access charges. The RA must make a very critical choice between, on the one hand, justified demands of the incumbent for a fair return on invested capital and, on the other, demands for supporting an effective competitive process.

We conclude:

1. Instead of using access charges, there is a strong case for the financing of all USOs from a USF into which all telecommunications network operators and service providers pay on the basis of their volume of activity in the market.
2. For the fixing of contributions to the USF (or an access charge), a very careful calculation fo the costs of USOs to be covered needs to be done.
3. Careful calculation of the local access loss would particularly be relevant. This would involve the establishment of the proper cost measure for the local loop (forward looking, based on current prices) as well as considerations regarding how much of these costs could be considered to be or have already been covered by the incumbent's supranormal profits from other services, currently and in the past.
4. If USOs have been measured and need to be covered by access charges they should be treated like common costs.

CONCLUSIONS

Theoretical models of welfare-optimal interconnection pricing call for use of the ECPR or for complicated types of Ramsey prices. We find these models are excessively demanding for regulatory use. The preconditions for the ECPR are too restrictive and exact Ramsey prices are too hard to implement in practice. Instead, we find that interconnection pricing policy can be based on reasonable lower and upper bounds within which negotiations for interconnection charges should settle and that would mark outer limits for a regulatory determination.

The lower bound is the LRAIC of the services used by the interconnector. The upper bound results from adding to LRAIC the minimum uniform mark-up that would result if the incumbent TO's legitimate revenue requirements were added proportionally to the LRAIC of all the TO's services. These revenue requirements, which would have to be justified by the incumbent, can include common and overhead costs, differences between historical and forward-looking costs, and local access losses or other USOs.

We prefer not to have access charges used to finance USOs, including local access losses. However, if access charges are deemed necessary, they should be imposed in the least distortionary manner. Capacity-based charges and fixed per-user charges, rather than per-minute usage charges, can most efficiently recover network and overhead costs of interconnection.

Over time, as the market for interconnection develops, regulation of interconnection should decline and give way to competition policy as a safety net. How can interconnection charges be gradually deregulated? In our view, the path toward deregulation will provide for pricing flexibility and the establishment of competition policy standards for interconnection charges.

Flexibility would first be introduced in the form of price caps. Over time, the price caps would be increasingly freed from rebalancing and restructuring constraints, and optional tariffs could be introduced as alternatives to capped interconnection tariffs. Then the scope of price caps would be changed. Paradoxically, both an increase and a decrease in their scope can add flexibility to a TO's pricing policy. An increase in scope allows the firm to do more restructuring, because now more different prices can be traded off against each other. A decrease in scope can increase pricing flexibility, because those prices outside the cap are constrained only by the market. The regulator will choose whether to deregulate by reducing the number of constrained services, or by constraining all services less tightly.

At the same time that pricing flexibility is increased, competition policy standards should be imposed for prices that are not considered compatible with competition—normally those below incremental costs or above stand-alone costs. If challenged under competition law, a price falling outside this range could only be justified by factual evidence of social benefits.

REFERENCES

Armstrong, M., & Doyle, C. (1993, December). *Network access pricing* (Report to HM Treasury).

Baumol, W. J. (1971). Optimal depreciation policy: Pricing the products of durable assets. *Bell Journal of Economics and Management Science.*

Baumol, W. J. (1983). Some subtle issues in railroad regulation, *International Journal of Transport Economics, 10,* (1–2).

Baumol, W. J. (1991). *Modified regulation of telecommunications and the public interest standard.* London.

Baumol, W. J. & Sidak, G. (1994). *Toward competition in local telephony.* Cambridge, MA: MIT Press & American Enterprise Institute.

Boiteux, M. (1956–64), Marginal cost pricing. In Nelson (Ed.), *Marginal cost pricing in practice*, Englewood Cliffs, NJ: Prentice-Hall.

Ergas, H., & Ralph, E. (1994). *The Baumol–Willig rule: The answer to the pricing of interconnection?* [mimeo]. Canberra, Australia. Trade Practices Commission.

Laffont, J.-J., & Tirole, J. (1993a, December). Access pricing and competition. Paper presented at Charges d'Acces, France Telecom and Institut d'Economie Industrielle, Universite des Sciences Sociales de Toulouse, Toulouse, France.

Laffont, J.-J., & Tirole, J. (1993a), *A theory of incentives in procurement and regulation*, Cambridge, MA: MIT Press.

Laughhunn, D. (1989). *Understanding the paradox of fully distributed cost.* Fuqua School of Business, Duke University, Durham, NC.

Littlechild, S. (1970). Marginal-cost pricing with joint costs. *Economic Journal.*

Mitchell, B. M., & Vogelsang, I. (1991). *Telecommunications pricing—Theory and practice.* Cambridge, UK: Cambridge University Press.

Park, R. E. (1989). *Incremental costs and efficient prices with lumpy capacity: The single product case.* (Research paper.): Rand.

WIK/EAC. (1994, November). *Network interconnection in the domain of ONP, Study for DG XIII of the European Commission* (Final report). Bad Honnef.

Willig, R. D. (1979). The theory of network access pricing. In H. Trebing (Ed.), *Issues in public utility regulation,* (pp. 109–152). East Lansing: Michigan State University.

6

Local Exchange Competition: Alternative Models in Maryland

Richard L. Cimerman
Geoffrey J. Waldau[1]
Public Service Commission of Maryland

This chapter examines alternative models for local exchange competition and interconnection rates. In July 1993, Metropolitan Fiber Systems–Intelenet, Inc. (MFS–I) applied for authorization to provide and resell local exchange and interexchange telephone service to business customers in Maryland.

The incumbent local exchange carrier (Bell Atlantic–Maryland) did not oppose competition but argued that the interconnection rates that entrants pay should cover direct incremental costs plus foregone contribution to the incumbent's shared and common costs. This foregone contribution would include lost revenue from toll and access (but exclude lost revenue from enhanced vertical services) when a business customer switches to MFS–I. The theoretical basis for this position is the efficient component pricing model (ECPM) developed by Willig, Baumol, and Sidak, presented on behalf of Bell Atlantic–Maryland (BA–MD) by Kahn and Taylor (1993). This model recommends that the optimal input (access) price should equal the input's direct per-unit incremental cost, plus the opportunity cost to the input supplier of the sale of a unit of input.

The Maryland Commission staff argued that interconnection rates should include long-run incremental costs (LRIC) plus make some minimal contribution to incumbent shared and common costs. Universal service support funding should not be included in interconnection rates but instead be targeted and effectuated through a separate mechanism. Requiring entrants to compensate the incumbent for large amounts of lost contribution would be onerous on new competition, low-

1. The views expressed in this chapter are solely those of the authors and do not necessarily reflect views of the Public Service Commission of Maryland. There is an abridged and extended version of this chapter. In the extended version, local exchange competition alternatives are discussed in more detail (e.g., network configurations, unbundling, collocation, number portability, traffic routing, and compensation).

er the probability of entry, and not incent the incumbent toward greater efficiency. The theoretical basis for Staff's position is the dynamic firm interconnection rate model (DFIRM), a series of examples that establishes the interrelationship between interconnection rates, contribution to common costs, efficiency improvement, and probability of entry. These two competing theoretical models are compared and contrasted.

Entrants and incumbents will have relative advantages and disadvantages. For example, entrants will be able to offer interLATA toll. Staff proposes that entrants temporarily be allowed some nondominant competitor advantages (e.g., customer targeting and pricing flexibility). The incumbent will have advantages in name recognition, a ubiquitous right-of-way network, and the ability to spread overhead costs over a larger customer base. They will also have the disadvantage of some obsolete or overvalued plant. These relative advantages and disadvantages would be difficult to quantify precisely. Regardless, the interconnection rate is an important factor in the competitive parity equation. Rough competitive parity exists when neither entrant nor incumbent has a significant net competitive advantage over the other.

In April 1994, the Maryland Commission laid out the initial terms and conditions for business local exchange competition.[2] MFS–I was granted co-carrier status. They will pay a 6.1 cent local access charge per call to BA–MD for calls originating from MFS–I customers that terminate on BA–MD's network, and they will receive a cost-based local access charge (established in April 1995 to be 6.1 cents) for reverse traffic. A proceeding is underway to consider opening the residential market to local exchange competition from a cable company. A future proceeding will consider a support mechanism for ensuring universal service.

INTERCONNECTION ARRANGEMENT MODELS

MFS–I could interconnect with BA–MD in many ways. There are important technical and operating differences between the alternative forms of interconnection. MFS–I can function as either a reseller or as a co-carrier. As a reseller, BA–MD would treat MFS–I's switch as a shared tenant service (STS) provider. As a co-carrier, BA–MD would treat MFS–I's switch in the same manner as an end office operated by an independent telephone company.[3]

Commission's Order: Reseller and Co-Carrier Models

The Maryland Commission found "that it is in the public interest to grant operating authority to MFS–I to provide local exchange service to business customers as a co-carrier, and as a reseller of BA–MD's local exchange service in Maryland"

2. Order No. 71155, Case No. 8584, April 25, 1994.

3. New York also permits true local exchange competition whereby a competitive carrier may compete for local access and collect an access charge for local originating and terminating calls. Proceedings for true local competition are underway in Pennsylvania, Michigan, and Illinois. Several other states permit resale of local services, but not true competition for local access.

(1994, p. 23). The Commission ordered BA–MD to offer MFS–I (a) interconnection to tandem; (b) where justified by the volume of traffic, central office switches; and (c) at such time as expanded interconnection is established for interexchange access, expanded interconnection for local exchange access. The Commission recognized MFS–I's market niche strategy as a legitimate form of competitive strategy.

Commission Order: Unbundling and Number Portability

The Commission ordered, "Conceptually, we approve of the idea of unbundling links and local exchange ports. However, we will defer any decision regarding unbundling until Phase II, which will permit us to examine more detailed information about implementation of the unbundling and to set appropriate rates for the unbundled elements" (1994, p. 36). The Commission further ordered that MFS–I may implement Flex-DID utilizing BA–MD's existing DID and PBX tariffs. The Commission indicated that any long-term solution would likely be the result of industry-wide discussions at the national level, and therefore, BA–MD and MFS–I were not expected to devise a long-term solution on their own.

INTERCONNECTION RATES

In the Maryland proceeding, the most contentious issue was how to set interconnection charges that MFS–I would pay Bell Atlantic–Maryland (the LEC) for its connection to the BA–MD ubiquitous network. If incumbent and entrant are of equal size and equal bargaining strength, then possibly they would negotiate suitable interconnection arrangements and rates to each other's networks without regulatory intervention. If interconnection is authorized, but the terms are unreasonable or discriminatory, then competition could be foreclosed just as effectively as if interconnection were denied. Bell Atlantic–Maryland, Staff, and MFS–I proposed alternative interconnection rate methodologies.

Bell Atlantic–Maryland

Bell Atlantic–Maryland (BA–MD) witnesses Professor Alfred E. Kahn and Dr. William Taylor proposed using the Efficient Component Pricing Model (ECPM) as the economic model for setting interconnection charges paid by entrants to the incumbent.[4] With this general guidance, BA–MD proposed that the interconnection charge paid by MFS–I include foregone contribution from toll and access, but exclude contribution from enhanced vertical services.

Under BA–MD's proposal, the appropriate interconnection charge would be equal to the sum of BA–MD's direct cost of providing MFS–I with access to the network and the opportunity cost of the contribution foregone by providing access. BA–MD proposed recovering these costs using two types of charges. First,

4. For an extensive discussion of the ECPM, see Baumol and Sidak (1994).

BA–MD proposed a monthly per-line charge of $14.47 called the competitive contribution charge (CCC). The remaining $20.48 in foregone contribution would be expected to be recovered from MFS–I for completion of local calls over the BA–MD network (because MFS–I would pay switched access charges to BA–MD to terminate these calls). The actual amount collected by BA–MD would be greater or less than $20.48 depending on the actual number of calls terminated. Cost differences to terminate local or toll calls would affect these rates.

BA–MD backed into the CCC charge using the ECPM by starting with the average revenue BA–MD receives per business line for the dial tone line, local usage, intraLATA toll and intrastate access services (i.e., $46.70 per month).[5] From this, BA–MD subtracted the monthly amount it would save in direct incremental costs for these services if the line is disconnected ($12.81). The difference ($33.89) represents the monthly loss of contribution per business line served by MFS–I. BA–MD then added the estimated direct cost of access ($1.06) to the $33.89 to arrive at the total interconnection charge of $34.95. According to BA–MD, this represents both the alleged subsidy to residential service and to the network's shared and common costs.

BA–MD proposed raising the CCC charge for any interconnection charges MFS–I charges BA–MD. According to BA–MD, these increases would be necessary to maintain what BA–MD identified as the "historical contribution levels" that BA–MD foregoes when it provides MFS–I interconnection.

In the New York arrangement, MFS–I will pay a $50 universal service element (USE) per trunk, which supports numerous individual lines, plus a small cost-based charge for the physical facilities to connect with New York Telephone. The New York USE differs from the CCC in that USE includes the lost contribution from only the specific element the LEC no longer provides. It does not include any other revenue that may have been generated from the lost lines.

BA–MD argued that unbundled links are comparable to voice-grade private line local channels that are already offered under tariff at a monthly rate of $16.75, but if required to provide such channels to MFS–I, BA–MD should be permitted to deaverage their prices to reflect variations in cost. BA–MD generally contended that its outside plant costs are lower (on a per-line basis) in high-density areas such as downtown Baltimore than they are in suburban and rural areas.

BA–MD agreed that if MFS–I were to assure (through an auditable procedure) that local and interexchange traffic would be routed on separate trunk groups so that the correct rates could be applied to each, then BA–MD would not assess the Carrier Common Line Charge on local traffic and would adopt a modified access rate of 6.1 cents per message for this traffic. The 6.1-cent rate would be the per-message equivalent of BA–MD's per-minute switched access rates.

5. Revenue per business line for all sources including vertical services, hi-cap, private line, and inside wire (but excluding Yellow Pages) is approximately $71.

Maryland Commission Staff

Staff proposed an alternative interconnection rate model, the DFIRM. This model describes the interrelationship between interconnection charges, contribution to common costs, efficiency gains, and probability of competitive entry. The main assumption of DFIRM is that competition will drive out excessive monopoly costs. The main conclusion is that a high interconnection charge as indicated by ECPM would discourage entry and jeopardize the beneficial effects of competition. Therefore, the interconnection charge should cover long-run incremental costs, which would include direct costs and make some minimal contribution to incumbent shared and common costs. Universal service should be supported through explicit subsidy mechanisms, and not included in interconnection rates.

Staff proposed that BA–MD be required to offer tariffs for unbundled links and ports. Under Staff's proposal, the combined port and link rates would equal the existing business dial tone rate plus the federal subscriber line charge. The business dial tone line rate is cost based in Maryland, thus, compensatory port and link rates could be established using the business dial tone rate. Staff also proposed that BA–MD file tariffs for physical (or virtual) interconnection between BA–MD switches and the MFS–I switches. Compensation for these interconnections should be calculated using the same methodology used by the Federal Communications Commission (FCC) in its special access and switched access collocation orders. The FCC methodology uses the incremental cost of collocation plus a reasonable allowance for overhead loadings. MFS–I would be charged for interexchange switched access at the same rates and in the same manner as those charged to interexchange carriers (IXCs).

Finally, Staff proposed a 4-cent per call charge for terminating local calls based on BA–MD direct cost data plus an appropriate mark-up. Because all of Staff's proposed rates would be set by way of a tariff, the rates would apply equally to all newly entering local exchange carriers, as authorized by the Commission.

Staff argued that BA–MD's proposal for interconnection charges that include foregone contribution continued the extraction of monopoly profits. Monopoly profits can exist in two forms: excessive earnings that benefit company shareholders, or excessive costs that benefit company employees. Although the Commission can limit excessive earnings, it is difficult for the Commission to see or limit excessive costs and inefficiencies. Regulatory agencies are simply not equipped to discover and ferret out all inefficiencies.

MFS–I

MFS–I proposed unbundled link and port tariffs where the sum of the unbundled link and port rates equals the previously bundled dial tone rate. For interexchange calls, switched access charges should apply based on meet-point billing arrangements similar to the agreement between BA–MD and Armstrong Telephone Company, the only small LEC in Maryland.

MFS–I proposed a "bill and keep" arrangement for terminating calls. Each carrier would bill its local usage charges to its own customers and keep the resulting revenues. Hence, each carrier competing within the local area would be required to terminate each other's calls at no charge (as opposed to Staff's recommended 4-cent charge). The originating carrier would be responsible for the cost of transporting the traffic to the terminating carrier's network. MFS–I suggested administrative savings and fewer billing disputes. Under the MFS–I proposal, there would be no tariffed interconnection rate for terminating local calls.

MFS–I would lease transmission facilities used to establish interconnection with BA–MD at tariffed rates if MFS–I chose not to build these facilities itself. Interexchange calls would be terminated at tariffed rates.

Allowing MFS–I to use its interconnected facilities for local and intrastate access traffic would have little benefit if MFS–I still had to pay BA–MD bundled rates for access services that included the cost of the connecting trunks that BA–MD was no longer providing. Thus expanded interconnection would have little value if BA–MD did not also restructure its switched access tariff to the extent necessary to unbundle the cost of dedicated transport facilities from other elements of switched access. The rates resulting from such a tariff provision need not be established using the same methods that the FCC approved for the interstate transport structure, as long as the rates for dedicated transport circuits are unbundled from other elements of transport.

MFS–I proposed that the price for access to a local co-carrier's network should depend on whether the traffic being transmitted is local (i.e., within the boundaries of a local calling area) or interexchange. For interexchange traffic, the same interconnection access rates should apply that are paid by IXCs who terminate this type of traffic on the local exchange network. Therefore, MFS–I should pay BA–MD's switched access charges (including the carrier common line charge) for all interexchange traffic it terminates on BA–MD's network, and BA–MD should pay MFS–I a tariffed access charge (which should be identical to the charge that MFS–I applies to IXCs) for all interexchange traffic it terminates on MFS–I's network. These rates should be unbundled so that MFS–I does not pay twice for the same thing. For example, if MFS–I provides its own transmission facilities to a BA–MD tandem, it should not also pay local transport charges designed to recover costs of BA–MD transmission facilities. MFS–I suggested that the Commission could require changes to MFS–I's access tariffs if market forces were not effectively constraining rates.

Commission's Order: Interconnection Rates

The Commission (1994) ordered BA–MD to file tariffs for expanded interconnection:

> to be based on the methodologies established by the FCC for interstate access. Since Phase II of this proceeding will deal primarily with local exchange issues, we expect to review BA–MDs tariffs for expanded interconnection for intrastate interexchange access at an Administrative

> Meeting where they will be accepted or set for hearing if the reasonableness of these tariffs is contested by MFS–I or any other party. (p. 10)

The Commission order does not rule on switched access charges for interexchange traffic.

The Commission ruled that MFS–I should pay BA–MD 6.1 cents per call for local calls that originate from MFS–I customers and terminate with BA–MD customers. MFS–I should provide justification for the amount to be collected from BA–MD for local calls that terminates with MFS–I customers. The CCC was not accepted. The Commission stated tariffs will be subject to "further consideration" or fine-tuning in Phase II:

> We are not persuaded that BA–MD's application of the Kahn/Taylor theory, which is aimed at efficient entry, rather than determining any cost support necessary to maintain universal service, reasonably will achieve this goal, at least as presently applied by BA–MD. Instead, other avenues of determining the amount of contribution needed, and methods of collecting it, should be explored. (p. 57)

Interpretation of Commission's Order

There are several possible interpretations of the Commission's Order. It appears that the Commission intends the interconnection rate of 6.1 cents to include some funding for BA–MD's fixed and common costs and support for universal service, but not to the extent requested by BA–MD. Hence, the Commission does not compensate BA–MD for all lost revenues, some of which could be monopoly excess. The Commission believes that the ordered interconnection rate would satisfy 60% of BA–MD's claim for contribution from MFS–I.[6] Both MFS–I and BA–MD stated to the press that the Commission's decision was a good one.

The Commission considered both models and established an interconnection rate in between the two. Staff's proposal to develop a separate universal service support mechanism is to be considered in Phase II. One interpretation of the Commission's order is that the 6.1 cents is to provide all contributions for fixed and common network costs and universal service. Any future implementation of a separate funding mechanism to support universal service would be coupled with a decrease in the 6.1 cents. Another interpretation is that any mechanism would be in addition to the 6.1 cents. The imposition of additional monetary flows from entrant to incumbent could be critical to entrant viability.[7]

---

6. The 60% figure, calculated as approximately $20/($20 + $14), is closer to Staff's DFIRM recommendation. The Commission's 6.1 cents gives BA–MD 30% of the difference between what Staff recommended (4 cents) and what BA–MD recommended (6.1 cents plus CCC). These percentages do not reflect offsetting charges for termination of BA–MD's customers' calls on MFS–I's network. MFS–I must file cost-based tariffs with supporting cost information.

7. According to MFS–I, interconnection rates are much more critical than the form of interconnection (e.g., physical or virtual colocation).

Other modifications to the interconnection rate are possible. The interconnection rate revenue corresponding to 6.1 cents per call could be obtained instead from a two-part tariff on both calls and lines. The interconnection rates could be different for competitors that supply business and residential local exchange service.

A Railroad Example

BA–MD witnesses Professor Kahn and Dr. Taylor illustrated the ECPM with a railroad example whereby an entrant pays an interconnection rate that includes incremental costs plus lost contribution to shared traffic-sensitive costs.[8] They concluded that the interconnection rate is irrelevant to competitive parity as long as the incumbent's prices to end users exceed the interconnection rate. In their view, only the margin matters.

Staff developed an alternative model to ECPM called the DFIRM. To illustrate DFIRM, Staff built on the Kahn-Taylor example by including nontraffic-sensitive (NTS), fixed, and common costs. Staff concluded that the interconnection rate is relevant and affects the entrant's ability to compete. The higher the interconnection rate, the more efficient the entrant must be to compete. If the entrant cannot meet an extraordinary efficiency standard, it cannot compete successfully and will leave the market. The beneficial effects of competition are not obtained.

Model

Railroad A has nontraffic-sensitive (NTS), fixed, and common costs equal to \$50. Shared traffic-sensitive costs are \$5 per ton. Direct cost for each leg (i.e., Leg 1, Leg 2) is the sum of track wear costs plus operating costs (e.g., for Leg 1, wear costs = \$.75 and operating costs = \$2.25)

Economic costs for Railroad A (t = tons)
= (NTS + Fixed,Common) + shared + Leg 1 costs + Leg 2 costs
$= 50 + 5t + (.75 + 2.25)t + (.5 + 1.5)t$
$= 50 + 10t$

Railroad A ships 10 tons and sets its price at \$15 per ton so that it breaks even (e.g., revenues equal economic costs, including normal profit).

$$Pt = 50 + 10t$$
$$P = 15$$

Railroad B owns track on Leg 1 and must lease track for Leg 2 (a bottleneck asset) from Railroad A.

Assume Railroad B's NTS, fixed, and common costs are expected to be one half (i.e., \$25) of Railroad A's because B is smaller and owns only one leg. Assume

8. Bell Atlantic–Maryland witnesses Kahn and Taylor (1993).

Railroad B is more efficient than Railroad A. Railroad B's NTS, fixed, common costs, and operating costs are lower by a multiplier factor of $e < 1$. Assume both Railroad A and Railroad B are equally efficient in causing railroad wear.[9]

Economic costs for Railroad B (if it owns track on both legs and operates both legs) = $e(50 + 5t + 2.25t + 1.5t) + 1(.75 + .5)t$

Railroad B does not own track on Leg 2 and it is more economical to lease. Under the Kahn-Taylor proposal, Railroad B compensates Railroad A for direct wear costs ($.50 per ton) and shared costs ($5 per ton) for trackage rights on Leg 2.[10]

Costs for Railroad B (owns and operates Leg 1 and compensates A for trackage rights to operate on Leg 2) = $e(25 + 5t + 2.25t + 1.5t) + .75t + (5 + .5)t$

It is possible to compute the efficiency multiplierfactor e that is necessary for Railroad B to be able to compete for traffic (on both legs) under the Kahn–Taylor proposal.

Assume for simplicity that Railroad A determines the market price. Assume the market expands by 4 tons and Railroad B attracts 4 tons of the expanded market. Railroad B can charge the same market price as Railroad A (i.e., $15 per ton) and breakeven when:

Revenues for B = cost for B = B's own cost + payments to A

$$15t = e(25 + 5t + 2.25t + 1.5t) + .75t + (5 + .5)t$$

Solving for e when t(tons) = 4

$$e = .58$$

Under the Kahn–Taylor proposal, Railroad B can compete with Railroad A when Railroad B is much more efficient (i.e., when $e = .58$). Railroad B cannot compete with Railroad A unless Railroad B has extraordinary efficiency. Hence, a more efficient Railroad B could have difficulty competing with Railroad A, which must be substantially compensated for lost contribution, unless Railroad B is substantially more efficient.

DFIRM proposes lesser compensation—direct incremental costs plus some minimal contribution.[11] It is possible to compute the efficiency multiplier factor e that is necessary for Railroad B to be able to compete for traffic (both legs) under

9. The model can include a more generalized cost function, more cost components, and a separate efficiency factor for each component. The main conclusions remain unchanged.

10. The Kahn–Taylor ECPM railroad example assumed shared traffic-sensitive costs but did not explicitly assume NTS or common costs. In the modified railroad example presented here, ECPM-based interconnection rates could range from $5.50 to $15. The exact rate is not important to the main conclusions.

11. In the modified railroad example, DFIRM-based interconnection rates could range from $.50 to $5.50. The exact rate is not important to the main conclusions

Staff's proposal, whereby Railroad B compensates Railroad A for direct wear costs ($.50 per ton) on Leg 2 plus some minimal contribution (e.g., $.50 instead of $5) to shared and common costs. Railroad B can charge the same market price ($15 per ton) and break even when:

Revenues for B = cost for B = B's own cost + payments to A
$15t = e(25 + 5t + 2.25t + 1.5t) + .75t + (.5 + .5)t$

Solving for e when t(tons) = 4
$e = .88$

Now Railroad B must only be slightly more efficient than Railroad A to compete.

In Staff's example, the absolute level of the trackage rights or interconnection charge—that is, $(5+.5)t$ or $(.5+.5)t$—is relevant to competitive parity because of the presence of NTS, fixed. and common costs. Both the interconnection charge and the entrant's efficiency affect the entrant's ability to compete.

If Railroad B can be expected to have no NTS, fixed, and common costs (i.e., $0 instead of $25), then competitive parity is more closely achieved under the Kahn–Taylor proposal.[12]

Revenues for B = cost for B = B's own cost + payments to A
$15t = e(0 + 5t + 2.25t + 1.5t) + .75t + (5 + .5)t$

Solving for e (which now does not depend on t)
$e = 1$

Hence, if Railroad B has no NTS, fixed, and common costs, then Railroad B can be as efficient as Railroad A on traffic-sensitive costs, pay a Kahn–Taylor interconnection rate that includes Railroad A's direct cost plus foregone contribution to shared costs, and be at competitive parity with Railroad A.

Dynamic Model Scenarios. The model presented so far is a static model at one point in time. In contrast, a dynamic model considers what happens over time. For example, suppose in response to competition, Railroad A sheds its old monopoly excess and reduces its price below $15. Over time, Railroad B also must increase its efficiency and/or volume to remain competitive. The tracing of these dynamic interactions and efficiency gains over time is missing from the static model. Dynamic models can be easily seen with the use of graphs to show change over time. Figures 6.1 through 6.9 show market dynamics for entrants and incumbent.

---

12. Other commentors on the ECPM reach a similar conclusion. See Tye, (1994): "The authors [Baumol and Sidak] explicitly or implicitly make a number of specific assumptions in reaching these conclusions ... .The nonintegrated competitor is assumed to have sunk no costs [footnote omitted], (or is assumed to have no need to recover them through efficient pricing of access)" (p. 206).

Fig. 6.1.
Entrant–incumbent relative efficiency ranges.

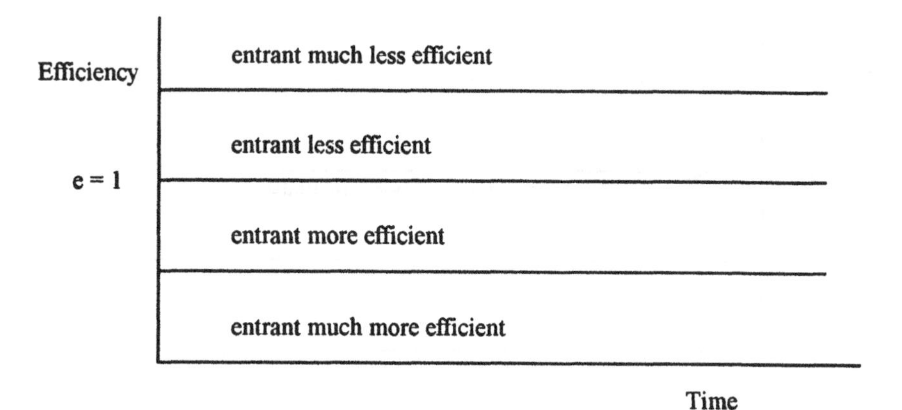

Figure 6.1 shows the four levels of entrant efficiency relative to incumbent.

Figure 6.2 shows how with a low interconnection charge, the entrant can compete and improve its efficiency. The entrant's initial efficiency is low because it must work out operating bugs. Also, initial capital suppliers require a higher return to compensate for start-up risk. Over time, the entrant improves its efficiency. As the entrant improves efficiency, so does the incumbent. The entrant begins with low output, spreads its overheads over few units of output, and loses money. Over time, the entrant builds up market share, works out inefficiencies in its operations, and begins to make money. The incumbent, sparked by competition and the desire

Fig. 6.2.
Low interconnection charge—Efficiency entry.

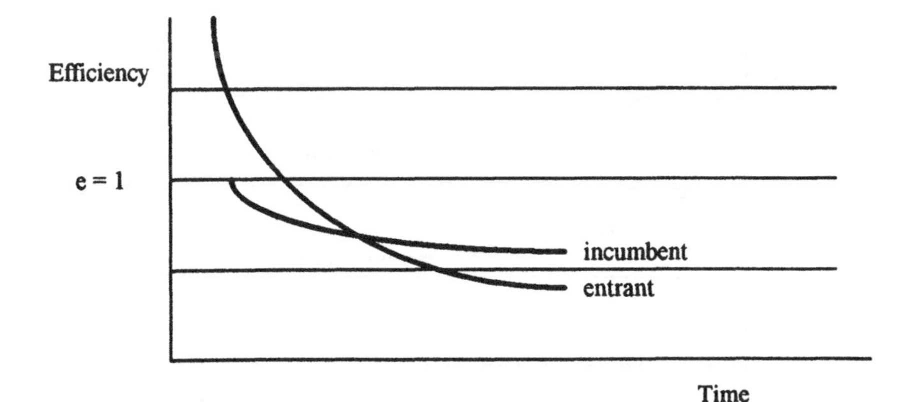

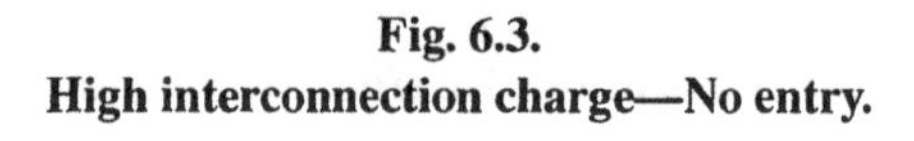

Fig. 6.3.
High interconnection charge—No entry.

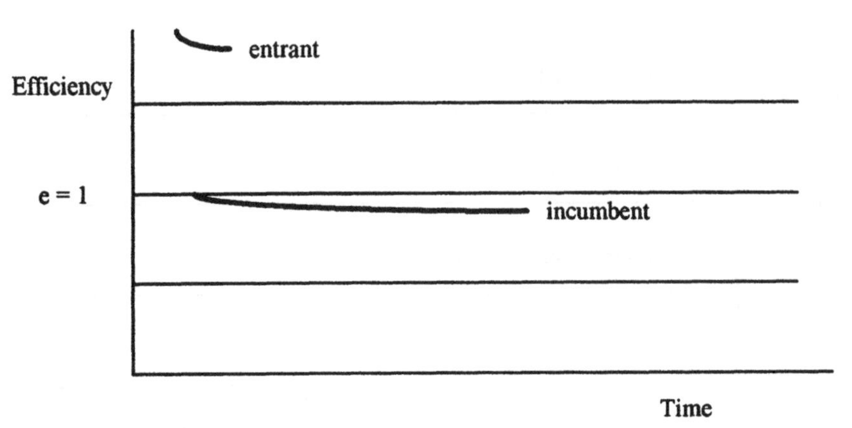

to prevent further loss of market share, finds more ways to improve its efficiency. In this scenario, although the efficiency of both entrant and incumbent improve, the entrant's efficiency is eventually greater than the incumbent's efficiency.

Figure 6.3 shows how with a high interconnection charge, an entrant either does not attempt to enter or leaves the market. The entrant can not attract sufficient market share. The time necessary for the entrant to break even is too long. The entrant loses money and leaves the market. The incumbent, without competitive pressures, either does not improve its efficiency or improves at a lesser rate than with competitive pressures.

Fig. 6.4.
Low interconnection charge—Inefficient entry.

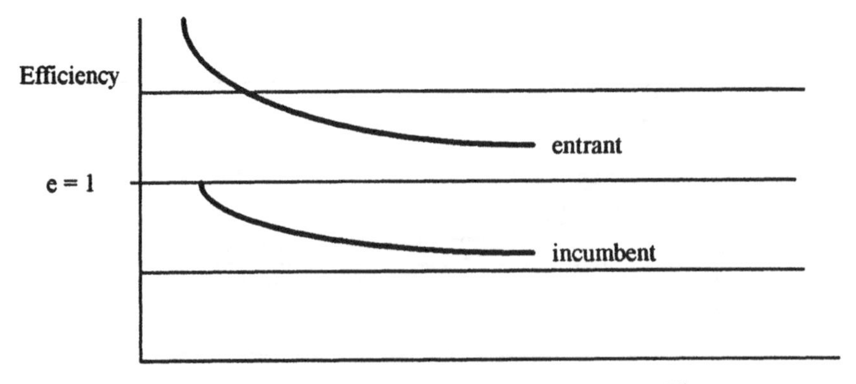

Figure 6.4 shows how with a low interconnection charge a less efficient entrant is able to capture market share. The incumbent, sparked by competition and the desire to prevent further loss of market share, finds ways to improve its efficiency at a greater rate than if there is no entry. The incumbent improves its efficiency, but the entrant never becomes more efficient than the incumbent.

The Fig. 6.2 scenario has positive net benefits. The Fig. 6.3 scenario has zero net benefits. The Fig. 6.4 scenario has indeterminant net benefits.

Probability-Weighted, Discounted Cost–Benefit Analysis

Setting an interconnection rate may be interpreted in terms of an investment. If entry is encouraged, and does occur, some fixed and common costs will be duplicated in the "network of networks."[13] Investment in duplicate costs could include duplicate facilities such as the drop wires leading into individual customer locations and duplicate management structures. However, the "investment" in these fixed costs could be more than made up by the probability-weighted expected benefits of greater overall managerial and infrastructure efficiency, variety, and innovation. Hence, a DFIRM-based low interconnection rate could be viewed as a seed investment made by a regulator in future benefits from competition.

If the data was available, one could conduct a probability-weighted, discounted cost–benefit analysis. Such data would be difficult to obtain. However, even without the data to conduct such an analysis, it is still possible to reach preliminary conclusions about likely scenarios.

If the interconnection rate is set low, the probability of the positive net benefit scenario (Fig. 6.2) or the indeterminant scenario (Fig. 6.4) is high. If the interconnection rate is set high, the probability of no entry or early exit is high (Fig. 6.3) and the benefits of competition are not obtained. If the "positive" and "indeterminant" scenarios are given equal probability, then the low interconnection rate policy could be optimal.

Retail and Wholesale Competition

The ECP theory was perhaps developed for resellers (Baumol & Sidak, 1994). Resellers can enter the market quickly with little sunk investment and can act as retailers. Retailing of final products to final customers is considered a "component" under the ECP theory. The ECP theory is concerned that the outcome of the competition be determined by the relative efficiency of competitors performing component functions. Relative marginal costs are important. Common costs of entrants should not matter because these costs are assumed to not exist or be miniscule. The assumption that entrants have little or no common costs may be true for resellers of interexchange services.

The assumption would not be true for a facility-based local exchange carrier like MFS–I. MFS–I provides switches, electronics, and fiber cable facilities. MFS–I

13. The duplication of costs can be minimized through unbundling.

performs both retail and wholesale functions. MFS–Telecom provides dedicated access to end users and to carriers that may in turn sell to end users. MFS–I proposes to provide switched access to end users and to carriers that may in turn sell to end users. MFS–I may someday sell local exchange services to other companies that in turn may resell these to residential customers. Hence, MFS–I may also be involved in wholesale functions in this application. In not considering fixed entrant costs, the ECP model is perhaps misapplied to "co-carrier" local exchange competition.

BA–MD Criticism of DFIRM and Reply

In testimony, Kahn and Taylor criticized Staff's model as creating "infant industry protection" that is unwarranted. In brief, BA–MD criticized DFIRM because it does not calculate a precise interconnection charge. Staff pointed out that economic models are used for different purposes. Some models are descriptive, others prescriptive, some provide precise estimates of parameters of interest, others provide general direction and guidance. The DFIRM model is of the latter type. It explains the interrelationship between interconnection rates, contribution to common costs, probability of entry, and efficiency. The main conclusion of DFIRM is that the high interconnection charge proposed by BA–MD would discourage entry and jeopardize the beneficial effects of competition. DFIRM does not recommend infant industry protection or favoritism because long-run incremental costs are covered by the interconnection rate. The model provides general policy guidance and is not intended to calculate a precise number. Staff's interconnection rate estimate of 4 cents was developed in accordance with that guidance.

Economic Welfare Analysis

ECPM and DFIRM are not based on an economic welfare analysis. Under the ECP theory, incumbent common costs become "opportunity costs incurred by the supplier in providing the product (access)" (Baumol & Sidak, 1994). The claim of efficiency and optimality for the ECP "input-pricing rule" is not based on an economic welfare analysis such as Ramsey pricing or optimal tariff pricing. DFIRM does not address allocative efficiency.[14]

ECPM and DFIRM are essentially theories about the allocation of common costs between firms. It is well accepted by economists that cost allocation is arbi-

14. For an allocative-efficiency model, see Armstrong and Doyle (1994). The Armstrong–Doyle (A–D) model derives a three-term optimal access charge that maximizes social welfare. The first two terms correspond to the Baumol–Willig ECPR. However, a third term indicates that when (incumbent and entrant) products are imperfectly substitutable (i.e., differentiated), the optimal access charge should be lower than the Baumol–Willig rule. Hence, the ECPR is only allocatively efficient under special circumstances. The A–D model also considers that when there is a fixed cost to entry, there may be insufficient incentive to enter (i.e., privately unprofitable but socially desirable) and some further entry assistance (e.g., lower access charge) may be optimal.

trary in the sense that many possible divisions of joint and common costs can be contemplated.

Dynamic Competition and Public Policy

Both ECPM and DFIRM may have strengths in particular circumstances (e.g., ECPM for single-component producers and resellers; DFIRM for multicomponent producers with substantial sunk costs).[15] The weakness of ECPM as a sole telecommunications policy tool is not the prescribed interconnection rate, but the assumptions necessary to make the prescription valid for co-carrier applications. ECPM essentially focuses on static interfirm cost allocation and not on the truly important dynamic economic issues of interconnection rate determination—probability of entry and the probable subsequent impact on efficiency and innovation. ECP's static focus does not consider fixed and common costs of the entrant, nor the role of competition in promoting dynamic efficiency over time.

In contrast, DFIRM has a dynamic focus and should be considered by policymakers in addition to ECPM. DFIRM argues that entry should be encouraged because inefficient firms can place competitive pressures on each other, drive out excessive costs, and become more efficient over time. Entry is in the public interest when regulated monopoly fails to achieve the lower costs that should be achieved.

A procompetitive telecommunications interconnection rate policy should recognize that co-carrier entrants will also have fixed and common costs, entrants cannot produce extraordinary efficiency in the short-term, and incumbents can and will achieve some of the efficiency gains needed to reduce overall costs. Public policymakers should conclude that ECPM-based interconnection rate proposals could be viewed by entrants as onerous and discourage entry. If interconnection rates are set to make the incumbent whole for opportunity costs and losses to competition, but entry occurs anyway, the incumbent could become an "indifferent"[16] competitor instead of a vigorous competitor. If the incumbent has less incentive to innovate and trim old monopoly excess, this could negate the beneficial results of competition.

In contrast, DFIRM is procompetitive by requiring lesser compensation from entrants to the incumbent. If entry occurs, an overall improvement in efficiency and cost is more likely. In the medium and long term, efficiency gains sparked by competition could come from two sources, both the incumbent and the entrant. The sum of these efficiency gains could more than offset the additional entrant fixed and common costs.

15. The ECPM could have application in pure reseller situations where the reseller only acts as an arbitrageur with little fixed or sunk costs.

16. The incumbent supplier would become indifferent as to whether components of final product are provided by itself or others. See Baumol and Willig (n.d.).

COMMISSION'S ORDER: UNIVERSAL SERVICE AND RATE DEAVERAGING

The Commission (1994) ruled that "This proceeding is not the forum for the Commission to address whether or how the provision of financial support for universal service should be modified in the future" (p. 22). Pricing policies should guard against an erosion of universal service and the broader issue should be taken up in a future proceeding. The Commission also stated:

> In Phase II, the Commission anticipates examining DTL [dial-tone line] and usage costs of serving customers in the various service areas, and the revenues received for those services. In addition, the Commission will examine revenues from basic service customers for all other-than-competitive services, including the contribution from Yellow Pages. In other words, the Commission is interested in determining in Phase II more precisely the extent of contribution that is needed to keep basic telephone service affordable for Maryland customers. (p. 58)

In addition, the Commission stated:

> We do not prejudge future BA–MD requests for further subdivision of the rate classes, in order to provide dial tone rates which more closely trace costs. While in testimony BA–MD witnesses maintained that regulatory changes, such as deaveraging of rates, should be authorized concomitant with MFS–I's entry into the Maryland market, on brief BA–MD seemed to drop pursuit of this course of action. (p. 69)

Instead, BA–MD proposed a rate cap plan. The Commission ruled, "Importantly, we emphasize our conclusion is that the Commission should have the option to consider other forms of rate setting, such as price cap plans, in future evidentiary hearings. We have not made any determination that any so-called price-cap plans should be adopted now or in the future (p. 71).

CONCLUSION

Maryland is the second state to authorize local exchange competition based on the co-carrier model. An entrant has nondiscrimintory access to Bell Atlantic's local network and may complete calls between its own customers using its own switching system rather than Bell Atlantic's. Some interconnection rates were decided as a compromise, others are still pending tariff review. Much telecommunications policy is yet to be decided. The Maryland Commission has deferred many policy issues to future proceedings (e.g., unbundling, support mechanisms for universal service, further rate deaveraging, price caps).

REFERENCES

Armstrong, M., & Doyle, C. , (1994, October). *Access pricing and the Baumol–Willig rule*, Paper presented at the Telecommunications Policy Research Conference.

Baumol, W., Bailey, E., & Willig, R., (1977, June). Weak invisible hand theorems on the sustainability of prices in a multiproduct monopoly. *American Economic Review, 67.*

Baumol, W., & Sidak, G., (1994). Toward competition in local telephony. Cambridge, MA: MIT Press and American Enterprise Institute.

Baumol, W., & Willig, R., (n.d.). *Brief of evidence: Economic principles for evaluation of the issues raised by Clear Communications Ltd. on interconnection with Telecom Corporation of New Zealand Ltd.*

Comanor, W. S., & Leibenstein, H., (1969, August). Allocative efficiency, X-efficiency, and the measurement of welfare losses. *Economica*

Kahn, A., & Taylor, W., (1993, November). *Direct testimony, Case No. 8584, before the Public Service Commission of Maryland.*

Leibenstein, H., (1966, June). Allocative efficiency vs. X-efficiency. *American Economic Review.*

Public Service Commission of Maryland, (1994, April 25). Case No. 8485, Order No. 71155, In the Matter of the Application of MFS Intelenet of Maryland to Provide and Resell Local Exchange and Interexchange Telephone Service; and Requesting the Establishment of Policies and Requirements for the Interconnection of Competing Local Exchange Networks. In the Matter of the Investigation by the Commission on its Own Motion into Policies Regarding Competitive Local Exchange Service.

Shin, R., (1988). *Econometric estimation of telephone costs for local exchange companies.* Unpublished doctoral dissertation, University of California, Berkeley.

Shin, R., & Ying, J., (1992, Summer). Unnatural monopolies in local telephones. *Rand Journal of Economics, 23.*

Tye, W. B., (1994). The pricing of inputs sold to competitors: A response. *Yale Journal on Regulation, 11.*

III

INTERNATIONAL TELECOMMUNICATION

7

The U.S. Stake in Competitive Global Telecommunications

John Haring
Strategic Policy Research, Inc.

Jeffrey H. Rohlfs
Strategic Policy Research, Inc.

Harry M. Shooshan III
Strategic Policy Research, Inc.

INTRODUCTION

Foreign impediments to free trade in international telecommunications services substantially harm the U.S. economy. These impediments take two forms: (1) anticompetitive protectionist regulation of telecommunications by foreign governments; and (2) excessive settlement rates paid to foreign governments or carriers by U.S. carriers for the completion of international calls by U.S. customers. This study provides a quantitative and qualitative assessment of the gains that would accrue to the U.S. economy if these impediments were abolished; *i.e.,* if foreign markets were fully opened to competition by U.S. telecommunications service providers and settlement rates paid by U.S. carriers to foreign carriers were set at cost. These gains will be lost if U.S. telecommunications policymakers fail to bargain tough to remove foreign protectionist regulations and excessive settlement rates.

FOREIGN IMPEDIMENTS TO FREE TRADE IN INTERNATIONAL TELECOMMUNICATIONS SERVICES

Supply Restrictions

Almost all foreign countries impose restrictions on the supply of telecommunications services. These restrictions serve to protect the national incumbent carrier. In most foreign markets, laws and regulatory rules prohibit any firm from offering basic services, either within the country or from that country to the rest of the world, in competition with the incumbent national carrier. To the extent any competition is permitted, most countries prohibit any competing carrier from constructing facilities; only resale is permitted, and even then regulatory rules often do not provide adequate safeguards to protect resellers from unreasonable pricing or other anticompetitive actions of the incumbent carrier.

In the few countries where facilities-based competition has been permitted, regulatory rules generally limit the scope of competition. New carriers are often not afforded the competitive safeguards necessary to facilitate effective competition. In particular, no major foreign economic power requires its incumbent carrier to offer its competitors equal access (including dialing parity). Thus, the new entrants must overcome expensive, inferior connections that require customers to dial extra digits and marketing disadvantages that inhibit customers from leaving the incumbent. Even where a modicum of competition has been permitted, laws and regulatory rules often leave opportunities for the incumbent to use its market power to forestall competition.

Excessive Settlement Rates

A second and significant impediment to free trade in international services is the excessive settlement rates paid to foreign governments or carriers for the completion of international calls made by U.S. customers.[1] Excessive settlement rates inflate the prices of international services and restrict demand. They are a form of subsidy to foreign carriers.[2] Because settlement rates exceed the costs incurred by foreign carriers to complete international calls made by U.S. customers, over $2 billion of U.S. income is siphoned each year to benefit foreign telephone carriers—most of which are government-owned monopolies.[3]

Settlement rates are widely acknowledged to be above cost. In Western European countries, settlement rates[4] for calls from the U.S. range from $0.35 per minute

1. When a U.S. carrier establishes facility arrangements with a foreign carrier to provide U.S. international service, the parties negotiate an *accounting rate*. For each minute of traffic delivered by the originating carrier, the originating carrier is obligated to pay to the terminating carrier one half of the accounting rate, which is called the *settlement rate*. In practice, the payments are netted and the settlement rates are paid only on the imbalance of traffic. Because U.S. carriers usually deliver more traffic to foreign carriers than foreign carriers deliver to the United States, U.S. carriers usually make net outpayments of settlements to foreign carriers, rather than vice versa.

2. Although the rates are negotiated by the U.S. and foreign carriers, the monopoly position of the foreign carrier leaves little room for negotiation. Despite the efforts of U.S. carriers, settlement rates with all countries remain above costs.

3. The total net settlement outpayment by U.S. carriers to foreign carriers in 1992, including the true economic costs of terminating international calls, was approximately $3.5 billion.

4. Settlement rates for calls from the United States to Western Europe, Japan, and Canada are derived from settlements to foreign carriers divided by minutes. See FCC, Industry Analysis Division, Common Carrier Bureau, *Preliminary 1992 Section 43.61 International Telecommunications Data*, Table A1, September 1993.

to $1.11 per minute—2 to 7 times actual costs.[5] The settlement rate to Japan is $0.70 per minute—3 to 4 times cost.[6] The settlement rate to Canada is $0.15 per

5. There are very few publicly accepted studies of the costs of terminating international calls in foreign countries. The Office of Technology Assessment (OTA) has concluded that the cost of terminating a call in Germany is no more than $0.15 per minute. (See U.S. Congress, OTA, *U.S. Telecommunications Services in European Markets* [OTA-TCT-548]. Washington, DC: U.S. Government Printing Office.) Most of the cost incurred for the termination of a U.S.–Germany call is the cost associated with the German domestic network. Transoceanic cable costs have now declined to approximately $0.04 per minute of calling. The U.S. and German carriers share the cost of the transoceanic facility, so only $0.02 per minute is incurred by the German carrier for the transatlantic transport.

The OTA estimate applies to actual costs in a monopoly environment and overestimates the economically efficient costs that would arise in a fully competitive environment. A reasonable surrogate for an economically efficient level of settlement rates would be prices to consumers in the competitive U.S. market for calls between the U.S. mainland and Hawaii. Such calls involve local access charges on each end, transmission costs within the mainland United States and submarine fiber optic cables between the U.S. mainland and Hawaii, and profit margins. AT&T's basic tariff rates for such calls range from $0.13 to $0.33 per minute. The corresponding estimate of settlement costs would be half of this price ($0.065 to $0.165) at each end. Numerous discount plans are available with even lower prices.

6. Longer transoceanic cables are required to complete U.S.–Japan calls, but cable costs are a relatively small part of the total costs incurred to terminate calls. Therefore, there is probably only a few cents per minute difference in costs between U.S.–Germany and U.S.–Japan calls.

minute, which exceeds costs by over 90 percent.[7] Further, some countries insist on a full rate 24 hours per day, failing to offer economically justified, appropriately discounted off-peak settlement rates.

There is absolutely no justification for this subsidy. The subsidy is a windfall that foreign carriers have received as a result of competition in the U.S. Prior to the advent of competition in the U.S., traffic streams between the U.S. and foreign locations were approximately in balance and settlement rates were set substantially above cost. As competition took hold, substantial price reductions for U.S. interexchange and international calling followed. These lower prices, in turn, stimulated calling by U.S. customers, causing increasing imbalances in the amount of U.S. originated traffic *vis-à-vis* foreign originated traffic to the U.S. The lack of competitive pressures in foreign markets, however, did not create any desire on the part of foreign carriers to reduce their own rates for international calls or the artificially high settlement rates. As a result, foreign carriers began to experience substantial inpayments, as U.S. carriers paid artificially high settlement rates on ever-increasing traffic imbalances. The net U.S. outflow grew from $40 million in 1970[8] to $3.5 billion in 1992. More than $2 billion per year (in excess of costs) of U.S. income now flows from the U.S. to foreign carriers. This subsidy in 1992 represents approximately one-third of the total U.S. budget for foreign aid.[9] But, a significant part of that subsidy is paid to industrialized nations with high standards of living, not to lesser developed or developing nations.

QUANTIFICATION OF BENEFITS

In this section, we quantify some of the gains to the U.S. economy from removing foreign impediments to free trade in telecommunications services. We project the respective gains in U.S. gross domestic product, employment and overall balance of trade (including services). The projection reflects the potential gains that would accrue by lowering settlement rates to competitive, cost-based rates and abolish-

7. The costs to terminate calls in Canada are considerably less than the costs to terminate calls in Germany. The U.S. and Canadian networks are physically integrated and transoceanic cables are not needed. The costs of U.S.–Canada traffic are probably comparable to those of U.S. domestic long-distance traffic. A cost-based settlement rate in that case would be less than $0.08 per minute. This estimate is based on AT&T's price for domestic switched services of $0.16 per minute, which includes access charges that are above cost. (See Richard Schmalensee and Jeffrey H. Rohlfs, *Productivity Gains Resulting from Interstate Price Caps for AT&T*, prepared for AT&T, September 3, 1992, for definition of *switched services*.) Because the settlement rate should cover only half the cost of the call (which is presumably less than the price of the call), the settlement rate for U.S.–Canada traffic should be no more than $0.08 per minute. In all three regions of the world, these above-cost settlement rates represent discrimination against U.S. carriers.

8. FCC, In the Matter of Regulation of International Accounting Rates, CC Docket No. 90-337, Notice of Proposed Rulemaking, adopted July 12, 1990, released August 7, 1990.

9. For fiscal year 1992, U.S. foreign aid—*that is,* bilateral assistance by the U.S. Agency for International Development (AID)—totaled $6.530 billion. U.S. AID, *FY 1994 AID Congressional Presentation,* Table 2.

ing protectionist restrictions *in every foreign country.*[10] The potential gains are so large that achieving only a modest portion of them would be a great boon to the U.S. economy. The gains to U.S. GDP, employment and overall balance of trade each consist of several parts, which we discuss separately and accumulate at the end of the section.

Removing foreign impediments to free trade would directly stimulate demand for U.S. production of telecommunications services between U.S. and foreign locations, and between pairs of foreign locations. This demand stimulation translates into jobs for U.S. workers.

The income of the U.S. would also increase, over and above the gains in production. The nation would pay out less subsidy to foreign carriers for the international telecommunications services it currently consumes, as foreign carriers would be motivated by market forces to reduce their prices. Price reductions by foreign carriers would stimulate new calling, thereby improving the current imbalance of traffic volumes between U.S. and foreign-originated calling. International facilities also would be used more efficiently. The direct gains in production and other gains in income both contribute to the U.S. GDP.

In the macro economy, these direct gains would generate further indirect gains. For example:

- Workers who produce the additional output will earn more income and therefore spend more on consumption. This stimulates the general economy.
- Firms which produce the additional output will invest more in plant and equipment. This also stimulates the general economy.

The overall U.S. balance of trade would also improve. The subsidy currently paid to foreign telephone carriers would be eliminated, and exports would be stimulated.

All of the above gains are initially estimated for a single year. The effects of removing foreign impediments to free trade in international telecommunications, however, would not be limited to a single year. They would persist indefinitely. Furthermore, they would grow rapidly over time, as international telecommunications is a growth market.

To estimate the long-term effects, we sum the annual impacts over 10 years. Long-term gains are estimated at the end of this section. The estimated long-term gains include all the components discussed in this section.

Direct Effect on U.S. GDP of Lowering Settlement Rates

For the purposes of this quantification of benefits, we estimate the level of costs that would be incurred by an efficient carrier operating in a competitive market. We have postulated that the competitive level of costs associated with terminating

10. There are other, potentially significant gains that would be achieved if competition were effective in foreign markets. These other gains, not quantified here, are discussed later. They include: (a) the offering of end-to-end global seamless services to U.S. customers by multiple U.S. carriers; and (b) higher call completion rates (and U.S. revenues) for calls to international points, as foreign carriers are driven to improve the quality of their networks.

international telecommunications calls is $0.147 per minute, averaged across all markets. This assumption is based on the estimated cost of approximately $0.08 per minute for calls to/from Canada[11] and $0.15 for calls to Western Europe.[12] For Latin America/Caribbean and Mexico, we have assumed a $0.15 cost, based on the view that, in a competitive market, these countries would experience efficiencies similar to Western Europe.

We assume, again conservatively, that the competitive level of costs for terminating traffic elsewhere in the world averages $0.20 per minute. As compared to Western Europe, this allows a few extra cents per minute for longer cables. That should be more than adequate because the average cost of international cables is only $0.02 at each end. A cost of $0.20 per minute also allows a few extra cents per minute for distribution costs within the foreign country. That should also be more than adequate, because many of our large trading partners (*e.g.*, Japan, Taiwan and Singapore, and Hong Kong) have small geographic areas.

We now estimate the direct impact if settlement rates fell to these cost-based levels in all markets. Reducing settlement rates would cause prices for international telecommunications services to decline. Indeed, if effective competition were permitted to develop in foreign markets, prices in both directions would rapidly fall to approximately twice the settlement rate plus an increment to cover the cost of marketing, billing and collection, and bad debts.[13] We estimate the increment to be approximately $0.08 per minute, mostly for customer operations, including marketing and customer service expenses. AT&T's expenses for customer operations amounted to 16.4 percent of revenues in 1992.[14] Applying this ratio to international service revenues, we estimate that customer operations for international calls cost about $0.059 per minute. Billing and collection costs amount to about $0.007 to $0.012 per minute.[15] We add an additional $0.009 to $0.014 per minute to allow for other factors, such as bad debts and fraud. Bad debts and fraud are significant for international calls, but that is largely a consequence of the high price levels. As international prices decline toward the level of domestic prices, the frequencies of bad debts and fraud should also decline toward domestic levels.

It follows from the above assumptions that prices for calls originating in the U.S. would fall an average of 63 percent, down to 37 percent of their current values. Prices for calls terminating in the U.S. would fall an average 88 percent, down to 12 percent of their current values.

The huge price declines would vastly stimulate demand for calls in both directions. To estimate stimulation, we assume a price demand elasticity of -0.5. This

---

11. See footnote 7.

12. The OTA study, cited earlier, concluded the cost of terminating a call in Germany is no more than $0.15 per minute. In our quantitative analysis, we use that estimate of German costs as a surrogate for Western Europe. See footnote 5 for further discussion of this estimate.

13. We assume that this would be the average price per minute, net of discounts, that would proliferate under effective competition.

14. FCC Form M, *Annual Report of AT&T Communications to the Federal Communications Commission For the Year Ended December 31, 1992*, pp. 21 and 21.2.

15. Bridger Mitchell, *Incremental Costs of Telephone Access and Local Use,* July 1990, p. 43.

assumption is conservative, based on estimates that AT&T has provided to the FCC in the past. The assumption implies that demand for international services is less elastic than demand for domestic long-distance services. We also did a sensitivity analysis with an assumed price elasticity of -0.8. Under the latter assumption, the elasticity of international calls would be a bit higher than that of domestic long-haul, long-distance calls.[16]

Our estimates of demand stimulation are as follows:

- Demand Elasticity of -0.5:
  —International calls originating in U.S.:
  Demand stimulation: To 160 percent of current level.
  —International calls terminating in U.S.:
  Demand stimulation: To 290 percent of current level.
- Demand Elasticity of -0.8:
  —International calls originating in U.S.:
  Demand stimulation: To 220 percent of current level.
  —International calls terminating in U.S.:
  Demand stimulation: To 550 percent of current level.

These calculations are shown in Table 7.1.

Effect on U.S. Production. The additional calls in both directions would be partially produced in the U.S. and partially produced abroad. The additional demand for U.S. production would stimulate the U.S. economy.[17] To estimate the dollar impact on the U.S., we value the additional calls at the new prices, rather than current prices.[18] The new prices are cost-based and more accurately reflect the effect of the additional calls on the real economy.

The direct effect of the additional calls on GDP (value-added in the economy), is the dollar value of the vast increase in demand. To calculate it, we simply multiply the increase in output by the value-added in the U.S. This calculation is shown in Table 7.2.

Income Transfers. The decline in the settlement rates would eliminate the subsidy of $2.3 billion per year. The reduction in subsidy constitutes a direct reduction in income transferred from the U.S. abroad. The reduced transfers cause U.S. income to increase. The subsidy is calculated as current total settlements outpayment ($3.5 billion) less cost-based settlements associated with current traffic ($1.2 billion). The calculation of this subsidy is shown in Table 7.3.

16. *See* J. P. Gatto, et al., "Interstate Switched Access Demand Analysis," in *Information Economics and Policy, 3* (4) 1988, p. 344. Still higher estimates of demand elasticities would increase our estimates of the gains from removing foreign impediments to free trade in international telecommunications.

17. The price changes additionally affect income transfers from the United States abroad. Income transfers are discussed later.

18. The alternative of evaluating impacts at current prices would lead to higher estimates of the gains from removing foreign impediments to free trade in telecommunications.

Table 7.1
Direct Effect of Reduced Settlement Rates on Demand

	U.S. Billed	*Foreign Billed*	*Source/ Comments*
	(1)	*(2)*	*(3)*
(A) Current Price ($/Minute)	$1.01	$3.05	1992 FCC data
(B) New Price ($/Minute)	$0.37	$0.37	Discussed in text—weighted average of Canada, Europe/Americas (other than Canada) and rest of world
	Elasticity = -0.5		
(C) New Demand (Proportion of Current Demand)	1.64	2.90	[(B) ÷ (A)] ^ (-0.5)
	Elasticity = -0.8		
(D) New Demand (Proportion of Current Demand)	2.22	5.50	[(B) ÷ (A)] ^ (-0.8)

Table 7.2
Direct Effect of Reduced Settlement Rates on U.S. Production

	U.S. Billed	*Foreign Billed*	*Total Traffic*	*Source/ Comments*
	(1)	*(2)*	*(1)+(2)* *(3)*	*(4)*
(A) Minutes of International Traffic (Billions)	10.224	5.314	15.538	1992 FCC data
(B) U.S. Value-Added per Minute ($/Minute)	$0.23	$0.14	-	Discussed in text—weighted average of Canada, Europe/Americas (other than Canada) and rest of world
	Elasticity = -0.5			
(C) Demand Increase	0.64	1.90	-	See Table 7.1
(D) Increment to U.S. Production ($ Billions)	$1.505	$1.414	$2.919	(A)x(B)x(C)
	Elasticity = -0.8			
(E) Demand Increase	1.22	4.50	-	See Table 7.1
(F) Increment to U.S. Production ($ Billions)	$2.869	$3.348	$6.217	(A)x(B)x(E)

Table 7.3
U.S. Subsidy Paid to Foreign Countries

	Canada	*Europe & Americas (Other Than Canada)*	*Rest of World*	*Total*	*Source/ Comments*
	(1)	*(2)*	*(3)*	*(1)+(2)+(3)* *(4)*	*(5)*
(A) Current Settlement Rate ($/Minute)	$0.15	$0.63	$0.89	-	Derived from 1992 FCC data
(B) Cost-Based Settlement Rate ($/Minute)	$0.08	$0.15	$0.20	-	Discussed in text
(C) Minutes of U.S. Billed International Traffic (Billions)	2.226	5.562	2.436	10.224	1992 FCC data
(D) Minutes of Foreign Billed International Traffic (Billions)	1.511	2.625	1.178	5.314	1992 FCC data
(E) Subsidy ($ Billions)	$0.049	$1.419	$0.863	$2.331	[(A)-(B)]x [(C)-(D)]

Direct Effect of Global Telecommunications Market on U.S. GDP

If foreign barriers to competition were eliminated, telecommunications could graduate to a truly global market. In that market, the U.S. could provide transport capabilities and value-added functions on many calls that neither originate nor terminate in the U.S. International cables to/from the U.S. could be used during off-peak hours, when the cables have substantial idle capacity. On transatlantic cables, much of the additional usage could occur during the European morning, before U.S. business hours. The additional usage on the transpacific cables could occur between midnight and 4:00 a.m. in the U.S.

U.S. carriers would be able to make sales relatively easily in a competitive global market. Incremental capacity cost would be close to zero; so carriers could substantially lower prices and still make a profit. In addition, U.S. carriers would be able to offer useful ancillary services and features that are unavailable in many countries.

The amount of such transiting traffic would, however, be limited by the amount of international capacity—facilities that currently exist and those that will be built to handle the increase in traffic to/from the U.S. Building additional international capacity for the express purpose of handling transiting traffic might not be profitable.

We can conservatively assume that the facilities could handle twice their current usage since the incremental usage would occur primarily during off-peak hours. Incremental usage could include both retail (customer) and wholesale (carrier) demand. It would be spread out over several hours, during late night and early morning in the U.S. During all those hours, international facilities could handle the additional transiting traffic and could still have significant idle capacity. Thus, little incremental cost would be associated with the transiting traffic.

It follows from the above assumptions that the amount of transiting traffic would equal approximately half of international traffic that either originates or terminates in the U.S.[19] Additional capacity utilization would equal current capacity utilization. However, transiting traffic requires about twice as much international transmission capacity per unit of traffic. An international facility is used to transport the call from its origin to the U.S. and another international facility is used to transport it to its destination.

The current amount of international traffic that originates or terminates in the U.S. is approximately 16 billion minutes per year. If settlement rates decline, as discussed above, traffic will grow to 30 billion-to-50 billion minutes per year depending on the demand elasticity. (We assume that international capacity will grow as needed to meet the increase in demand.) It follows that approximately 15 to 25 billion additional minutes of transiting traffic could be carried on the facilities (at low incremental cost).

We next need to specify how much U.S. value-added will be provided per minute of hubbed traffic. We estimate that the average U.S. value-added will be roughly $0.05 per minute. On some calls—for example, those between Europe and the Far East—the U.S. would displace foreign ownership of two transoceanic cables. The value-added would exceed $0.10 per minute. On other calls—for example those between the U.K. and the European continent—the U.S. value-added would be much less. The U.S. service would displace only a cable between the U.K. and the Continent. The value of the service would be only $0.01 or $0.02 per minute, in addition to ancillary services provided by the U.S. (*e.g.*, detailed billing or private network features). We use $0.05 a minute as a reasonable rough estimate of the average value of this disparate traffic.

We estimate that roughly half of the $0.05 per minute value-added consists of U.S. production; *i.e.*, switching, transmission, and ancillary services. The remainder consists of additional returns from more efficient utilization of international transmission facilities. The entire amount represents an increase in U.S. income.

We can now estimate the direct effect of this transiting traffic on U.S. GDP. We simply multiply $0.05 by the number of additional transiting minutes. Half of the product is the direct effect on U.S. production; the other half is the direct effect on U.S. income. This calculation is shown in Table 7.4.

19. Current transiting traffic is only about 2% of the total. It is ignored in this analysis.

Table 7.4
Direct Effect of Global Telecommunications Market on U.S. GDP

	Transiting Traffic	*Source/ Comments*
	(1)	*(2)*
(A) Minutes of International Traffic (Billions)	7.769	Derived from 1992 FCC data
(B) Value-Added per Minute ($/Minute)	$0.05	Discussed in text
	Elasticity = -0.5	
(C) Increment to U.S. GDP ($ Billions)	$0.416	(A)x(B)x(Demand Increase) (See Table 7.2)
	Elasticity = -0.8	
(D) Increment to U.S. GDP ($ Billions)	$0.910	(A)x(B)x(Demand Increase) (See Table 7.2)

Direct Effect on Jobs

The direct effect on jobs derives from increased demand for U.S. production of international telecommunications. The effect of reduced settlement rates on production and the effect of increased transiting traffic on U.S. GDP were discussed previously and are shown in Tables 7.2 and 7.4. As discussed in that section, approximately half the gain in GDP corresponds to increased production. The increase in production has a direct effect on jobs, while the other gains in income do not.

We use a macroeconomic model to estimate the effect of the increase in production on jobs. In particular, we ran the model and observed that approximately one additional job is created for each $100,000 per year of additional production.[20]

Using this ratio, we estimate the number of additional jobs created by increased demand for U.S. production of international telecommunications services. The calculation is shown in Table 7.5.

Total Effect (Including Indirect Effects) on U.S. GDP

Macroeconomic models typically embody "multiplier effects," which transform direct effects to total effects, including indirect effects. We estimate the total effects by

20. This ratio represents the *marginal* ratio of GDP to jobs. The *average* ratio (*i.e.*, GDP/employment) is much less. The marginal rate is higher, because firms often respond to short-run demand increases by utilizing the existing labor force more intensively. Doing so incurs additional costs (*e.g.*, overtime pay) but it avoids the need to hire and train new employees. To meet long-run demand increases, however, the firm may find it more economical to hire and train new employees. For this reason, our study may underestimate the long-run impact on employment.

Table 7.5
Direct Effect of Increased U.S. Production on Jobs

	Amount	*Source/ Comments*
	(1)	*(2)*
(A) ΔGDP / ΔJobs ($/Job)	$100,000	Discussed in text
	Elasticity = -0.5	
(B) Direct Effect of Reduced Settlement Rates on U.S. Production ($ Billions)	$2.919	See Table 7.2
(C) Direct Effect of Global Telecommunications Market on U.S. Production ($ Billions)	$0.208	Half of effect on GDP in Table 7.4 (See text)
(D) Increment in Jobs (Thousands)	31	[(B)+(C)] ÷ (A)
	Elasticity = -0.8	
(E) Direct Effect of Reduced Settlement Rates on U.S. Production ($ Billions)	$6.217	See Table 7.2
(F) Direct Effect of Global Telecommunications Market on U.S. Production ($ Billions)	$0.455	Half of effect on GDP in Table 7.4 (See text)
(G) Increment in Jobs (Thousands)	67	[(E)+(F)] ÷ (A)

inputting the direct increases in demand into a commercial macroeconomic model. We included the effects of increased production from both lowering settlement rates and increasing transiting traffic. The multiplier turned out to be 1.4. That provides a reasonable estimate of total macroeconomic effects relative to direct effects.

The resultant multiplier is applied to the increments of U.S. production, as estimated in the previous sections. It is not applied to other increases in U.S. income. Other increases in income have indirect effects, as the beneficiaries of the increased income spend part of the money on goods and services, and thereby stimulate the macro economy. We do not, however, have an estimate of the appropriate multiplier. By disregarding the indirect effects of the increase in income not associated with increases in production, we *underestimate* the gains from removing foreign impediments to free trade in telecommunications.

The calculation of total effects and indirect effects is shown in Table 7.6.

Effect on Overall U.S. Balance of Trade (Including Services)

Removing foreign impediments to free trade in international telecommunications would improve the overall U.S. balance of trade. The reduction in U.S. income transferred abroad is the elimination of subsidy that the U.S. pays to foreign tele-

Table 7.6
Total Effect (Including Indirect Effects)

	Direct Effect	*Total Effect*	*Source/Comments*
		(1)x1.4	
	(1)	*(2)*	*(3)*
	Elasticity = -0.5		
	-----------*($ Billions)*-----------		
(A) Effect of Reduced Settlement Rates on U.S. Production	$2.919	$4.087	See Table 7.2
(B) Effect of Reduced Settlement Rates in Reducing Income Transfers from the U.S. Abroad	$2.331	$2.331[a]	See Table 7.3
(C) Effect of Global Telecommunications Market on U.S. Production	$0.208	$0.291	Half of effect on GDP in Table 7.4 (See text)
(D) Other Effects of Global Telecommunications Market on U.S. Income	$0.208	$0.208[a]	Half of effect on GDP in Table 7.4 (See text)
(E) Effect on U.S. GDP	$5.666	$6.917	(A)+(B)+(C)+(D)
	----------*(Thousands)*---------		
(F) Effect on Jobs	31	43	See Table 7.5
	Elasticity = -0.8		
	-----------*($ Billions)*-----------		
(G) Effect of Reduced Settlement Rates on U.S. Production	$6.217	$8.704	See Table 7.2
(H) Effect of Reduced Settlement Rates in Reducing Income Transfers from the U.S. Abroad	$2.331	$2.331[a]	See Table 7.3
(I) Effect of Global Telecommunications Market on U.S. Production	$0.455	$0.637	Half of effect on GDP in Table 7.4 (See text)
(J) Other Effects of Global Telecommunications Markets on U.S. Income	$0.455	$0.455[a]	Half of effect on GDP in Table 7.4 (See text)
(K) Effect on U.S. GDP	$9.458	$12.127	(G)+(H)+(I)+(J)
	----------*(Thousands)*---------		
(L) Effect on Jobs	67	94	See Table 7.5

a. Indirect effect is not calculated for income transfers.

phone companies. The entire amount constitutes an improvement of the overall U.S. balance of trade. In addition, lower prices would stimulate international calling. That would further improve the overall balance of trade, since calls terminating in the U.S. would be expected to increase more than calls originating in the U.S (because prices for calls terminating in the U.S. decline more than prices for calls originating in the U.S.).

In addition, the gain in U.S. GDP from additional transiting traffic consists solely of exports—both the effect on U.S. production and the other effects on U.S. income. Hence, the entire gain constitutes an improvement in the overall U.S. balance of trade.

These effects on the overall balance of trade are shown in Table 7.7.

Long-Term Effects

All the effects estimated in the preceding sections are for one year. In reality, however, the gains from removing foreign impediments to free trade (compared to the base case of not removing impediments) would persist indefinitely and grow over time, as the international telecommunications market expands. In this section, we estimate the long-term gains by accumulating over 10 years.

Table 7.7
Effect on Overall U.S. Balance of Trade

	Amount ($ Billions)	*Source/ Comments*
	(1)	*(2)*
	Elasticity = -0.5	
(A) Total Effect of Reduced Settlement Rates in Reducing Income Transfers from the U.S. Abroad	$2.331	See Table 7.6
(B) Total Effect of Global Telecommunications Market on U.S. Production	$0.291	See Table 7.6
(C) Total of Other Effects of Global Telecommunications Market on U.S. Income	$0.208	See Table 7.6
(D) Total Effect on Balance of Trade	$2.830	(A)+(B)+(C)
	Elasticity = -0.8	
(E) Total Effect of Reduced Settlement Rates in Reducing Income Transfers from the U.S. Abroad	$2.331	See Table 7.6
(F) Total Effect of Global Telecommunications Market on U.S. Production	$0.637	See Table 7.6
(G) Total of Other Effects of Global Telecommunications Market on U.S. Income	$0.455	See Table 7.6
(H) Total Effect on Balance of Trade	$3.423	(E)+(F)+(G)

Table 7.8
U.S. and Foreign Billed Minutes of International Traffic

Year	*U.S. Billed* (1)	*Foreign Billed* (2)	*Total Billed* (1)+(2) (3)	*Annual Growth Rate (Percent)* (4)
	(Billions of Minutes)			
1985	3.349	2.250	5.598	-
1986	3.907	2.482	6.390	14.13%
1987	4.480	2.722	7.202	12.72
1988	5.190	2.979	8.169	13.42
1989	6.109	3.449	9.558	17.01
1990	7.215	3.897	11.112	16.26
1991	9.072	4.769	13.841	a
1992	10.224	5.314	15.539	12.27

Note: Detail may not add to total due to rounding.
a. The annual growth rate from 1990 to 1991 is not calculated due to FCC reporting and definitional changes in 1991.
Source: Cols. 1 & 2: FCC, Industry Analysis Division, Common Carrier Bureau, "International Communications Traffic Data Report," 1985–1992, Table A1 and Appendix 1, Table A-1.
Cols. 3 & 4: Derived from data in Cols. 1 and 2.

In making these calculations, we take into account secular growth in the international telecommunications market. Total growth in international minutes has varied from 12 to 17 percent per year (see Table 7.8). We use 12 percent as a conservative estimate of future growth.

Price changes have been modest in the past—less than 1 percent per year. Consequently, virtually all the demand growth is exogenous growth—not stimulation from price reductions. Tables 7.9, 7.10 and 7.11 show growth factors associated with a growth rate of 12 percent per year. Final results are as follows:

- Demand Elasticity of -0.5:
  —Increase in GDP: + $120 billion (sum of impacts for next 10 years)
  —Increase in jobs in 10th year: +120,000 jobs
  —Improvement in overall balance of trade: $50 billion (sum of impacts for next 10 years)
- Demand Elasticity of -0.8:
  —Increase in GDP: + $210 billion (sum of impacts for next 10 years)
  —Increase in jobs in 10th year: +260,000 jobs
  —Improvement in overall balance of trade: $60 billion (sum of impacts for next 10 years)

Table 7.9
Long-Term Effect on U.S. GDP Over a 10-Year Period

	Growth Factor for Total International Minutes[a]	Elasticity = -0.5: Growth Factor Multiplied by One-Year Effect[b]	Elasticity = -0.8: Growth Factor Multiplied by One-Year Effect[c]
		($ Billions)	
		(1) x $6.917	(1) x $12.127
Year	*(1)*	*(2)*	*(3)*
1	1.00	$6.917	$12.127
2	1.12	7.747	13.582
3	1.25	8.674	15.207
4	1.41	9.718	17.038
5	1.57	10.887	19.088
6	1.76	12.188	21.368
7	1.97	13.654	23.939
8	2.21	15.293	26.813
9	2.48	17.126	30.026
10	2.77	19.181	33.628
Long-Term Effect		$121.385	$212.816

a. Discussed in text.
b. See Table 7.6, Row E, Column 2.
c. See Table 7.6, Row K, Column 2.

ADDITIONAL BENEFITS TO THE U.S.

The preceding section quantified some important benefits of removing foreign impediments to free trade in telecommunications. However, there are many additional benefits that cannot be easily quantified. They are discussed in this section.

Most importantly, U.S. firms are extremely well-positioned to provide competitive telecommunications services between and within foreign countries. U.S. interexchange carriers and suppliers of value-added services have had many years of experience in competitive markets. Their skills in developing and marketing new services and responding to customer needs have been honed, while foreign carriers have typically enjoyed stable monopoly markets. In addition, U.S. firms, unlike foreign telephone monopolists, have experience in competitive advertising and developing new competitive rate plans.

U.S. firms also have the edge in technology. The U.S. leads the world in telecommunications technology. U.S. firms have developed most of the technological advances in telecommunications since the industry began. Beginning with the invention of the telephone itself in Boston and the subsequent initial deployment of telephony, U.S. firms have played a leading role in each of the industry's major technological advances. These include automatic switching, long-distance direct dialing automation, cellu-

Table 7.10
Long-Term Effect on U.S. U.S. Jobs Over a 10-Year Period

		Elasticity = -0.5: Growth Factor Multiplied by One-Year Effect[b]	*Elasticity = -0.8: Growth Factor Multiplied by One-Year Effect*[c]
	Growth Factor for Total International Minutes[a]	*(Thousands of Jobs)*	
Year	*(1)*	*(1) x 43* *(2)*	*(1) x 94* *(3)*
1	1.00	43	94
2	1.12	48	105
3	1.25	54	118
4	1.41	60	132
5	1.57	68	148
6	1.76	76	166
7	1.97	85	186
8	2.21	95	208
9	2.48	106	233
10	2.77	119	261

a. Discussed in text.
b. See Table 7.6, Row (F), Column (2).
c. See Table 7.6, Row (L), Column (2).

lar, Ethernet LANs, intelligent networks, and common channel signalling. Especially important in this regard are enhancements to 800 services. U.S. interexchange carriers have many features currently in place for routing calls flexibly, depending on customer needs. None of our major trading partners have such services in place. Consequently, the U.S. would have a substantial advantage in marketing 800 services.

U.S. firms have also led in the development of virtual private networks. If foreign impediments to free trade in telecommunications were eliminated, multinational corporations could have seamless global virtual private networks that would provide calling capability to meet their total needs: within the U.S.; between the U.S. and foreign countries, and within and between foreign countries. These seamless services would have advanced features such as simplified, integrated corporate wide dialing plans, number portability inside the corporation and call forwarding inside the organization.[21] These features are readily available in the

21. Many features (including simplified integrated corporatewide dialing plans and number portability inside the organization) can be offered even in countries whose telecommunications infrastructure is not highly advanced. Indeed, the features may be most valuable in such countries. Other features do, however, depend on the infrastructure of the country. For example, DTMF capability (*e.g.*, touchtone) is required for call forwarding. As a consequence of differing infrastructures, today's bilateral VPN services do not offer the same feature sets in both countries. The new, global services would overcome this deficiency by employing technology platforms that would provide identical capabilities in multiple countries.

Table 7.11
Long-Term Effect on U.S. Overall Balance of Trade Over a 10-Year Period

	Growth Factor for Total International Minutes[a]	Elasticity = -0.5: Growth Factor Multiplied by One-Year Effect[b]	Elasticity = -0.8: Growth Factor Multiplied by One-Year Effect[c]
		($ Billions)	
		(1) x $2.830	(1) x $3.423
Year	(1)	(2)	(3)
1	1.00	$2.830	$3.423
2	1.12	3.170	3.834
3	1.25	3.549	4.292
4	1.41	3.976	4.809
5	1.57	4.454	5.388
6	1.76	4.986	6.031
7	1.97	5.586	6.757
8	2.21	6.257	7.568
9	2.48	7.007	8.475
10	2.77	7.848	9.492
Long-Term Effect		$49.663	$60.069

a. Discussed in text.
b. See Table 7.7, Row (D), Column (1).
c. See Table 7.7, Row (H), Column (1).

U.S., but are not offered by most U.S. major trading partners. U.S. interexchange carriers could be strong competitors in this area.

Value-added services are another area in which the U.S. excels. The U.S. led the world in opening these services to competition in the 1970s. While many countries have now copied this U.S. initiative, the U.S. gained a significant advantage from its head start. Electronic mail (E-mail) is an important example of U.S. leadership. The U.S. has always had an open market for E-mail services. As a consequence, the E-mail industry in this country has grown vigorously. Indeed, the E-mail systems in many other countries use software originally developed for the U.S. market. Similarly, the Internet was able to grow rapidly in the U.S. because the regulatory structure permitted such innovation. Currently, it is estimated that the worldwide Internet provides connectivity to more than two million hosts, and several times as many users worldwide[22] (most of which are in the U.S.).

The U.S. edge in technology could be even more important in the future. Indeed, the U.S. could become a hub for global intelligent network services. For example, suppose a call is going from the U.K. to the European Continent (or even within

22. SRI International, *Internet Domain Survey*, October 1993 (supplied by InterNIC Information Service, San Diego, California).

the U.K.). The first step might be to query a database in the U.S., which would provide instructions for routing the call. The instructions might depend on the number called, the number of the caller, the time of day, or specific instructions by the party being called. The U.S. would thereby provide the software and databases for flexibly routing calls throughout the world.[23] The technology required is closely related to technologies in which the U.S. already excels; *e.g.*, those used to provide enhanced 800 services.

Competition would also spur improvements in the quality of international telecommunications. At present, the number of international calls blocked is disproportionately higher than that experienced on calls within the U.S. because foreign carriers do not have adequate facilities. Competition would attract new entrants, eager to participate in the lucrative international market by providing better service than the incumbent. The incumbent carrier would likely respond to this competitive pressure by improving the quality of its network and its service.

U.S. business generally—not just the telecommunications sector—would benefit from quality improvements in international telecommunications. It would also benefit from the development of a seamless global network.

All of these developments are possible if the U.S. Government bargains tough to open up foreign telecommunications markets to competition. However, they are *not* possible with the current protectionist barriers to competition erected by most foreign governments.

PROBLEMS WITH ALLOWING FOREIGN ENTRY WITHOUT GETTING COMPARABLE ACCESS

The preceding sections described some of the economic benefits of removing foreign impediments to free trade in telecommunications. These results can be achieved only through tough bargaining by officials and agencies of the U.S. Government. Failure by the U.S. Government to bargain aggressively will perpetuate a *status quo* in which foreign markets move slowly to introduce competition and the U.S. citizenry continues to pay over $2 billion in subsidy to foreign governments and their monopolies. Failure to bargain aggressively will also forfeit the benefits of the Golden Age in international telecommunications.

Federal inaction can also lead to a bad result of another kind. Suppose that the Federal Government stands by, allowing foreign carriers to enter the U.S. market without demanding comparable access to foreign markets. What would the impacts be? The answer is: U.S. citizens would not enjoy the consumption benefits and employment opportunities of increased competition in a growing market; foreign countries would have a significant edge in competing for international traffic, and even domestic U.S. traffic; and, to the extent that occurs, foreign monopolies

23. Under this scenario, the calls themselves would not be routed to the U.S. Only signaling information would be exchanged between the U.S. and the foreign countries. The incremental cost of transmitting signaling information to and from the U.S. would be negligible.

would become even more powerful, and more determined to protect their home markets. Once having acquired the benefits of entry to the U.S. market without sacrifice, there is no incentive to open their home markets, and every incentive to keep them closed for as long as possible.

No Price Reductions in Foreign Countries

Foreign monopolistic telecommunications operators have no incentive to voluntarily allow competition in their markets. But, in most foreign countries, the most immediate pressure for competition could come from the U.S. Government as foreign operators seek to expand their operations to the U.S. Foreign operators have no incentive whatever to permit competition if they can obtain access to the U.S. market without making any concessions at all.

No Stimulation of Demand Resulting from Reduction of Subsidies

The reason that settlement rates are currently excessive (with respect to cost) is because foreign monopolistic telecommunications operators prefer them to be excessive and have not agreed to (commensurate) settlement reductions as costs have fallen. One effect of high settlement rates is that U.S. residents fund part of the cost of the foreign country's domestic telephone system. Another result is that U.S. customers fund the expansion plans of foreign carriers as they enter new markets, like the U.S. Without competitive pressures or additional pressure from the U.S. Government, the telecommunications operators are unlikely to reduce settlement rates; foreign governments also are not inclined to take steps to make them do so—even if the telephone system is privately owned.

Competitive Edge for Foreign Countries

Foreign carriers, by obtaining access to the U.S. market, while denying comparable access to U.S. carriers, gain a significant advantage in competing for international traffic.[24] The advantage would spill over to the domestic U.S. market. Many U.S. firms may prefer one-stop shopping and will select their carrier in the U.S. based on the customer's total calling needs: domestic interexchange, international and global. If only foreign carriers and their U.S. affiliates can provide seamless global network services, the foreign affiliate will have an advantage in the competitive battle in the U.S. domestic market, but not because it has a lower price or better quality. Rather, it wins because its foreign affiliate,

24. Multinational corporations today are denied the full technological benefits of the U.S. network when they procure international and global services. They can have corporate networks with uniform equipment and services in the U.S. and abroad, but they necessarily conform to the "lowest common denominator" of what is technically available from the various monopolist telecommunications administrations around the world. The alternative is to bear the cost and inefficiency of administering a corporate network with different suppliers, features and functions, billing and payment options, pricing plans, and service intervals.

with the support and protection of its government, maintains restrictive practices in the foreign market.

Future Bargaining Leverage Lost

Until foreign entry is permitted, the U.S. has substantial bargaining leverage. It can credibly threaten to deny entry unless the foreign country meets certain conditions. Denying entry causes no dislocations in the U.S. as the market is already competitive. U.S. customer needs are already being met better and more efficiently than in other countries.

After foreign entry is permitted, however, much of the bargaining leverage of the U.S. is lost. Theoretically, the U.S. could threaten to discontinue the foreign carrier's operations, but the threat may not be credible. Without a credible threat, the U.S. would have no bargaining leverage. The U.S., therefore, could not expect, after entry by the foreign carrier, to exert sufficient pressure to reduce barriers to competition in foreign markets or to lower settlement rates.

This loss of U.S. leverage is precisely what happened with regard to competition in terminal and network equipment. The U.S. granted unilateral entry to foreign suppliers without demanding comparable access. Afterwards, the U.S. had little leverage to open up foreign markets.

U.S. BARGAINING LEVERAGE

The U.S. can successfully use its leverage in bargaining for access by its carriers to foreign telecommunications markets. As the leverage is used, foreign governments will understand they have more at stake than the U.S. and they will likely yield. The trend is toward a global marketplace for most goods and services. No carrier/country can be a credible global player without a significant capability to enter and operate in the U.S. marketplace. Therefore, the U.S. market is a key requirement for services.

The U.S. is the largest, most lucrative market in the world. Table 7.12 illustrates this point by showing import, export, and GDP measurements for the major economic powers. It shows that the U.S. is the largest economy in the world, nearly twice the size of the next largest, Japan. Most of the countries shown rely far more on international trade than does the U.S., as shown by the column measuring exports as a percent of GDP. This puts the U.S. in a position of bargaining strength versus other countries. Table 7.13 shows the trade flows between the U.S. and the U.K. The U.K.'s trade with the U.S. represents nearly 10 percent of its entire trade, while the U.S. trade with the U.K. is less than 4 percent of its trade.

The U.S. also has great bargaining power with respect to trade in telecommunications services. The U.S. is by far the largest telecommunications market. Table 7.14 shows incoming and outgoing international calling for the major countries. Foreign carriers extract enormous profits from this unbalanced trade with the U.S., especially on calls originating in the U.S.

Table 7.12
Trade and Gross Domestic Product (GDP) of the Group of Seven Countries and Spain, 1991

	Imports	*Exports*	*GDP*	*Exports as a Percent of GDP (Percent)*
	(Billion U.S. Dollars)			
Country	*(1)*	*(2)*	*(3)*	*(2) ÷ (3)* *(4)*
Canada	$149.7	$144.0	$ 582.0	24.7
France	267.0	270.5	1,199.3	22.6
Germany	436.4	537.3	1,574.3	34.1
Italy	223.4	224.4	1,150.5	19.5
Japan	286.0	347.5	3,362.2	10.3
Spain	107.5	91.0	527.1	17.3
U.K.	248.7	239.6	1,009.5	23.7
U.S.	620.0	589.4	5,610.8	10.5

Source: Organization for Economic Co-operation and Development (OECD), "OECD in Figures, 1993."

Table 7.13
Trade of Commodities, 1991

U.S. with U.K.

U.S. Imports			*U.S. Exports*		
From U.K.	*Total*	*Percent of Total (Percent)*	*To U.K.*	*Total*	*Percent of Total (Percent)*
(Thousand U.S. Dollars)			*(Thousand U.S. Dollars)*		
(1)	*(2)*	*(1) ÷ (2)* *(3)*	*(4)*	*(5)*	*(4) ÷ (5)* *(6)*
$19,022,504	$507,255,488	3.75%	$20,301,956	$397,447,618	5.11%

U.K. with U.S.

U.K. Imports			*U.K. Exports*		
From U.S.	*Total*	*Percent of Total (Percent)*	*To U.S.*	*Total*	*Percent of Total (Percent)*
(Thousand U.S. Dollars)			*(Thousand U.S. Dollars)*		
(1)	*(2)*	*(1) ÷ (2)* *(3)*	*(4)*	*(5)*	*(4) ÷ (5)* *(6)*
$20,478,698	$210,002,608	9.75%	$18,268,295	$185,120,421	9.87%

Source: OECD, "Foreign Trade by Commodities, 1991," 1992.

Table 7.14
Total Outgoing and Incoming Minutes of Telecommunications Traffic (MiTT) of Group of Seven Countries and Spain, 1991

	Total Outgoing MiTT	*Total Incoming MiTT*
	(Millions)	
Country	*(1)*	*(2)*
Canada[a]	647	398
France	2,295	2,355
Germany	NA	NA
Italy[a]	239	281
Japan	NA	NA
Spain	719	737
U.K.	NA	NA
U.S.[a]	5,985	2,830

NA — not available
a. Intercontinental traffic only.
Source: IIC, "TeleGeography 1992, Global Telecommunications Traffic Statistics and Commentary," Table 3a.

Because other countries rely more on trade than does the U.S. and rely particularly on trade with the U.S., they stand to receive a far greater blow than the U.S. if international telecommunications trade was interrupted. The U.S. is, therefore, likely to prevail if it bargains tough. The best way to achieve this end is to establish an unambiguous policy that settlement rates must fall to nondiscriminatory, cost-based levels and that access to foreign markets must be open to the degree they are in the U.S. If the U.S. continues to deal with these issues in an *ad hoc* fashion, it risks sending conflicting or diluted messages that will not lead to needed changes.

CONCLUSIONS

Foreign impediments to free trade in telecommunications services cause the U.S. to pay out more than $2 billion per year in subsidy to foreign carriers and governments; deny U.S. customers the full benefits of competition in U.S. international services; and limit the growth of the U.S. economy. In this paper, we show that removing these impediments to free trade would usher in a "Golden Age" of international telecommunications for U.S. and foreign customers. Prices of international services would tumble; demand would grow enormously; and international facilities would be used far more efficiently. These developments could, over the next ten years:

- Create 120,000 to 260,000 new jobs;
- Add $120 to $210 billion to U.S. gross domestic product (GDP); and

- Improve the overall balance of trade (including services) by $50 to $60 billion.

In addition, the U.S. would reap additional gains that are not quantified here: In particular,

- The U.S. is well-positioned to compete for telecommunications services between and within foreign countries;
- The U.S. could also become a hub for future global intelligent network services;
- U.S. business would benefit from the development of a seamless global network; and
- Competitive pressures would cause foreign carriers to improve the quality of their international networks and thereby allow more U.S. calls to be successfully completed.

Achieving only a modest portion of these gains would be a great boon to the U.S. economy. The nation could get large payoffs in economic value-added and challenging, well-remunerated jobs.

To garner these benefits, U.S. policymakers will have to bargain hard to remove foreign impediments. All of these benefits will be forfeited, if foreign impediments to free trade are permitted to persist. Prospects will be even worse if the U.S. accedes to attempts by foreign telephone companies to enter the U.S. market and does not insist on comparable access for U.S. companies in foreign markets. In that case, the U.S. will lose its bargaining leverage and there will be little or no incentive for foreign telephone companies, or their governments, to remove the restrictions that limit both the supply of and demand for telecommunications services.

8

The Changing Face of Transatlantic Telecommunications

Richard A. Cawley*
European Commission

1994 saw the commencement of GATS negotiations on trade in telecommunications services, just as a number of international alliances and agreements are beginning to form in order to compete in the so-called global telecommunications services market, a market that has essentially appeared because a number of multinational companies with large telephone bills are beginning to assert their market power on the demand side. At the same time, the three major blocs or telecommunications services markets in the world, North America, Europe, and the Pacific rim are fundamentally reviewing their regulatory structures in light of the rapid development of wireless communications and the progressive, or prospective intermingling of voice, data, and visual signals on networks.

For all the talk of global information highways (and it may be that telecommunications can promote development rather than reflect it), the fact remains that the Organisation for Economic Co-operation and Development (OECD) countries generate nearly 80% of international public switched telephone (PSTN) traffic. Moreover, the United States and Western Europe between them are responsible for almost 70% of the total. Given that the growing international market is one of the major targets of competition and that, in any case, these large international markets reflect substantial domestic ones, the United States and Western Europe are bound to dominate the trade negotiations. They, therefore, share an important responsibility for helping to develop fair and open ground rules for the sector that are, above all, forward looking. This means taking into account the major developments alluded to, including the growth of wireless and personal communications and the changes in the market and regulatory situation.

In the last few years, much attention has been given to international accounting rates as a barrier to market growth and a distortion of trade. Although few would

* The views expressed in this chapter are those of the author and do not necessarily represent the policies of the European Commission.

disagree that accounting rates remain too high, comprehensive reform of a system based on traditional correspondent relations in which major differences of interest persist between different areas of the world was never going to be easy or fast.

Meanwhile, correspondingly high international prices have provided the incentive for some existing operators, users, and new operators to find various ways around the barriers. In recent years there has been substantial growth in private traffic, direct, and premium card-based service and various forms of resale, including voice refile. Estimates developed in the paper indicate that these submarkets now constitute at least 20% of the transatlantic telephone market and are growing at a faster rate.

Card-based service and virtual private networks constitute an important means of extending service beyond traditional geographic boundaries, given constraints on getting control over foreign infrastructure and/or buying into foreign infrastructure. In addition, the prospect of simple international resale spreading from 9% of global international traffic flows (essentially U.S.–Canada traffic) to between 40% and 50% over the next 4 years is a real one. Card-based direct and premium service, virtual private networks, and simple resale have interesting impacts on accounting rates. The first worsens the settlement or trade imbalance as domestic operators become more successful at expanding their business; the second ignores accounting rates; and the third undermines the system altogether. All this demonstrates that attention should focus on the efficient development of the new competitive markets rather than the levels of accounting rates in the old system.

The respective U.S. and U.K. authorities have now agreed to permit simple international resale on this route. This agreement depended in part on establishing that domestic interconnection regimes offered equivalent market opportunities. For international competition, including simple resale, to spread in an efficient manner, fair and transparent interconnection regimes will be needed elsewhere. This will require significant efforts by regulators and coordination between regulatory bodies.

Examples of such regulatory cooperation and coordination have occurred recently between the United States and United Kingdom and between the United States and the European Commission in the context of the approval of the BT–MCI agreement and shareholding. Cooperation between regulatory and competition authorities will become even more important in the next few years as international agreements and alliances occur in advance of the widespread lifting of current constraints on effective telecommunications services competition. Regulating the transition to more competitive market structures in the international arena will become just as important as regulating the transition to a more competitive situation in the domestic context.

The interplay between the two is already apparent. Telefonica has joined the Unisource equity partnership, which aims to provide a variety of telecommunications services to multinational clients. Unisource will also market its own telephone card. Will calls using the card between Spain and Holland or between Spain and Switzerland be based on traditional correspondent relations? At the same time, Telefonica tried during 1994 to reach an agreement with the Spanish government on a tariff adjustment or rebalancing package in preparation for telephone service liberalization in Europe. It may not be the first operator that is reluctant to unilaterally lower international prices and the accounting rates that go with them until it gets domestic regulatory approval of an overall package of tariff changes.

Until now the major focus of concern in the international arena has been on the excess of settlement rates with respect to cost, and differences of settlement rate according to correspondent country. As international competition grows, although unevenly, and international alliances develop, the sector will increasingly exhibit a mixture of arm's length and internal transfer prices. The agreement to liberalize services and now infrastructure by 1998 in the European Union, together with competitive developments on transatlantic routes means that operators, both old and new, will increasingly distribute services on a mixture of owned and rented infrastructure. Accounting rates will not disappear overnight but they will gradually be displaced by interconnection agreements.

The excessive attention to accounting rates is therefore misplaced. Focus should be put on the efficient development of fair competition at an international as well as a domestic level. That requires increasing cooperation between regulatory and competition authorities both to oversee international agreements and alliances, particularly in advance of the lifting of constraints on effective competition, and to ensure efficient and forward-looking interconnection regimes.

The agreement to liberalize telecommunications services and infrastructure in the European Union by 1998 and the moves to lift remaining local and long-distance restrictions in the United States mean that the two largest telecommunications markets in the world will be fully open to competition. This chapter examines some of the implications of this opening by focusing on the market that connects these two blocs, the transatlantic telecommunications market.

The chapter attempts to do three things. The first is to enumerate and set in context the volume and value of international telephone traffic and settlements, in particular transatlantic traffic. The second is to examine to what extent this international market is breaking up into substitutable but separate submarkets as indirect competition, resale, and private markets develop and as international alliances appear. The third is to examine the impact that these new markets are having on traditional correspondent relations.

One conclusion is that interconnection issues are now moving from the domestic situation into the international limelight with all that implies for regulatory oversight.

THE TRANSATLANTIC MARKET

Much of the background information for the first task is well documented in ITU and Telegeography (1994) and only salient features are presented here. Of the world's almost 600 million telephone lines, just over 70% are situated in the 24-member countries of the OECD. Of the world's approximately 50 billion annual minutes of outgoing international traffic[1] (42.5 billion in 1992), approximately

78% is generated by OECD countries. The United States on its own accounts for nearly 25% of outgoing international PSTN traffic. The whole of Europe generates about 47% of the total, of which the 12-member European Union is responsible for about 35% and the current European Economic Area generates about 43%. In other words, the United States and expanded European Union between them account for almost 70% of the international PSTN traffic market.

With respect to transatlantic routes, about 25% of total U.S. outgoing traffic is destined for the European Union and the European Free Trade Area (EFTA). Of the 43% of world international traffic generated in the Union and EFTA, about two thirds (29 percentage points), remains within EU and EFTA and another one tenth is destined for the United States. This means that nearly 10% of the world's international traffic consists of flows between the United States and EU/EFTA, almost 6% of the global total from the United States to Europe, and just under 4% in the other direction. Traffic totals and flows for North America and Europe are shown in Figs. 8.1 and 8.2 and are enumerated in the respective Tables 8.1 and 8.2.

It is difficult to calculate the exact value of this transatlantic telephone market because international operator revenues are not sufficiently desegregated by route, but about $5 billion seems a reasonable estimate. Moreover, as the previously mentioned report points out, reported international outgoing traffic grew by about 15% annually between 1983 and 1992, implying that international revenues are probably growing by between 5% and 10% per year.

However a number of additional remarks must be made concerning this transatlantic market. The first is that this estimate understates the size of the market. Traffic between the continents, particularly for business communications, travels on private circuits as well as on the public switched network.

Capacity, in terms of voice paths between North America and Europe (both cable and satellite) has grown from 100,000 in 1986 to about 800,000 in 1992 and a prospective 1.2 million in 1994, a 12-fold increase in 8 years, and PSTN traffic has risen just under threefold. Satellite-based capacity has increased about sevenfold during the period, and cable-based capacity by a factor of about 30.

Bearing in mind a number of features that could have influenced the relationship between capacity and traffic over this time, including most importantly a move from capacity shortages to capacity excess, in particular provisions for backup or rerouting, it would still appear that there has been a dramatic increase in the "private" transatlantic market that is much greater than the growth in the public market.

Such a conclusion is supported by other information that is available on international traffic flows. The ITU/Telegeography report highlights country pairs in the OECD where traffic in one direction is greater than twice the volume in the other direction. The most significant in absolute traffic terms reveal a distinct pattern. They are the United States to Germany, Germany to Turkey, and Belgium to Portugal. In other words, they all involve correspondence where significant num-

1. For the purposes of this paper, international switched traffic comprises traffic currently subject to international accounting rates and therefore also includes cross-border traffic within the European Union.

Fig. 8.1.
Outgoing international PSTN traffic flows, 1992 (percentage of global total)

Canada	5 %
USA	24 %
Mexico	2 %

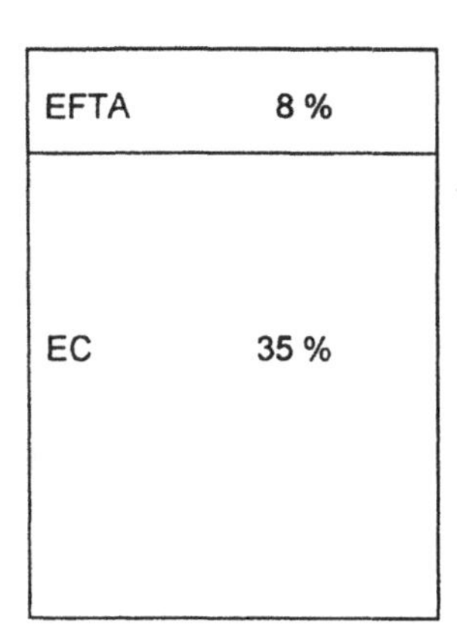

In addition outgoing totals from Japan (3%) and Australia/New Zealand (2%) meant that total OECD was 78% in 1992. Eastern Europe and other European countries (4%) meant that the total for Europe was 47%. South America constitutes 2% whereas Central and South America, including Mexico totals 5%. Global total was 42.5 billion minutes for 1992.

Fig. 8.2.
Major international PSTN traffic flows, 1992 (percentage of global total)

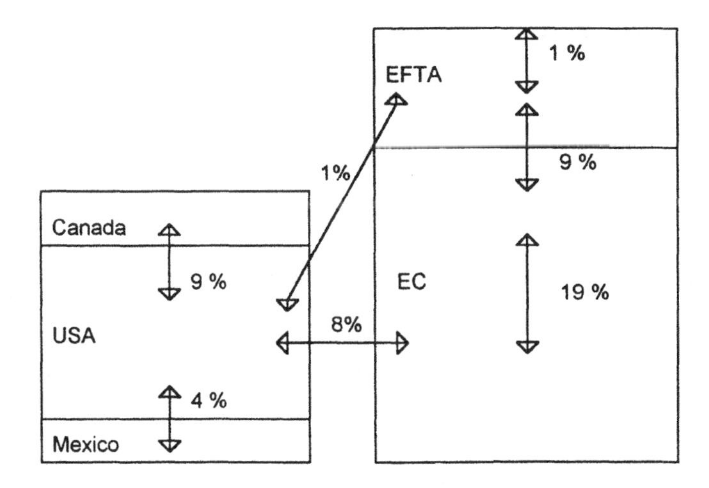

International PSTN traffic flows between and within these two blocs represent 51% of global traffic flow of 42.5 billion minutes.

Table 8.1.
USA, EC/EFTA, International PSTN Traffic Matrix, 1992 (Billion Minutes)

	Outgoing (minutes)					
Incoming	*USA*	*Canada*	*Mexico*	*EC/EFTA*	*Rest*	*World*
USA	–	1.5	0.6	1.4	1.8	5.3
Canada	2.2	–	0	0.2	0.3	2.7
Mexico	1.3	0	–	0	0	1.3
EC/EFTA	2.4	0.3	0	12.5	2.2	17.4
Rest	4.3	0.5	0.1	4.4	6.5	15.8
World	10.2	2.3	0.7	18.5	10.8	42.5

Table 8.2.
USA, EC/EFTA, International PSTN Matrix, 1992 (% of Global Traffic)

	Outgoing (%)				
Incoming	*USA*	*Canada/ Mexico*	*EC/EFTA*	*Rest*	*World*
USA	–	5.0	3.4	4.1	12.5
Canada/Mexico	8.2	–	1.2	0	9.4
EC/EFTA	5.7	0.7	29.3	5.4	41.1
Rest	10.1	1.2	9.6	6.1	37.0
World	24.0	6.9	43.5	25.6	100

bers of family members are living and working abroad and with the family member abroad doing the calling. What is it that leads these expatriate-type communications to have such an influence on the balance of international PSTN traffic flows, when one would expect business traffic to be greater than personal or residential traffic? The reason is that residential traffic is confined (largely although not exclusively) to the public market whereas much business traffic is increasingly passing via the private market.

The first conclusion, therefore, is that the U.S.–EU/EFTA transatlantic PSTN market comprises nearly 10% of global international traffic, is valued at about $5 billion annually and is growing about 15% a year in volume terms and between 5% and 10% a year in nominal terms, as tariffs fall.

However, the growth in this public market is being outstripped by the growth in the private market where traffic flows over private circuits. It is impossible to assess the traffic in this private market and it is also relatively difficult to assess the magnitude of the market in terms of number of circuits or capacity equivalent.

However, there is evidence to suggest that at least a quarter of transatlantic capacity in 1992 was accounted for by private circuits. Moreover, over half of this private capacity is concentrated on U.S.–U.K. routes. Private circuits (in excess of 9,6 Kbit/s) between the United States and Europe increased from about 1,500 to just over 1,800 between 1991 and 1992.

At the same time, U.S. carriers' revenue for private circuits to Western Europe increased from $90 million to $122 million (a 30% increase), implying a transatlantic private circuit market of about quarter of a billion dollars (counting the revenue from half-circuits at the other end), but growing much faster than the PSTN market.

THE TRANSATLANTIC SUBMARKETS

The first section highlighted the importance of taking account of traffic that flows over private circuits when examining the transatlantic telephone market, or for that matter, any international market.

One fundamental difference between the public and the private market is that settlement fees (the part of the accounting rate due to the terminating operator) are only payable in the public market. In the private international market, operators pay annual fees for the leased half-circuits in question, irrespective of the traffic that flows over them. For the moment, simple resale (where leased circuits may be connected to the public network at both ends) is not generally permitted on routes between the United States and Europe, although it is permitted (subject to certain conditions) between the United States and Canada and between the United Kingdom and Canada and has recently been approved on the United States–United Kingdom route. The private market, therefore, consists (in theory) of traffic passing over circuits that are either not connected to the public network or at one end only.

However, a number of developments are taking place that highlight the need to examine more closely the increasingly "grey" area between public and private markets. Demand from users in the grey area of the market is developing because service in the public market, although flexible, is expensive whereas service in the private market is insufficiently flexible or cost effective even for the larger users who can afford the choice.

The Prospect of International Simple Resale

One important prospect in the medium term is the spread of international simple resale. This already exists between the United States and Canada (a traffic stream representing 9% of the global total), and between the United Kingdom, Canada, Sweden, and Australia (and more recently between the United Kingdom and the United States); a number of international operators have been given licenses to offer service to the public on the latter group of routes. The recent agreement to allow simple resale between the United States and the United Kingdom adds a significant traffic stream (about 3% of global international traffic) to the "grey" area of the market and opens the possibility for U.S. and U.K. operators, if they so wish, to deliver traffic to each other's domestic networks at national or local bulk tariff or interconnection rates, as opposed to using accounting rates as the basis for transfer prices.

In addition the prospect of cross-border simple resale in the European Economic Area (accounting for 29% of global international PSTN traffic) from 1998 also adds a substantial chunk of international traffic to the potential resale area. It would also increase pressure for the whole of the transatlantic traffic stream (10% of global international PSTN traffic) to fall under simple resale. There is the prospect, therefore, of something between 40% and 47% of current global traffic flows falling under international simple resale arrangements.

No wonder then that many operators are taking steps to try to increase their control over some of these international markets. At one level it is motivated by the desire and the need to follow larger corporate customers further abroad as their traffic grows in terms of volume, geography, and application (voice, data, image)—the new so-called global telecommunications markets. At another level, it is motivated by the realization that the traditional international market, where traffic is passed to a correspondent operator at some notional midpoint for a fee based on accounting rates, is gradually breaking down. It is therefore necessary for operators to go beyond simply exporting to something akin to international merger or establishment abroad.

Indirect Competition Occurring Ahead of Simple Resale or Facilities Competition

Even in advance of the prospect of significant international simple resale, a number of submarkets are developing. The two most interesting questions seem to be whether these submarkets are based on the traditional settlements of international accounting rates and to what extent these submarkets are open to competition.

The fastest growing markets appear to be card-based direct and premium telephone service, various forms of resale, and services marketed as virtual private networks. Direct telephone service began as a convenience service, predominantly for business travellers abroad and was pioneered by the U.S. international operators. It has grown into a form of global indirect telephone competition, which on some routes has put considerable downward pressure on traditional outbound public telephone tariffs.

It is difficult to estimate the value of the direct and premium telephone service market or even the levels of traffic involved, although it is already running to hundreds of millions of dollars for some operators. An indication of the development and extent of the market is that nearly 70 million calling cards were in circulation worldwide in mid-1993, of which about 60 million were issued by AT&T, MCI, and Sprint. In Europe, the estimated card base in mid-1993 was about 10 million cards, of which about one third were issued by the three U.S. operators (European Commission, 1993).

Both in Europe and worldwide, this card-based market is dominated by the traditional infrastructure-based operators. However, resale telecom operators, such as Viatel, IDT, ACC, and Mastercall, and other entities that are increasingly active in international telecommunications services, such as Visa, Mastercard, Diners Club, and so on, use cards as a basis for offering service to individual customers on the move. In the main, however, the current business of resale operators is directed at international outbound traffic from business sites.

It is important to emphasize that the majority of traffic for direct and premium card-based telephone service passes over the PSTN, and is therefore subject to accounting rate settlements. However, settlements relations are reversed. With traditional PSTN use, the outgoing operator collects monies from the subscriber and pays settlement to the destination (and if necessary transit) operator. With direct and premium service, the operator issuing the card does the collecting and pays settlement to the operator in the outbound country (as compensation for network facilities and provision of the green number arrangement), and if necessary to additional destination countries where calls are handed on. Such settlement arrangements over the PSTN network could in principle be replaced by leased circuit arrangements without settlement but this would be tantamount to simple international resale.

The growth of direct and premium card-based services may, therefore, have a number of impacts on prices, traffic, and settlement flows. Increased use of card-based service in a country either by visiting nonnationals or by domestic residents will increase outgoing international traffic. To the extent that the prices of card-based service by foreign operators compete with the prices of the domestic operators, domestic international prices may fall and lead to further increases in amounts of outgoing traffic, possibly at the expense of incoming traffic. However, where increased outgoing international traffic normally raises both receipts to the domestic operator and increases settlement outpayments to foreign operators, more card-based traffic will raise the level of settlement inpayments from foreign operators.

One means of estimating the size of the card-based market is to examine the difference between the balance of actual traffic flows and the balance of traffic flows for determining the accounting rate settlement. Table 8.3 illustrates the situation for traffic between the United States and Germany in 1991. The traffic balance pertinent to accounting rate settlement was 352 million minutes giving rise to payments by U.S. carriers of about $258 million. However, about 147 million minutes were due to direct telephone calls from Germany to the United States. The underlying balance of traffic flow was therefore 58 million minutes, with a direct service market representing 20% of the combined traffic flows and worth in the region of

Table 8.3.
The Impact of Card-Based Traffic, US–Germany 1991 (Millions of Minutes)

	U.S.–Germany	*Germany–U.S.*	*Balance*
Recorded minutes	561	209	352[a]
Direct traffic	n.a.	147	–
Actual traffic flow	414	356	58

a. This is the balance for settlement purposes and gave rise to settlements of about $258 million in 1991.

Respective figures for 1992 were U.S.–Germany 551 million minutes; Germany–U.S. 233 million minutes; settlement balance 318 million minutes; direct traffic 135 million minutes implying actual traffic flow imbalance of 48 million minutes.

Source. Comments of DBP Telekom to Dept. of Commerce, NTIA, Washington, DC March 27, 1993.

$150 million. In 1992, the traffic balance for settlement purposes and the effective traffic balance narrowed slightly to 318 and 48 million minutes, respectively.

Information concerning direct service traffic flows is not available on a more systematic basis. However, country-direct services from all U.S. carriers in 1991 accounted for about 10% of total U.S. settlements or about 900 million minutes in that year, implying a global country-direct market well in excess of $1 billion dollars. In addition, about 25% of outgoing settlement from the United States to the United Kingdom in 1992 is estimated to be for country-direct service, that is just over 180 million minutes of the total of 730 million minutes.

Given this background and the size of the underlying card base, and making conservative assumptions about the proportion of cards used for direct and premium service and average value of card usage, the transatlantic market is estimated to be in the region of half a billion dollars annually.

With respect to international resale services, this market is still relatively small in value terms, even if considerable growth potential exists over the next few years. Callback and related resale services were estimated to be worth about $200 million in 1993.

The third market area, virtual private networks, has been thriving in the United States for some time. One recent key development in Europe has been the award of contracts for international virtual private network services within Europe by a group of companies, known as the European Virtual Private Network Users Association (EVUA). The combined telecommunications services expenditure of the 18

Table 8.4
International Telecommunications Services Submarkets by Region, 1992

Region		*Transatlantic (U.S.–EC/EFTA)*	*EC-EFTA*
Market			
PSTN	• Volume	3.9 billion minutes	12.5 billion minutes
	• Growth rate	15%	10%
	• Value	est. $5 bn.	est. $8 bn.
	• Growth rate	5–10%	5%
	• Settlements	$615 million[a]	–[b]
Private	• Volume	approx. 2000 circuits	n.a.
	• Growth rate	25 %	n.a.
	• Value	$250 million	n.a.
	• Growth rate	30%	n.a.
Card-based	• Value	est. $500 million	est. 10 million cards
	• Settlements	est. $250 million	n.a.
Resale	• Value	est. $200 million	n.a.
VPN	• Value	n.a.	$1 billion

a. Net settlements from U.S. carriers to EC/EFTA; total settlements (both directions) approximately $2 billion (i.e., average of about 50 cents per minute).

b. Net settlements within EC/EFTA are zero; gross settlements are estimated at about $ 3 billion (i.e., average of about 25 cents per minute).

former members of EVUA is almost £ 1 billion annually and the initial VPN contract is expected to be worth about £ 500 million per year. There are approximately 3,000 companies in Europe with more than 100 access lines, although the present contractual arrangements are limited to a relatively small group of companies.

The various markets and submarkets that have been alluded to and their estimated values are summarized in Table 8.4.

THE IMPACT OF INTERNATIONAL ALLIANCES

The previous sections have identified the major components of the transatlantic telephone market including the traditional PSTN market, the private circuit market, subject to increasing entry by third-party operators and resellers, and the card-based direct and premium service market, still based mainly on the public switched network.

It has also been pointed out that the prospect of international simple resale initially on United States–United Kingdom routes and later within Europe will lead to increasing pressure for competitive service provision between North America and Europe generally, albeit subject to the pricing of underlying infrastructure.

Although, therefore, the current spate of international alliances is motivated in the short term by the desire to more effectively offer a number of advanced services to multinational companies and their employees, it may also be motivated by the desire to be positioned for more substantial market openings as regulatory restraints are relaxed. In the same way that international alliances in the airline sector are motivated by difficulties in acquiring trading slots, the need to placate national regulators (including ownership restrictions) and the advantage of offering an identifiable brand name, similar parallel considerations apply in the telecommunications sector. Card-based services constitute a very powerful instrument for building a customer base abroad.

The North American market, including the domestic U.S. and Canadian markets as well as the cross-border market, is already subject to significant competition and tariffs on higher priced routes are in the region of 25 cents per minute. In contrast, competition is only beginning to occur on transatlantic and intra-European routes and the "peak" price ranges per minute in the former (Europe-originated calls) are $0.65 to $1.50 and in the latter 0.4 to 0.8 ecu (45–90 cents).

However, the previous section emphasized that the prospect of international resale in these markets (current PSTN value of over $15 billion) developing in the next 4 years is a real one, even if the current inroads into the market by resellers or card-based competition are still relatively marginal.

A number of implications follow. First of all, considerable downward pressure will be put on accounting rates, as has been the case between the United States and Canada. Second, the alliances that are beginning to take shape to compete for large international accounts will be able to use this base to compete for profitable PSTN traffic, particularly given that constraints on access to competitively priced infrastructure in foreign markets are likely to be slower to disappear. Unisource, the joint venture between KPN, the Swiss PTT, Telia, and now Telefonica (the operator in the country that U.S. executives *could* find on the map) has not only won part of the European virtual private network contract, but is launching a card that can be used for direct and premium service throughout Europe and the rest of the world.

The current proposed major international alliances involve equity participation or potential equity participation between 10 of the largest 17 international operators in the world (see Table 8.5 and Fig. 8.3). Between them, these 10 companies account for about 55% of global outgoing international PSTN traffic (see Table 8.6). Inevitably, these alliances or potential alliances raise significant regulatory concerns at a world level. The BT–MCI grouping has involved considerable regulatory scrutiny in both the United States and in Europe.

At the European end, the proposed stake by BT in MCI was initially notified and examined as a concentration under European Community Merger Control Regulation. In fact, the case was finally examined under Article 85 of the EC Treaty (dealing with agreements between undertakings that may restrict competition) and under Article 53(3) of the EEA Agreement, which is a legal agreement on trading relationships, including the application of competition rules, between the European Union and the majority of EFTA countries.

In giving an exemption or clearance to the agreement, a number of points were taken into account and a number of safeguards required. Among these was the fact that the joint venture company created, "Concert," intended to offer a portfolio of global products, such as data services, value-added services, calling card services, IN services and outsourcing, and so on, which were increasingly subject to third-

Table 8.5
Major Operator Traffic Shares, 1992

Country	*Global Traffic Share (%)*	*Operator*	*Company Share (%)*
U.S.	24.0	AT&T	16.4
		MCI	4.9
		U.S. Sprint	2.2
Germany	9.6	DBPT	9.6
U.K.	6.7	BT	5.4
		Mercury	1.3
France	5.4	France Telecom	5.4
Canada	5.3	Stentor	3.6
		Teleglobe	1.7
Switzerland	3.7	Swiss PTT	3.7
Netherlands	2.7	KPN	2.7
Spain	1.9	Telefonica	1.9
Sweden	1.5	Telia	1.2
Italy	3.5	Telecom Italia	3.5
Japan	3.0	KDD	2.1
Hong Kong	2.7	Hong Kong Telecom	2.7
Mexico	1.6	Telmex	1.6
South America	1.6	Various	

Fig. 8.3.
Traffic flows and major internatonal alliances, (based on 1992 traffic flows).

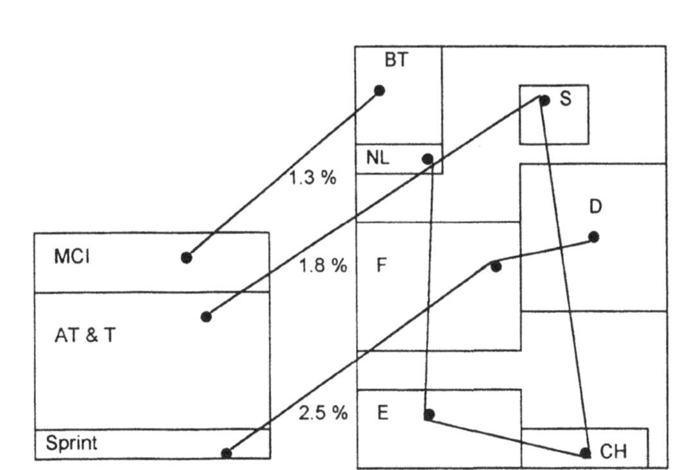

The percentages indicated for the three groupings are the current PSTN traffic flows within the groupings handled by the operators concerned expressed as a percentage of global international traffic. Traffic flows on routes where more than one operator is responsible for outgoing traffic are weighted by operator shares; for AT&T, MCI and Sprint these are 68%, 20% and 9% respectively, for BT a figure of 78% is used.

party competition as well as the bargaining power of purchasers at an international level. Also, safeguards were sought to avoid exclusive provision in the medium and long term in Europe by BT rather than MCI, in order that future competition was not restricted. In addition, arm's length accounting arrangements remain in place for PSTN services.

The question of accounting arrangements for PSTN services in the context of international alliances or agreements is a key issue in the examination of possible distortions to competition, given that alliances are occurring ahead of increasing resale or direct competition on international routes. On the relevant United Kingdom–United States route, BT and MCI do not have exclusive licenses at either end of the relationship. Nevertheless, if they were able to pass traffic to each other at rates significantly below arm's length accounting rates agreed with other operators, then this could distort competition. The same principle applies in the relevant

Table 8.6
Operator Groupings—Combined Global International Traffic Shares, 1992 (%)

AT&T, Unisource	25.9
DBT, France Telecom, Sprint	18.2
BT, MCI	10.3
TOTAL	54.4

United States–Germany–France Telecom submarkets for the proposed Sprint–Deutsche Telekom–France Telecom link and in other submarkets for the potential AT&T–Unisource agreement; all the more so there, because of the potential reinforcement of an existing dominant and exclusive position in some of the countries concerned, where access to the domestic market is still highly restricted.

Regulatory decisions, therefore, regarding the legitimacy and scope of these international alliances depend on the assessment and balancing of the potential benefits of increasing competition in relatively limited but growing segments of telecommunications services to multinational companies, against the potential cost of restricting the potential for more substantial competition to emerge as current restrictions on both services and infrastructure are gradually lifted over the remainder of the decade. The fact remains that the majority of the telecommunications services market will remain oligopolistic, simply because of the substantial resources required to fund and have control over underlying infrastructure and also due to the licence limitations that will remain in place.

CONCLUSIONS

The first, rather obvious, conclusion is that North America and Western Europe have a considerable weight in global telecommunications traffic flows. Together, they account for well over 70% of outgoing international PSTN traffic. Moreover, international traffic flows between and within these two blocs account for about 50% of the global total. Inevitably, the United States and the European Union share an important responsibility for the development of fair and open trading and competition rules in international telecommunications.

The transatlantic market is a large one (10% of global international PSTN traffic and in excess of $5 billion) and growing fast. Net accounting settlements from U.S. to European carriers form a declining proportion of total U.S. settlement outpayments, (in absolute terms it decreased to $646 million in 1992 from totals of $571 million and $686 million in 1990 and 1991, respectively), despite the development of country-direct service that tends to raise net outpayments by U.S. operators (see Fig. 8.4).

More importantly, the transatlantic market can be broken down into a number of overlapping submarkets. In addition to traditional outbound traffic on the PSTN there are thriving markets in private traffic, country-direct service, and other forms of card-based traffic and various forms of resale. Estimates developed in the paper indicate that these sub-markets constitute at least 20% of the transatlantic telecommunications market.

In addition, following agreement on international simple resale between the United States and United Kingdom plus widespread liberalization in the expanded European Union from 1998, the possibility of simple international resale spreading from 9% of global internation traffic flows to between 40% and 50% over the next 4 years is a real one.

Fig. 8.4.
Net settlement outpayments by U.S. operators

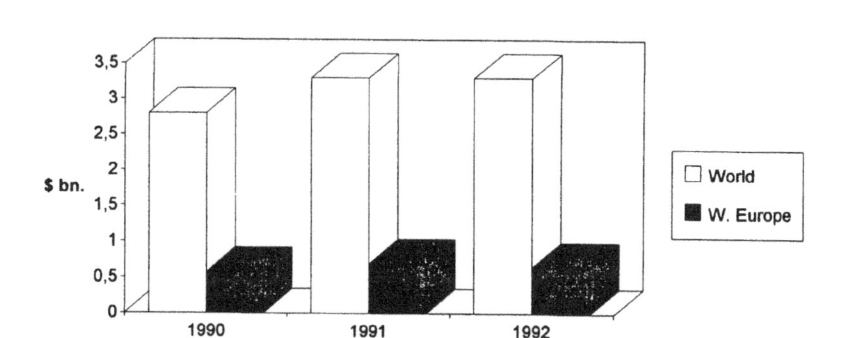

Card-based direct and premium service, virtual private networks, and simple resale have interesting impacts on accounting rates. The first worsens the settlement or trade imbalance as domestic operators become more successful at expanding their business; the second ignores accounting rates; and the third undermines the system altogether. All this demonstrates that attention should focus on the efficient development of the new competitive markets rather than the levels of accounting rates in the old system.

It is not surprising, therefore, that many operators are taking steps to increase their presence in international markets, either via alliances or by taking equity participation or by constructing networks outside their geographical base. It may be motivated at one level by the need to follow larger corporate customers abroad, but it is also motivated by the expectation that the traditional international correspondent relations between operators will gradually break down. This makes it doubly imperative that national regulators and competition authorities increasingly cooperate.

The excessive attention to accounting rates is, therefore, misplaced. Smoothing the transition to more competitive market structures in the international arena will become just as important as managing the transition to a more competitive situation in the domestic context. That requires increasing cooperation between regulatory and competition authorities both to oversee international agreements and alliances, particularly in advance of the lifting of constraints on effective competition, and to ensure efficient and forward-looking interconnection regimes.

ACKNOWLEDGMENT

I am indebted to Dr. Ken Stanley at the FCC who provided publicly available material on transatlantic traffic.

REFERENCES

European Commission, (1993). *New forms of competition in voice telephony services in the European Union.*

ITU and Telegeography Inc., (1994, June). *Direction of traffic.*

9

The Anatomy and Regulatory Repercussions of Global Telecommunications Strategies

Johannes M. Bauer
Michigan State University

The past decade marks the beginning of a transformation of the telecommunications services industry from a predominantly national structure to a structure in which major service providers operate on an international or global level. The emerging regime deviates significantly from the long established framework of the industry, underscoring the fundamental character of the change. Influenced by national security and strategic concerns of governments, the very early years of electronic telecommunications were characterized by more or less separated national systems. The high costs of such an arrangement were recognized soon by business and government leaders and led to several regional and later international agreements to interconnect these national systems, including the foundation of the International Telecommunication Union (ITU) in 1865. In the political framework of the late 19th century, the emerging international arrangements were inspired by the maintenance of national autonomy and sovereignty. This was accomplished by a collaborative system of service provision by national carriers. Thus during this second era, beginning in the mid-19th century and lasting until the present, domestic telecommunications services were provided by national carriers, frequently equipped with far-reaching monopoly privileges, and international services were provided in a system of *joint provision* by these national carriers.[1] Currently, this institutional arrangement is gradually being overlapped and superseded by a multifaceted expansion of telecommunications service providers into the formerly exclusive service territories of other carriers.

1. Only a handful of exceptions existed to this general pattern. For instance, Cable & Wireless and France Cable et Radio were both established as telecommunications carriers for colonial empires and continued to provide services on an international basis. American Telephone and Telegraph (AT&T) until the 1920s or International Telephone and Telegraph (ITT) before it divested its telecommunications activities in the 1980s served foreign markets. Unlike telecommunications service provision, the telecommunications equipment market, although fragmented by numerous trade barriers, was characterized by a much higher degree of international investment and trade (Neu & Schnöring, 1989).

Several factors are commonly identified as contributing to this reorganization (Hausman, 1993; Johansson, 1994). Most prominently among them ranks technological change, supporting the development of a wide array of innovative and specialized equipment, services, and applications, for which the monopoly environment did not seem to provide a conducive framework. Multinational and globally operating enterprises, whose emergence was facilitated by previous advances in communication technologies, in turn demand customized and specialized global telecommunications services provided by one rather than a multitude of vendors, requiring a global presence of carriers to retain large customers. Also, constrained by emerging domestic competition and regulatory restrictions, large equipment and service providers started to reach for profitable investment projects abroad. Last, but not least, the traditional system of service provision through state-owned monopolies frequently suffered from weak performance and did not meet the expectations and needs of their increasingly diverse customer base. As a response, many countries adopted policies of privatization and liberalization of former state-owned monopolies and began to open national telecommunications markets for foreign competitors.[2]

Not surprisingly, the gradual opening of entry into foreign markets after a period of more than a century unleashed a transformation process that is fractured, uneven, and in a state of disequilibrium. Major telecommunications service providers began to transcend their predominantly domestic orientation and embark on a multitude of international business strategies. These strategies differ with respect to the markets and regions targeted, the importance of international activities in the overall portfolio of activities, or the market entry strategies chosen for these ventures. The complexity of approaches makes it particularly difficult to analyze the emerging trends and patterns in a coherent way. Most of the research published so far is of a predominantly descriptive character (Johansson, 1994) or focuses more narrowly on single countries or companies (Cave & Sharma, 1994; Fransman, 1994; Kramer & NiShuilleabhain, 1994; Massari, 1994; Tucker, 1994; Warwick, 1994; Wasden, 1993; Williams & Taylor, 1994). In this paper I attempt to develop a conceptual framework that allows us to organize the empirical evidence and to trace the main currents of the multifaceted inroads of telecommunications service providers into new foreign markets more systematically. Given the rapid pace of change and the early stage of the process, such an undertaking will necessarily be of a tentative and incomplete nature. Nevertheless, we hope to provide a framework that may be useful to derive further questions for future research. Because the current international reorganization takes place within an asymmetric

2. It needs to be noted, though, that the general judgment that the public and/or monopoly provision of telecommunications services always failed is not supported by the historical evidence, nor is the argument advanced by Noam (1992, chap. 3) that it is the success of the model of monopoly provision that finally leads to its abandonment. Moreover, privatization policies are frequently pursued for other reasons than to increase the efficiency of the public telecommunication organizations (PTOs; Bauer, in press).

regulatory and institutional environment, attention will be paid to the implications of the current transformation for national and international regulatory policies.

The next section of the chapter develops a stylized model of a horizontally and vertically integrated firm attempting to maximize profits on a global level and derives some results for its investment and pricing strategies. It also discusses some implications of uncertainty for the specific organizational form chosen for a venture. Following that, I briefly review some of the main strategies of 20 of the largest telecommunications service providers and the emerging regional and service patterns of their international expansion. Next, I discuss some regulatory repercussions and welfare effects. I conclude the chapter with hints for further research.

TOWARD A MODEL OF TELECOM GLOBALIZATION

Telecommunications services to a final customer can be provided in a variety of ways. For instance, a telephone call can be transported via a wire-based or wireless network, in a carrier-owned network or via leased lines, in a service-integrated or specialized network, using digital or analog technology, routed by decentralized traffic control software located in terminal equipment or by centralized software in main switches, and so on. As a result, many strategies are available for carriers attempting to enter a new market. Service providers may decide to invest in their own network infrastructure or just lease transmission capacity, they may attempt a sequential entry strategy by first targeting the terminal equipment market and subsequently value-added services and basic services, or they may follow a broad approach, targeting all those areas at once. Entry can also take place in a multitude of organizational forms including stand-alone projects, alliances, or joint ventures with local or nonlocal partners.[3] The optimal entry strategy is dependent on the specific competitive and regulatory framework of a particular country or market.

A conceptual framework for the study of the international activities of telecommunications service providers should help our understanding of the level, geographic patterns, and the organizational forms of international projects. This leads to the related issue of explaining why a firm engages in foreign direct investment activities or other forms of collaborative ventures instead of exporting goods or services. Furthermore, such a framework should provide a basis to assess the welfare consequences of global strategies in both the source and host countries.

Several of these issues have been addressed in the extensive debate on multinational enterprises (Caves, 1982, is an excellent survey) and the related research on

3. In a stand-alone approach a firm seeks to establish a presence in a foreign market without any involvement of local equity capital. Alliances are agreements between firms that largely maintain the autonomy of the participating organizations although they may be solidified by equity swaps like in the BT–MCI alliance or by mutual participation in the boards of the companies. Joint ventures typically lead to the foundation of a new organizational entity capitalized by the partner firms.

corporate diversification (Montgomery, 1994).[4] Early research by economists and political scientists studied foreign direct investment and diversification processes out of a distress about the market power exerted by large firms and conglomerates and its detrimental effects on competition. Concerns raised included the threat of cross-subsidization and predatory pricing, mutual forbearance (the recognition of interdependence of competitors meeting each other in multiple markets leading to less vigorous competition between them), and the reduction of competition through strategies of reciprocal buying (Hughes & Oughton, 1993; Montgomery, 1994; Shepherd, 1985). In a similar vein are approaches that attribute the emergence of multinational corporations to oligopolistic interaction between suppliers (Knickerbocker, 1973) or the desire to price discriminate (Dunning, 1988).

International economics suggested an arbitrage approach to foreign direct investment, stating that capital would flow from countries where its marginal product is low to countries where it is high. Whereas some of the flows of foreign direct investment are compatible with this argument, at least in the aggregate it has limited power to explain reciprocal investment between countries (Bagchi-Sen & Das, in press; Caves, 1982). More promising than the latter are approaches that emphasize transactional advantages as an explanation of international investment and diversification. Such advantages may emanate from intangible assets of a firm such as know-how, managerial and organizational skills, research and development results, or a specific corporate culture that cannot easily be traded (Buckley & Casson, 1976; Caves, 1982; Dunning, 1988). However, Hallwood (1994) showed that even if no failure in the arm's-length market for firm specific intangible assets exists, it may be advantageous for the profit-maximizing firm to directly invest in foreign operations if an independent firm suffers from a cost and hence price disadvantage (i.e., if a subsidiary can pay a higher price for services rendered by the parent company).

Evolutionary approaches go beyond these market failure and imperfect information approaches. Based on historical evidence, Chandler (1992) argued that the "basic technological characteristics of an industry in which the firms operated at the time of overseas investment were more important than imperfect information costs in determining the number and location of plants built abroad" (p. 89). In this context, he interpreted *technology* as encompassing "the coordinated learned routines in production, distribution, marketing, and improving existing products and processes" (p. 93). A firm is more likely to internalize its expansion if its operations are sufficiently capital intensive to require extensive expertise and skills for success, the minimum efficient plant size is small enough to eliminate the cost advantage of producing from a single plant, and the host market is favorable. An expansion into foreign markets is, furthermore, more likely, if the "extension of a known technological basis is relatively easier than an expansion into new product

4. Throughout the chapter, I interpret *diversification* rather broadly including the provision of services in new geographic markets (*market expansion*), the provision of products or services related to core activities of a firm, or the provision of products or services totally unrelated to the core activities of a firm (*conglomerate diversification*).

markets that requires investment in new technologies, both physical and human" (Scaperlanda, 1993, p. 608).

Another evolutionary approach was proposed by Penrose (1959) and may be of some explanatory power to understand the international expansion of firms. She argued that firms diversify in response to excess capacity in productive factors, such as factors purchased from the market, services created from those factors, and knowledge accumulated over time. An equilibrium of the firm is precluded by three factors: (a) the indivisibility of resources, (b) the fact that resources can be used differently under different circumstances and in a specialized manner, and (c) in the course of the "ordinary processes of operation and expansion" by which "new productive services are continually being created" (p. 68).

Yet another approach to the study of international diversification is provided by modern agency theory. In their seminal work, Berle & Means (1932) pointed to the efficiency implications of the separation of ownership and control in the modern corporation. Agency theory emphasizes the potential conflicts between manager and owner interests, monitoring problems, and the corporate decision-making dynamics resulting from the separation of ownership and control. In the absence of significant ownership stakes, management may face incentives to reduce the value of the firm if such a strategy promotes its own goals (e.g. Mueller, 1972; Shleifer & Vishny, 1989). For instance, managers may pursue strategies of corporate mergers and diversification to reduce the overall risk of operations (and increase the security of their own position) even if shareholder value is not maximized with such a strategy. Diversification and market expansion may also serve the self-interest of managers if they maximize the demand for their particular skills (Montgomery, 1994).

These approaches lead to very different predictions as to the expected behavior of internationally expanding companies and the resulting welfare implications. The market power, the transaction cost, and the evolutionary approaches are compatible with profit maximization behavior. However, only the latter two are also consistent with an efficient use of resources. As Montgomery (1994) pointed out, the agency view is consistent with neither profit maximization nor efficiency. Given the widely divergent performance of telephone companies worldwide, the welfare implications of such diversification strategies are even more complex to assess. It may well be the case that, compared to the status quo ante, even a strategy that is based on an expansion of market power may lead to short-run efficiency improvements.

The empirical evidence as to the effects of diversification on performance is likewise inconclusive. In a study of regulated utilities, Sherman (1989) concluded that the average market value to book value ratio for diversified firms is about the same as for undiversified utilities in very favorable regulatory climates and better than the ratio for undiversified utilities in adverse regulatory circumstances. He interpreted these results as evidence in favor of diversification. Diversification strategies seem to be more successful if targeted to core competencies instead of unrelated products and services. Furthermore, an optimal degree of diversification seems to exist beyond which the complexity of the organization reduces the potential benefits of diversification (Hoskisson & Hitt, 1994; Teece, Rumelt, Dosi, & Winter, 1994). Hoskisson & Hitt (1994) argued that international diversification may stimulate innovation and performance due to the larger market volume to recover research and development (R&D) and other joint and common costs. They concluded that international diversification in core areas may contribute more effectively to corporate performance than domestic product diversification.

Whereas those frameworks provide meaningful ways to organize thought and inquiry into the global strategies of telecommunications operators, it is necessary to recognize some particular features of the industry that may call for modifications of these approaches. First, many telecommunications services are only partially tradeable. Furthermore, some features of telecommunications services may be "embedded" in the delivery infrastructure. Hence, no full substitution of trade for foreign direct investment is possible and some local investment may be a condition for presence in foreign markets. Second, the economic and technological characteristics of networks, such as high investment and perhaps sunk cost, economies of scale, scope, density, and coordination may constitute significant barriers to entry and exit. In addition, the heterogeneous institutional and regulatory framework as well as the political sensitivity of telecommunications impede a free flow of resources between countries and regions. All these factors contribute to discontinuities and strategic actions in the global diversification of telecommunications carriers that need to be accounted for.

With these caveats in mind I propose, as a starting point, a rather simple model of the investment and pricing decisions of an internationally, horizontally, and vertically diversified firm (see Bauer, 1994a, for a more detailed and formal treatment). The globalized firm possesses two interrelated sets of choice variables in addition to variables such as the price and quality strategies for its services. These are the location, level, and structure of investment as well as the pricing strategies for transactions between operations in different locations. In making these decisions, a profit-maximizing firm has to take into account its competitive advantage compared to incumbent suppliers and other competitors, differences in regulatory policy, and also tariff barriers, differences in corporate tax rates, and the specific risks of international activities, such as risks emanating from exchange rate fluctuations or cultural differences affecting the business environment. Hence, although regulatory differences matter for the decisions of the firm, they have to be analyzed in a more general context.

In a simplified but generalized example, assume that Firm 1 is the parent firm located in source Country 1, and Firms 2 and 3 are subsidiaries located in host Countries 2 and 3 (such models have been studied by Copithorne, 1971; Eden, 1985; Horst, 1971). Firm 3 will produce inputs (goods or services) for sale to the two other firms (vertically integrated internal trade). Firms 1 and 2 sell an identical or at least overlapping bundle of goods and services in different locations and may engage in intrafirm horizontal trade. In the case of telecommunications operators, a vertical trade relationship can be the supply of switching, transmission, or terminal equipment by Firm 3. This subsidiary could also provide information like R&D results or consulting expertise to the other firms. Due to the character of telecommunications as a service that is to a large degree dependent on local facilities, horizontal trade is at present a less frequent phenomenon and limited to components of a service. However, with the advancement of technology, it is possible that unbundled telecommunications service elements such as switching or traffic control are provided in the form of horizontal trade between subsidiaries. A similar case is the provision of database and information services for simple resale by an affiliated firm.

Given the present market structures of telecommunications, we are able to make use of some simplifying assumptions. First, it seems justified to assume that the firm, due to its unique tangible and intangible assets, faces a downward-sloping demand curve and hence has some pricing flexibility. However, this flexibility might be reduced or even eliminated by the degree of competition in a market and/or regulatory oversight. For the sake of simplicity, we assume that effective regulatory oversight (if it exists at all) imposes an upper boundary on the prices the firms can charge and thus limits the degree of market power the firms can exert. To simplify the analysis of the investment and pricing decisions, we assume that effective regulation has similar effects as a tax on profits.[5] The global net profits of the firm are the sums of the profits of the individual firms expressed in the currency of the source country. Second, we assume that there are barriers between markets that enable the firm to price discriminate between its markets. Third, we assume that the firm has a clear nationality and aims to maximize its profits in the currency of this source country (see also Caves, 1982).

Assuming first conditions of certainty and an absence of barriers to entry, the decision of the firm can be envisioned as an interrelated two-stage process. In a first step, the production capacity for each location is selected and in a second step optimal policies to allocate production between those locations and policies for the internal transfer prices of transactions are implemented. Both decisions are dependent on the specific market and production conditions, the regulatory and tax conditions (in our model described by the proportion of profit left for the firm after the effects of regulation and taxes are taken into account), the exchange rates between the different countries involved, tariffs levied on transactions between locations in different countries, and the possible transfer prices between different locations.

By totally differentiating the first-order conditions, important comparative static results of the impact of marginal changes in the exogenous variables can be derived. Other things being equal, an increase in the cost advantage of a firm or an increase in tariffs will increase the incentive to invest in another market. On the contrary, an increase of the exchange rate of the foreign currency or an increase in the intensity of regulation in the host market will reduce the incentives to invest abroad (Bauer, 1994a). For our problem, particularly interesting are the effects of asymmetric regulatory policies on the price and output decisions of the multinational telecommunications operator. A relaxation of the regulatory conditions in a

5. Regulatory price control and taxes may differ significantly with respect to their allocative effects. For the purposes of this analysis, the suggested simplification seems justified.

host country (or a tightening of regulatory control in the source country) will provide incentives for a profit-maximizing telecommunications operator to increase investment in the country with more relaxed regulatory oversight and vice versa. If regulatory entry barriers exist, the globalization may take the form of alliances or joint ventures to accomplish a similar effect (Llerena & Wolff, 1994).

Likewise, regulatory differences between countries provide incentives for a globalized telecommunications operator to use internal transfer prices to shift profits to jurisdictions with looser regulation.[6] For instance, let us assume a situation in which the combined effect of regulation and taxation in the host country is tighter than in the source country. Other things equal, a relaxation of regulation in the host country would result in the expected shift of production to the country with the now improved profitability conditions and vice versa. An increase in the stringency of regulation in the source country would induce a relocation of production to other locations. In the case of changes in one or more of the exogenous variables, the overall net effect is dependent on the specific circumstances. For instance, the production relocation effect of relaxed regulatory conditions in the host country can be (more than) offset by a reduction in tariffs levied on exports from the source to the host country.

Although our stylized model allows us to derive some important conclusions as to the behavior of profit-maximizing globalized telecommunications providers, the model needs to be augmented to take account of some real-world characteristics. For instance, to simplify matters, the model does not directly take account of the risks associated with foreign direct investment. Depending on the specific circumstances, foreign investment might carry a higher or a lower degree of risk than domestic investment. The risk of a foreign investment project depends on the specific market situation and the regulatory framework in the host country, the prospective course of future economic development in the host country, exchange rate developments, and cultural and political risks emanating from uncertainty related to different business practices or institutional frameworks.

The overall risk of international operations, very much like a diversification of activities at the national level, not only depends on the risk of each individual project but also on the covariance of risks between different activities. Indeed, an international expansion of activities can contribute to the reduction of the overall entrepreneurial risks if the developments of major market parameters in different host countries are not perfectly correlated. In this case an international portfolio of investments can reduce the risk of shareholders and at the same time stabilize or even increase the rate of return on the invested capital.

One important determinant of the risk of direct foreign telecommunications investment is the market structure in the host country as well as the established regulatory framework. Telecommunications operators have diversified into several

6. From the model analysis it also becomes clear that under conditions of asymmetric regulation the multinational enterprise chooses a transfer price for intrafirm trade that is either above or below the efficient shadow price depending on the existing differential between the effective regulation, taxation, and the existing tariffs.

submarkets of the telecommunications sector including markets with intense competition, but also markets with a relatively low level of rivalry. To the first category belong the "upper end" services such as specialized data communications services for high-volume users. To some extent the bidding processes for franchises for the provision of cellular services (Central and Eastern Europe, Central and Latin America), international switching (Russia), or cable TV services show relatively intense competition, but after the award of the franchises most companies operate in monopoly or duopoly market environments. Frequently, only weak regulatory oversight is established. A similar argument may hold for investment in foreign telephone companies as in the cases of Argentina, Chile, Mexico, New Zealand, or Australia. Indeed, in the competition for foreign investment capital and expertise, national authorities may be tempted to use favorable regulatory conditions as a tool to attract foreign investors with perhaps unintended efficiency and welfare consequences. At present, and given the anticipated future growth in the targeted foreign markets, most of the present globalization projects seem to fall into the category of medium- to low-risk endeavors.

International asymmetries in regulation, tax policy, and tariffs create incentives for a multinational firm to choose profit-maximizing internal transfer prices that deviate from the free trade shadow price. This way, the firm can "relocate" earnings into areas allowing a higher net profit on revenues. The direction of such transfers depends on the differential between the combined tax and regulation rate and the tariff rate between countries. As Bauer (1994a) demonstrated, a profit-maximizing firm will attempt to set transfer prices above marginal cost (of the exporting firm) if the net return to the exporter is higher than the net return of the importer per dollar of profits of imports or vice versa. The optimal direction of profit transfers, therefore, may be either way.

Due to the particularities of telecommunications as a services sector (which somewhat limits the ability of multinational providers to trade horizontally between subsidiaries), vertically integrated firms may enjoy strategic advantages to engage in profit-shifting transfer pricing. The extent to which firms engage in profit-shifting transfer pricing may also depend on differences in the performance level between source and host countries. Technologically more sophisticated firms may possess a significant cost advantage over incumbent suppliers. This may lead to the perverse effect that a technologically superior firm operating in a host country with a low performance level earns part of its profits in a "covert" (i.e., undeclared and undetected) way and retransfers those profits to the source country (or another host country with a higher performance level) to cross-subsidize activities there (Hallwood, 1994, presents a related argument).

The incentive to engage in foreign direct investment and transfer pricing may depend to some extent on the specific ownership structure of firms. For instance, it may not be justified to assume that public telecommunications operators, especially if they are organized as a public administration, are acting as pure profit maximizers. Public enterprises may face a limited incentive to achieve profits if the management cannot fully dispose of those profits (as is frequently the case when profits are used to supplement the general public sector budget).

Under conditions of risk and uncertainty as well as regulatory asymmetry, the organizational form of a project becomes an important strategic variable. Alliances or joint ventures may be important tools to create organizational synergies and induce learning effects among the partners that lead to sustained competitive advantages (Ioannides, 1994; Kanter, 1994; Lawrence & Vlachoutsicos, 1993). However, they may also be important vehicles to overcome regulatory and other entry barriers into markets (Barbet & Benzoni, 1994; Brewer, 1992; Llerena & Wolff, 1994). The formation of alliances and other forms of collaboration also impact the patterns and intensity of the emerging forms of competition in global telecommunication markets (Gomes-Casseres, 1994).

THE ANATOMY OF TELECOM GLOBALIZATION

The sample of carriers studied for this paper differs widely with respect to several variables, such as the importance of global ventures within the overall corporate strategy; the specific forms of global expansion through foreign direct investment, joint ventures, and alliances; and markets that are targeted. Several segments and approaches to international expansion can be distinguished. The market for *genuine global services*, such as virtual private networks, is growing fast but is relatively small compared to the totality of domestic and bilateral services.[7] Only a few carriers decided to explore this market in a stand-alone fashion and even fewer decided to pursue a facilities-based strategy. Frequently, companies form alliances or pursue a strategy of service provision via leased lines, assembled in bilateral negotiations with foreign PTTs and governments. Many new firms have entered specialized subsegments of this market and have contributed to a high intensity of competition. The recent wave of privatizations of PTTs and openings of cellular markets has led to an international expansion into national and regional markets in *multidomestic* strategies. In this area, firms more frequently attempt to get a hold in a foreign market through investment into network facilities. The specific features of inroads into foreign markets are shaped by regulations in both the source and the host countries, which I briefly review now.

The International Institutional Framework for Global Strategies

The strategies for global expansion of telecommunication service providers are shaped and constrained by a diverse set of national legal and regulatory frameworks linked by a few supplementary international agreements. The deregulation movement of the past two decades has somewhat homogenized the previously more fragmented national frameworks for the industry. Common international trends are the liberalization of the terminal equipment market and the market for

7. In 1993, the worldwide market for telecommunications was estimated to be $560 billion, of which about 78% are generated by services and 22% by equipment. The market for genuine global services such as virtual private networks was comparatively small, with a total revenue of approximately $2 billion to $4 billion.

value-added services. However, significant cross-national differences continue to exist in key areas of the industry.

Most significantly, the extent of monopoly privileges established for the provision of basic services, such as voice telephony or packet-switched data communication services, as well as the provision of network facilities, varies greatly among countries. For instance, policy in the United States and a few other countries including the United Kingdom, Sweden, New Zealand, or Japan has attempted to open market access for the provision of basic services and network facilities. Most other countries maintain monopolies for network provision and basic services and, perhaps, envision a gradual transition to a more open framework. Despite this largely monopolistic framework, a growing number of countries have partially or fully privatized the dominant service provider or are considering such a measure. This is frequently being done without the establishment of an effective regulatory framework designed to prevent extortionist pricing, unfair price discrimination, or the cross-subsidization of competitive activities from monopoly revenues by the privatized monopolies.[8]

National rules governing the horizontal and vertical integration and diversification of telecommunications service providers, such as those that deal with the cross-ownership of cable television systems by telephone companies, also differ. Last but not least, national frameworks differ with respect to the stringency of regulatory control and antitrust oversight exerted over dominant providers of telecommunication services. Countries in the developing world and former Communist countries in transition sometimes are forced to create very favorable conditions for infrastructure service providers including exclusive franchises and lax price control to succeed in the attraction of foreign investment capital and technology.

These diverse national settings are complemented by a somewhat rudimentary international framework of technical and trade rules. Key provisions about technical standards are arranged through the International Telecommunication Union and, more recently, regional standards organizations such as the European Telecommunications Standards Institute and industry groups. The latest round of General Agreement on Tariffs and Trades (GATT) talks has outlined provisions for the international trade in services that directly and indirectly affect the global offering of telecommunications services. Both technical and trade rules do not establish a coherent framework to oversee the corporate strategies of service providers. This institutional diversity, together with technological and managerial advantages of telecommunications service providers from industrialized countries, creates profitable investment opportunities outside of the established service domains of telephone companies.

---

8. In recent years, international organizations such as the World Bank, the International Telecommunication Union (ITU), or the Organization for Economic Cooperation and Development (OECD) have accentuated the need for a stable (but not necessarily efficient) regulatory framework as a precondition for increased private-sector participation in telecommunications (ITU, 1994; O'Neill, 1994; Wellenius & Stern, 1994; World Bank, 1994).

Corporate Approaches

The ongoing global diversification of carriers is not an entirely new phenomenon. Firms such as AT&T, ITT, or Cable & Wireless had extensive investments outside of their source countries around the turn of the century. Likewise, the telecommunications equipment industry was operating on an international level from its very beginning (although it was for a long time constrained by numerous entry barriers into national markets such as procurement codes favoring national suppliers). Bohlin & Granstrand (1991) estimated the degree of internationalization from around 5% for AT&T to almost 95% for Philips. By the same token, the loss of a protected home market for many manufacturers has resulted in a global wave of mergers, acquisitions, joint ventures, and alliances. These include the cooperation between AT&T, Philips, and STET, the fusion of CGE and ITT Europe into Alcatel NV.; the merger of GEC/Siemens and Plessey; the cooperation between Fujitsu and GE; the purchase of Rolm by Siemens; or the partial takeover of STC by Northern Telecom. Several of the large manufacturers have recently attempted to diversify into the provision of (specialized) telecommunications services. Ericsson, Alcatel, and Siemens provide VANS, and Motorola is part of several consortia to build up personal communications services. By pursuing such strategies of vertical forward integration into niche markets, manufacturers and service providers build future entry opportunities for other, related service markets.

In the telecommunications services markets, three generic strategies are visible and are often pursued in parallel. As conditions to lease lines and provide value-added services were gradually liberalized by an increasing number of countries, carriers from industrialized nations took advantage of the new conditions by entering bilateral agreements with the local PTOs to either provide international voice or value-added services. Examples of such a strategy of *service diversification* are GEISCO or MCI, which has built a service presence in more than 60 countries (Johansson, 1994). In markets where the facility basis is important to provide high quality service, such as the markets for corporate services, carriers such as AT&T, BT, Cable & Wireless, MCI, or Sprint have embarked on a second strategy of building their own (global) backbone networks and securing a presence in important international markets. Similar approaches are being taken by the various satellite-based consortia such as Iridium or Inmarsat. To facilitate such a time-consuming and daunting task as well as to share risks, many carriers, again including AT&T, MCI, or BT, but also many of the (former) PTTs, have joined forces in regional or global partnerships (Table 9.1). These include the "tighter" alliances of EuNetCom, and Unisource, the "looser" alliances of AT&T World Partners or the Nordic alliance, and the cooperations between BT and MCI and Cable & Wireless and Bell Canada Enterprises (BCE), each solidified by a partial takeover of stock. These strategies have, so far, led only to a limited cross-penetration of foreign markets via foreign direct investment, maintaining the predominantly national structure of the industry.

Table 9.1
Selected Transnational Alliances (Status October 1994)

Alliance	*Partners*	*Plans*
EuNetCom	Deutsche Bundespost Telekom and France Télécom	European backbone
Financial Network Association (FNA)	Collaboration by MCI and including Belgacom, France Télécom, Hongkong Telecom, Italcable, Kokusai Denshin Denwa, Mercury Communications, Singapore Telecom, Stentor, Telecom Australia, and Telefónica	Global backbone, initially leased lines
Global European Network (GEN)	Cooperation between BT, Deutsche Bundespost Telekom, France Télécom, STET, and Telefónica	Backbone between member countries, not directly marketed to end-users
Infonet	Collaboration between 11 national service providers	VANS
Managed European Transport Network (Metran)	Collaboration involving 26 European PTTs	Synchronous Digital Hierarchy backbone planned for 1996
Nordic Alliance	Tele Denmark, Telecom Finland, Norwegian Telecom	
Temanet A/S	Collaboration between Telecom Denmark and Maersk Data A/S	Leased lines
Unisource N.V.	PTT Telecom Netherlands, Swedish Telecom, Swiss PTT, Telefónica, Telia	European backbone interconnected with U.S. and Japan; leased lines, data, messaging, network management
AT&T World Partners	By AT&T, includes Deutsche Bundespost Telekom, France Télécom, Kokusai Kenshin Denwa, Unitel, Singapore Telecom, Telstra	

Sources: Own research; ITU (1994); Heywood (1993).

More recently, and based on the continuing gradual liberalization of data communications, cellular and satellite-based markets, and the privatization of state-owned telephone companies, more direct strategies of foreign direct investment, joint ventures, and alliances are being pursued and mark the beginning of a more significant cross-penetration of national markets. Unlike the firms participating in the earlier stages of internationalization, in this latest surge carriers that are, due to their monopoly or dominant status in all or part of their source operations, subject to national price control and other forms of social regulation, play a much more visible role.

On a global level, investment funds flow from industrialized countries to developing countries especially in Latin America and Asia (but barely to Africa) as well as to the countries emerging from the former Communist bloc. A similar movement of carriers to cross-penetrate markets within the industrialized nations is ob-

Table 9.2
The Regional Structure of Foreign Direct Investment Projects
(Number of Projects as of October 1994)

	North America	*South/Central America*	*Western Europe*	*Eastern Europe/ NIS*	*Middle East*	*Africa*	*Asia*	*Australasia Oceania*
AT&T	1	8	1	4	1	–	17	2
Ameritech	–	–	2	2	–	–	–	2
Bell Atlantic	–	1	1	2	–	–	–	2
BCE	4	2	3	1	–	1	–	1
Bell South	–	5	8	–	1	–	2	4
BT	3	1	2	–	–	1	–	1
DBP Telekom	–	–	–	8	–	–	–	–
France Télécom	1	2	2	2	–	–	2	–
MCI	1	2	–	–	–	–	1	1
NTT	2	–	–	–	–	–	2	1
NYNEX	–	–	2	2	–	1	1	1
Pacific Telesis	–	–	7	–	–	–	7	–
PTT Nederlands	–	–	–	3	–	–	–	–
SW Bell	–	1	1	–	1	–	1	1
Sprint	1	–	2	3	–	–	2	–
STET	1	3	1	1	–	–	–	–
Telefónica	–	7	1	1	–	–	–	–
Telstra	2	–	–	3	1	–	7	1
US West	–	–	2	9	–	–	2	–
Total	15	32	35	41	4	3	44	17

Source: Azoulay, Sarkar, & Bauer (1994).

servable. Of the 190 recent projects surveyed in this third category of approaches,[9] 23.2% target Asia, 21.6% target Eastern Europe and the Newly Independent States (NIS), 18% focus on Western Europe, 16.8% target South and Central America, 8.4% target Australasia and Oceania, 7.9% focus on North America (mostly Canada), and only 3.7% of the projects focus on Africa and the Middle East (Table 9.2). Fifty-seven percent of the projects use wire-based and 43% use wireless technology. Basic services are provided by 56% of the projects and 44% of the projects focus predominantly on enhanced services.

The only carrier with a presence in all regions is Cable & Wireless. Geographically, AT&T is the second most diversified firm followed by Bell Canada Enterprises. Most other carriers have targeted a few regions, not surprisingly with a strong presence in the OECD countries, the fast growing markets of Asia, and the prospective growth markets of Central and Eastern Europe. Several carriers have focused on one or two core regions. For example, Deutsche Bundespost Telekom and PTT Nederlands have a strong presence in Eastern and Central Europe, and Pacific Telesis (through Air Touch) has targeted Western Europe and Asia.

Based on the types of network technologies employed by different carriers, another pattern becomes visible. The majority of carriers are pursuing a somewhat balanced expansion with a dominance of wire-based over wireless communications projects. Some carriers, including AT&T, Bell Canada Enterprise, BT, MCI, and Sprint focus their activities mostly on wire-based concepts. On the other hand, U.S. West, Pacific Telesis, and in particular Bell South have predominantly penetrated the markets for wireless services.

These tentative patterns emerge out of the coincidence of several factors. With the mentioned caveat of the limited tradeability of telecommunications services, the data are consistent with the hypothesis that technological and transactional advantages are the main driving forces of the international expansion of carriers. These may be based on a record of rapid network and services improvement (even if the domestic performance is weak, such in the case of Telefónica) or on an overall history of excellent performance. Such advantages may also stem from a competitive advantage based in a geographic or "cultural vicinity" such as the expansion of Telefónica into the South American markets, of Southwestern Bell into Mexico, or of Deutsche Bundespost Telekom into Eastern Europe and the Newly Independent States.

9. The carriers included in the research for this chapter are AT&T, Ameritech, Bell Atlantic, Bell Canada Enterprises, Bell South, BT, Cable & Wireless, Deutsche Bundespost Telekom, France Télécom, MCI, NTT, NYNEX, Pacific Telesis (Air Touch), PTT Nederlands, Southwestern Bell (now SBC Corporation), Sprint, STET, Telefónica de Espana, Telstra, and US West. For most of the analysis, Cable & Wireless, due to its unique historical evolution, is treated separately to gain more precise insights into the more recent trends. Foreign equity investment projects were included in the survey if they had resumed operations or were at least close to resuming operations. Local bureaus or other forms of representation that many carriers have established worldwide, were not included. Also, cooperation in the area of research and development was not included. For this reason, the sample is smaller than the one studied by Johansson (1994), but a better indication of the foreign direct investment activities of telcos.

The strong presence of the North American carriers (74.2% of all projects) and the Regional Bell Operating Companies (RBOCs) (37.4% of all projects) is probably due to the earlier opening of the telecommunications markets in the United States and the investor-owned status of U.S. companies as opposed to the still predominantly public ownership status of many foreign PTOs. It is furthermore consistent with strategies to brace for more competitive domestic market conditions. In addition, these carriers have technological and managerial advantages that can constitute a significant competitive advantage in entering a foreign market, especially in areas such a wireless communications and digital network concepts in which the technology is becoming increasingly "homogenized" and thus transferable to other markets with low adjustment costs (Hausman, 1993). In the case of the RBOCs, which are still subject to several of the line-of-business restrictions imposed by the Modification of Final Judgment (MFJ) of 1982, some of the projects may have the character of field trials for an anticipated opening of similar markets at home. The cable ventures in the United Kingdom may fall into this category. It is interesting to note, though, that some RBOCs have consolidated or divested their inroads into cable. Furthermore, the RBOCs did not enter the manufacturing business as vigorously as the argument that their international expansion is a reaction to MFJ restrictions would predict (McCormick, 1993). The significant cash flows generated by shortened service lives and accelerated depreciation of equipment may also have contributed to the search for investment opportunities abroad (Economics and Technology [ETI], 1993).

The observation period is not yet sufficient to discriminate empirically between the various conceptual scenarios presented. The scattered evidence does not favor any approach over another. For instance, whereas some of the international ventures, especially in the provision of basic services are profit making, many ventures still have to reach a significant profit-generating stage. From a strategic perspective, many ventures may serve as gatekeepers to secure future market access and contribute to an expansion of market power beyond national borders (Urey, in press). Last but not least, the agency theory perspective, which would predict revenue maximizing and risk diversification behavior, cannot be refuted by the existing empirical evidence.

REGULATORY AND ANTITRUST IMPLICATIONS

Starting from a status quo ante of monopoly or joint provision of services, global networks in telecommunications evolve in an only partially competitive business environment. It is difficult to predict whether telecommunications is indeed going through a transition period toward a truly global and competitive industry (as many vivid supporters of rapid deregulation believe) or whether the ongoing process will lead to oligopolistic structures, continued fragmented networks and service availability, only weak stimuli for competition in many submarkets of telecommunications, and limited ubiquity in service diffusion (Mansell, 1994, terms these alternative visions as "idealist" and "strategic" scenarios). From a policy perspective, some short-term implications may deserve careful analysis. Regulatory concerns may emerge in two related areas. First, the implications of the globalization of networks for the existing approaches of national regulatory control as well as the repercussions of these regulatory approaches on the globalization strategies need to be assessed. In addition to such a predominantly domestic perspective, the design of an international, or perhaps global, framework that would maximize the efficiency gains from the global restructuring process of the industry needs some thought. Under conditions of very heterogeneous deployment of telecommunications infrastructures worldwide, particular interest needs to be paid to the tension between regulation as a promotional tool to speed up infrastructure development and regulation as a restraining tool to avoid welfare-reducing effects of market power.

As we demonstrated in our conceptual analysis, the international investment strategies of profit-seeking firms depend on the prices of inputs and outputs in the source and host countries, the relevant exchange rates, and tax policies in the source and host countries. In the case of the differentially regulated telecommunications industry, the strategies will also be influenced by the specific regulatory approaches in source and host countries, especially if they lead to differing profitability of the capital invested. If all risks are accounted for properly, global diversification may stabilize and increase the profitability of a firm. However, an asymmetric regulatory framework may create efficiency distorting incentives.

If firms do not operate in perfectly competitive markets (and other things being equal), profit-seeking firms will have incentives to reallocate their investment portfolio from jurisdictions with tighter regulatory control to jurisdictions with less rigid explicit or implicit regulatory constraints. However, the intensity of this effect will depend on the specific regulatory instruments employed. It will be less articulate under a regime of price regulation (which allows profits to fluctuate) than under a regime of profit regulation (which holds profitability constant; Braeutigam & Panzar, 1989). In network-based industries, the quality of service is dependent on the condition of the network infrastructure. Hence, if such a reallocation of funds reduces domestic network investment as some of the time series for investment budgets seem to indicate, a negative local welfare effect may result through a reduction in the quality of service. Due to the limited tradeability of telecommunications services, consumers may not have the ability to substitute for the deteriorated service. Such a strategy may improve the overall profitability of a holding company's operations. From a welfare perspective an assessment is more complicated and clearly shows the tensions between public policy and private behavior. Such a policy by the parent company may interfere with a "social compact" aiming at the provision of a continuous expansion and modernization of the local infrastructure. A telecommunications service provider may postpone investment projects in an attempt to negotiate more favorable regulatory conditions such as a higher rate of return and thus a transfer of rent from consumers to shareholders. However, such a negative local effect need not occur if foreign expansion is financed with extra funds or, alternatively, if other providers are permitted to enter the local market. This last option is frequently obliterated by the remaining monopoly framework in the local service areas.

Another way to evade domestic regulatory constraints open to diversified firms is the strategic use of transfer prices for affiliate transactions to "shift" profits from more stringent to less stringent or unregulated jurisdictions and markets. Such transactions include the purchase of equipment from an affiliate, or the provision of services such as research and development, planning expertise, administrative services, finances and so on. Affiliate transactions can be a vehicle to shift costs to a regulated arm of a holding and, by the same token, to shift profits to a less stringent regulatory jurisdiction, thus increasing the overall profitability of a holding operation. Effective monitoring of those transactions between a local and a foreign affiliate may be rather complicated and safeguards may be difficult to devise.

Last but not least, the risk assessment of foreign ventures and, as a consequence, the cost of capital of the firm may be influenced by the regulatory regime in the source and host countries. A globally diversified firm may attempt to "hedge" investment risks with revenues from (domestic) regulated operations and enter ventures that it would not enter without its partially protected market position. In the case of failure, the captive customers of a firm may have to bear part or even all of a bailout. For the same reason, a firm with operations in regulated markets may be perceived as an organization with "deep pockets" and thus a more attractive partner in alliances and joint ventures.

These potential effects of regulatory asymmetries raise the question of whether the emerging global networks operate under sufficiently competitive conditions to accelerate deregulation. A meaningful delineation of the relevant markets has to acknowledge that some limited substitutability exists between wired and wireless telecommunications services and within those categories perhaps between cable and telephone companies or cellular and microcell-based systems. For the better part of the spectrum of services offered, however, these services are complements rather than substitutes.

For the purposes of this chapter it is sufficient to distinguish between the markets for residential services, small businesses, and medium and large businesses. It is especially the last group that has a high demand for international and global services. Most markets for medium and large business telecommunications services are highly competitive and would justify regulation with a light touch supplemented by antitrust review. On the other hand, at present, residential and small business customers typically face either a monopoly provider or a dominant firm with a fringe of smaller competitors. Typically, these markets are characterized by a fairly high concentration of market power. In the presence of significant legal and/or economic entry and exit barriers as well as the flexibility of incumbents for retaliatory pricing, most of these markets do not meet the criteria for contestability either (Bauer, 1994b). Thus, continued regulatory oversight seems justified in those cases. However, in some more specialized submarkets the fast pace of technological change may create sufficient dynamic entry opportunities to render them at least imperfectly contestable.

It has become standard procedure to initiate a bidding process for wireless or other specialized telecommunications licenses (and sometimes in the process of privatizing state-owned enterprises). Whereas this mechanism introduces some competition for the market, the efficiency effects are doubtful due to the possible dissipation of expected future monopoly profits in the bidding process. In all, at this point in time both actual and potential competition seem powerful but, nevertheless, insufficient forces to justify total deregulation of the emerging global telecommunications market (Trebing, in press).

AN ASSESSMENT AND CONCLUSION

In this chapter, I established a simplified framework to analyze how the existing international asymmetries in market entry, price, and conduct regulation influence the globalization decisions of telecommunications operators. These asymmetries may distort the incentives for international diversification and create incentives to engage in international transfer pricing and profit shifting. On the other hand, these asymmetries also pose obstacles to efficiency-increasing internationalization and diversification strategies. Existing entry barriers into the provision of telecommunications services might prohibit strategies to utilize economies of international joint production or limit market access for more efficient suppliers. The welfare effects of globalization strategies under such asymmetric regulatory conditions need to be assessed on a case-by-case basis but may have both welfare-increasing and welfare-decreasing consequences. Compared to the status quo ante, welfare-enhancing effects of the current restructuring process in one country may coincide with welfare-reducing effects in another country. In the absence of an agreed on framework for the evaluation of such welfare trade-offs, an overall assessment of the global restructuring of telecommunication networks may come to different conclusions depending on the adoption of a specific national or a global perspective.

An assessment also depends on the quality of existing regulation. Nowell & Tschirhart (1993) argued that regulation in general will serve both public and special interest aspects. The dynamics resulting from the enlarged set of strategies of globalized service providers may help improve poor regulation but also contribute to the evasion of sound regulation. Although local improvements in the efficiency of telecommunications service provision can be expected, this analysis has demonstrated that there is a continued need for regulatory oversight and antitrust control. This task is complicated by the "mismatch" between the global operations of telecommunications service providers and the largely national institutional framework of regulation and antitrust enforcement. The challenge of the years to come will be to overhaul this institutional framework as well as the regulatory instruments employed to take advantage of possible efficiency-enhancing strategies and avoiding efficiency-reducing effects. It may be an efficiency improving approach to homogenize the framework for international investment in telecommunications and to establish some minimal oversight over competitive practices, perhaps along the lines proposed by Scherer (1994).

The level of international diversification and foreign direct investment of telecommunications operators is still relatively limited, although a steady increase may be expected over the coming years. For the telecommunications research and policy-making community a number of important issues need to be addressed in this context. Once more observations are available, the issues raised in this paper will be accessible for more detailed empirical study and provide more precise quantitative information of the overall effect of internationalization on the profitability, efficiency, and performance of telecommunications suppliers. There is a clear need to assess the global and local welfare effects emerging from the globalization of telecommunications operators. Last but not least, more research as to the proper international regulatory framework that facilitates-efficiency enhancing globalization while it reduces the probability of efficiency-reducing distortions, will be an important task.

ACKNOWLEDGEMENTS

Part of the research for this chapter was funded by an All University Research Initiation Grant of Michigan State University. The author wishes to thank Martin Cave for helpful comments. Pierre Azoulay and Mitrabarun Sarkar provided valuable assistance in collecting empirical information. Discussions with Patricia McCormick helped clarify some of the arguments.

REFERENCES

Azoulay, P., Sarkar, M., & Bauer, J. M. (1994). *Foreign direct investment of telecommunications service providers. A research compendium* (Working paper). East Lansing: Michigan State University, Institute of Public Utilities.

Bagchi-Sen, S. & Das, P. (in press). Foreign direct investment by U.S. telecommunication companies. In B. Mody, J. M. Bauer, & J. D. Straubhaar (Eds.), *Telecommunications politics: Ownership and control of the information highway in developing countries*. Hillsdale, NJ: Lawrence Erlbaum Associates.

Bauer, J. M. (in press). Alternatives to private ownership. In B. Mody, J. M. Bauer, & J. D. Straubhaar (Eds.), *Telecommunications politics: Ownership and control of the information highway in developing countries*. Hillsdale, NJ: Lawrence Erlbaum Associates.

Bauer, J. M. (1994a). Globalization of telecommunications operators under conditions of asymmetric national regulation. In G. Pogorel (Ed.), *Global telecommunications strategies and technological changes* (pp. 315–331). Amsterdam: North-Holland.

Bauer, J. M. (1994b). The emergence of global networks in telecommunications: Transcending national regulation and market constraints. *Journal of Economic Issues, 28*, 391–402.

Barbet, P., & Benzoni, L. (1994). Asymmetries, competition and internationalization: Telecommunications operators and the future Single European Market. In E. Bohlin & O. Granstrand (Eds.) *The race to European eminence* (pp. 401–409). Amsterdam: Elsevier Science.

Berle, A. A., & Means, G. C. (1932). *The modern corporation and private property.* New York: Macmillan.

Bohlin, E., & Granstrand, O. (1991). Strategic options for national monopolies in transition: The case of Swedish Telecom. *Telecommunications Policy, 15,* 464–482.

Braeutigam, R. R., & Panzar, J. C. (1989). Diversification incentives under price-based and cost-based regulation. *Rand Journal of Economics, 20,* 373–391.

Brewer, T. L. (1992). Effects of government policies on foreign direct investment as a strategic choice of firms: An expansion of internationalization theory. *The International Trade Journal, 7,* 111–129.

Buckley, P., & Casson, M. (1976). *The future of the multinational enterprise.* London: Macmillan.

Cave, M., & Sharma, Y. (1994). Foreign entry and competition for local telecommunications service in the UK after the duopoly review. In E. Bohlin & O. Granstrand (Eds.), *The race to European eminence* (pp. 39–50). Amsterdam: Elsevier Science.

Caves, R. E. (1982). *Multinational enterprise and economic analysis.* Cambridge, MA: Cambridge University Press.

Chandler, A. D. (1992). Organizational capabilities and the economic history of the industrial enterprise. *Journal of Economic Perspectives, 6,* 79–100.

Copithorne, L. W. (1971). International corporate transfer prices and government policy. *Canadian Journal of Economics, 3,* 324–341.

Dunning, J. H. (1988). *Explaining international production.* London: Unwyn-Hyman.

Eden, L. (1985). The microeconomics of transfer pricing. In A. M. Rugman & L. Eden (Eds.), *Multinationals and transfer pricing* (pp. 13–46). New York: St. Martin's.

ETI. (1993). *Patterns of investment by the Regional Bell Holding Companies.* Boston, MA: Economics and Technology, Inc.

Fransman, M. (1994). AT&T, BT, and NTT: Vision, strategy, corporate competence, path-dependence, and the role of R&D. In G. Pogorel (Ed.), *Global telecommunications strategies and technological changes* (pp. 277–314). Amsterdam: North-Holland.

Gomes-Casseres, B. (1994). Group versus group: How alliance networks compete. *Harvard Business Review, 72,* 62–74.

Hallwood, P. C. (1994). An observation on the transaction cost theory of the (multinational) firm. *Journal of Institutional and Theoretical Economics, 150,* 351–361.

Hausman, J. A. (1993). The Bell Operating Companies and AT&T venture abroad and others come to the United States. In S. P. Bradley, J. A. Hausman, & R. L. Nolan (Eds.), *Globalization, technology and competition. The fusion of computers and telecommunications in the 1990s* (pp. 313–334). Boston, MA: Harvard Business School Press.

Heywood, P. (1993, April). Global carriers: Fresh air for cross-border networking. *Data Communications*, 83–89.

Horst, T. (1971). The theory of the multinational firm: Optimal behavior under different tariff and tax rates. *Journal of Political Economy, 79*, 1059–1072.

Hoskisson, R. E., & Hitt, M. E. (1994). *Downscoping. How to tame the diversified firm*. New York: Oxford University Press.

Hughes, K., & Oughton, C. (1993). Diversification, multi-market contact and profitability. *Economica, 60*, 203–224.

ITU. (1994). *World telecommunication development report 1994. Globalisation, technology, restructuring, development*. Geneva: Author.

Ioannides, D. (1994). The internationalization of telephone operators: Survival in an integrating world. In E. Bohlin & O. Granstrand (Eds.), *The race to European eminence* (pp. 391–400). Amsterdam: Elsevier Science.

Johansson, O. (1994). Internationalization and diversification of technology-based services: A comparative analysis of 25 telecommunication service corporations. In E. Bohlin & O. Granstrand (Eds.), *The race to European eminence* (pp. 471–500). Amsterdam: Elsevier Science.

Kanter, R. M. (1994). Collaborative advantage. *Harvard Business Review, 72*, 96–108.

Knickerbocker, F. T. (1973). *Oligopolistic reaction and multinational enterprise*. Cambridge, MA: Harvard University Press.

Kramer, R., & NiShuilleabhain, A. (1994). Cable & Wireless: Services, investments and prospects. In E. Bohlin & O. Granstrand (Eds.), *The race to European eminence* (pp. 79–87). Amsterdam: Elsevier Science.

Lawrence, P., & Vlachoutsicos, C. (1993). Joint ventures in Russia: Put the locals in charge. *Harvard Business Review, 71*, 44–54.

Llerena, P., & Wolff, S. (1994). Inter-firm agreements in telecommunications: Elements of an analytical framework. In G. Pogorel (Ed.), *Global telecommunications strategies and technological changes* (pp. 257–275). Amsterdam: North-Holland.

Mansell, R. (1994). A networked economy: Unmasking the 'globalisation' thesis. *Telematics and Informatics, 11*, 25–43.

Massari, F. (1994). International telecommunications operations: STET and the Argentina experience. In B. Wellenius & P. A. Stern (Eds.), *Implementing reforms in the telecommunications sector. Lessons from experience* (pp. 403–408). Washington, DC: The World Bank.

McCormick, P. K. (1993). *The RBOCs diversify. International joint ventures and investments*. Unpublished paper, Department of Telecommunication, Michigan State University, East Lansing.

Montgomery, C. A. (1994). Corporate diversification. *Journal of Economic Perspectives, 8*, 163–178.

Mueller, D. C. (1972). A life cycle theory of the firm. *Journal of Industrial Economics, 20*, 199–219.

Neu, W., & Schnöring, T. (1989). The telecommunications equipment industry. Recent changes in its international trade pattern. *Telecommunications Policy, 15*, 25–39.

Noam, E. M. (1992). *Telecommunications in Europe*. New York: Oxford University Press.

Nowell, C., & Tschirhart, J. (1993). Testing theories of regulatory behavior. *Review of Industrial Organization, 8*, 653–668.

O'Neill, J. (1994). Privatization of telecommunications enterprises: The viewpoint of foreign operators. In B. Wellenius & P. A. Stern (Eds.), *Implementing reforms in the telecommunications sector. Lessons from experience* (pp. 385–390). Washington, DC: The World Bank.

Penrose, E. T. (1959). *The theory of the growth of the firm*. Oxford, UK: Basil Blackwell.

Scaperlanda, A. (1993). Multinational enterprises and the global market. *Journal of Economic Issues, 27*, 605–616.

Scherer, F. M. (1994). *Competition policies*. Washington, DC: The Brookings Institution.

Shepherd, W. G. (1985). *The economics of industrial organization*. Englewood Cliffs, NJ: Prentice-Hall.

Sherman, R. (1989). Efficiency aspects of diversification by utilities. In M. A. Crew (Ed.), *Deregulation and diversification of utilities* (pp. 43–63). Boston, MA: Kluwer Academic.

Shleifer, A., & Vishny, R. W. (1989). Management entrenchment: The case of manager-specific investments. *Journal of Financial Economics, 25*, 123–139.

Teece, D. J., Rumelt, R., Dosi, G., & Winter, S. (1994). Understanding corporate coherence. Theory and evidence. *Journal of Economic Behavior and Organization, 23*, 1–30.

Trebing, H. M. (in press). Privatization and the public interest: Is reconciliation through regulation possible? In B. Mody, J. M. Bauer, & J. D. Straubhaar (Eds.), *Telecommunications politics: Ownership and control of the information highway in developing countries*. Hillsdale, NJ: Lawrence Erlbaum Associates.

Tucker, H. (1994). Evaluating investment opportuinities: Bell Atlantic's approach and the New Zealand experience. In B. Wellenius & P. A. Stern (Eds.), *Implementing reforms in the telecommunications sector. Lessons from experience* (pp. 391–401). Washington, DC: The World Bank.

Urey, G. (in press). Telecommunications investment and global capitalism. In B. Mody, J. M. Bauer, & J. D. Straubhaar (Eds.), *Telecommunications politics: Ownership and control of the information highway in developing countries*. Hillsdale, NJ: Lawrence Erlbaum Associates.

Warwick, W. (1994). A review of AT&T's business history in China. The Memorandum of Understanding in context. *Telecommunications Policy, 18*, 265–274.

Wasden, C. D. (1993). *A descriptive compendium of the international activities of major U.S.-based utility holding companies*. Columbus, OH: The National Regulatory Research Institute.

Wellenius, B., & Stern, P. A. (Eds.). (1994). *Implementing reforms in the telecommunications sector. Lessons from experience*. Washington, DC: The World Bank.

Williams, H., & Taylor, J. (1994). Competencies and diversification: The strategic management of BT. In E. Bohlin & O. Granstrand (Eds.), *The race to European eminence* (pp. 67–78). Amsterdam: Elsevier Science.

World Bank, (1994). *Infrastructure for development. World development report 1994*. Washington, DC: Author.

IV

UNIVERSAL SERVICE

10

Universal Service: Prosaic Motives and Great Ideals

Harmeet Sawhney
Indiana University

Since the divestiture of AT&T, universal telecommunication service has been discussed from a number of angles. Some researchers have focused on the historical development of this concept (Dordick, 1990; Hills, 1989; Lavey, 1990; Mueller, 1993). Others have studied the impact that unavailability of universal service would have on society, focusing on the stratification of society between the "information rich" and the "information poor" (Bowie, 1990; Hudson & Parker, 1990; Pressler & Schieffer, 1988). Recently, the focus has shifted toward creating a new definition of universal service (National Telecommunications and Information Administration [NTIA], 1991; Pacific Bell, 1988).

The problem of formulating appropriate universal service policies was relatively simple when the telephone networks provided only voice communication. The challenge then was to extend the network in such a manner that the entire population was served. The biggest problem was the cost involved. The solution was found in an elaborate set of internal subsidies between long-distance and local rates that enabled the extension of telephone service to almost the entire U.S. population. However, the recent proliferation of telecommunications-based services has made the issue of universal service very complicated. In a situation where there is a wide range of services it is extremely difficult to develop a consensus about what should constitute a "bare essential" package. Although there is very little disagreement on the necessity of providing enhanced 911 emergency services on a universal basis, the discord becomes severe with other services such as home banking, distance learning, remote medical diagnostics, surveillance, energy management, special services for the hearing impaired, automatic language translation, voice mail, computer conferencing, and access to databases.

This chapter was originally published in the *Journal of Broadcasting & Electronic Media* (1994, Fall), *38*(4), 375–395.

There is general agreement that the definition of universal service needs to be extended beyond basic voice communication (Gillan, 1986; Hadden, 1991a, 1991b; Information Infrastructure Task Force, 1993; NTIA, 1988, 1991; O'Connor, 1991; Office of Technology Assessment [OTA], 1990; Parker, Hudson, Dillman, & Roscoe, 1989; F. Williams, 1991). Scholars and policymakers have accordingly focused their attention on the redefinition of universal service. The National Telecommunications and Information Administration (NTIA) suggests that expanded basic service should include touch-tone, emergency communication services (e.g., 911), services for the hearing impaired, and equal access to competitive long-distance carriers (NTIA, 1991). Others have sought to include voice, video, and data in the redefined universal service (O'Connor, 1991; Pacific Bell, 1988; F. Williams, 1990). Some have even suggested that redefined universal service should include access not only to communications networks but also to information services (Dordick, 1991; F. Williams & Hadden, 1991). Perhaps the most comprehensive attempt at redefining universal service was made by California's Intelligent Network Task Force (Pacific Bell, 1988). The Task Force redefined universal service as affordable access for virtually all citizens to: (a) the intelligent network,[1] and (b) a package of essential services that includes touch-tone service, access to emergency services, access to public information services, access to educational services, services for customers not fluent in English, and services for persons with disabilities.

The Task Force recommendations are fairly detailed and specific, and they have a ring of definiteness about them. However, the definitions proposed by the Task Force and others are rather arbitrary in nature, because they are not based on any theoretical rationale for including some services and excluding others. The selections are essentially made on the basis of good judgment. At best, the Task Force recommendations and other such attempts serve an illustrative purpose by providing a sense of what should be included in redefined universal service. They have failed to facilitate the emergence of a consensus. Yet, the lack of success to date has not dampened efforts. A new definition is still considered to be the key piece of the universal service puzzle.

There is little indication that a more intensified effort in the current direction will lead to a resolution of the universal service problem. We need to reassess the basic assumptions underlying the current debate and explore alternate ways of framing the issue of universal telecommunication service. The analysis in this chapter approaches the issue within telecommunications by looking at how similar

---

1. The Task Force defined an *intelligent network* as a telecommunication system that offers the following services to all the subscribers: a transparent gateway to databases and other information services provided from a variety of sources; network protocol conversion between unlike computer systems; assured privacy for communications and transactions handled via the network; simultaneous voice and data services; store-and-forward services such as voice mail, software delivery, some forms of videotex and audiotex, and advanced 976 services; transmission and routing for such home-oriented services as household security, health care monitoring, and remote environmental control; provision for network access by disabled persons and those not fluent in English; and, as technology advances, such services as automatic language translation.

situations were tackled in the case of universal education, universal telephone service, and other such "universal" services. The study also includes universal suffrage, which, although not a "service," is a principle that created the foundation for the provision of other services on a universal basis. This study is based on the logic that it is not the first time that society has been faced with the problem of providing a service to all its citizens. The repertoire of past experience is a useful resource and elements drawn from there can have a heuristic value for devising creative solutions that might provide a conceptual breakthrough. This chapter adopts an innovative approach to overcome the shortcomings of the past attempts that have suffered from a narrow focus on telecommunications technologies. Limited experience with the emerging technologies and the resulting paucity of data have constricted the development of an appropriate conceptual framework for analyzing universal service issues. In this study, the net has been cast very wide.

The main thesis here is that the development of universal service is primarily a function of politics, economics, and social values; the specific characteristics of a particular technology or service are of secondary importance. The problem is fundamentally the same whether the service under consideration is education, electricity, or telecommunications. Therefore, there is a great deal of consistency in the way a society resolves the question of providing a service on a universal basis. In other words, although the specifics of each individual situation are different, there is a pattern that underlies the development of universal service within a society. An understanding of this pattern can aid the formation of a conceptual framework that would be most appropriate for analyzing universal service issues in the telecommunications arena.

The chapter is organized into four sections. The first part of the analysis deals with the public posturing by the different players. The discussion highlights the themes that have recurred in the rhetoric on universal service. The second part of the analysis examines their actual motives and reveals a wide chasm between rhetoric and reality where different groups supported universal service only when it furthered their private agendas. At the same time, the hypocritical "talk" has served a useful purpose, because it was a combination of "talk" and motives that made the achievement of the ideal—universal service—possible. The third part focuses on the role of competition and the impetus it provides for making the service ubiquitous. Finally, the implications of this analysis are discussed to shed some light on the current debate over universal service for telecommunications.

THE "TALK" ABOUT UNIVERSAL SERVICE

The road to the provision of a service on a universal basis is a contentious one. The main point of disagreement is not the ideal but the price the society is willing to pay for it. These debates reflect a culture's struggle to come to terms with its internal inconsistencies—the ideal of equal participation and the lack of resolve to make it possible. As Bellah, Madsen, Sullivan, Swidler, and Tipton (1985) pointed out, "cultures are dramatic conversations about things that matter to their participants" (p. 27). The United States has had many such conversations before about universal service. Only the topics have been somewhat different—suffrage, education, rural electrification, telephones, and other such rights and services. These conversations have been dramatic, because universal service has always mattered to Americans, although in a very complicated way.

The discourse about universal service takes place at three levels—individual, social system, and humanity. At the individual level, the discussion is on the welfare and rights of each individual human being. The discourse on the social system level deals with the benefits to the society as a whole. The relationship between universal service and humanity is complex and therefore the discourse is not well articulated. The discussions range from the religious to secular hopes for human unity. The recurring themes in these conversations are discussed in the following.

Universal Service and the Individual

One of the arguments for universal service is that access to many modern services, including telecommunications, is a basic human right. Every person has a right to these services by the mere virtue of being a citizen. "The entitlement argument asserts that in a modern society telephonic communication, like education, basic medical care, and postal service, is an inherent attribute of citizenship" (Pool, 1984, p. 115). The moral basis of this claim is that telecommunications services have now become so important that an individual without access to them is not equipped for everyday life. The telephone is no longer a luxury. It is a necessity in a modern society. Therefore, no one, including the poorest individuals, "should be denied the opportunity to phone for help in an emergency or be denied the participation in the life of the community that the telephone provides" (Pool, 1984, p. 115). The concern for potential isolation has led to policies that include telephone subscription as a part of welfare assistance. This argument has also gained some legal force. For example, the Montana Supreme Court ruled in a 1987 case that lack of a telephone is a significant "barrier to employment" (Hadden, 1991b, p. 65).

We find similar appeals for the provision of education, electricity, and medical coverage on a universal basis. In the mid-1800s the masthead of the *Working Man's Advocate* read, "all children are entitled to equal education; all adults to equal privileges" (Binder, 1974, p. 33). The argument here was that universal education is a necessary requirement for modern life. In 1925 this sentiment reappeared in a speech by L. J. Taber, master of National Grange, who saw electricity as a basic right and therefore implored the electric utilities "to render conspicuous service to humanity and to bring Electrical Sunshine to all American homes, and with it the confidence that the rights of the humblest citizens are being protected" (Nye, 1990, p. 301). Today, the concern about universal medical coverage is generating similar pleas. The individual rights argument was well articulated by Pope John XXIII who, in his 1963 encyclical, wrote that all humans had "the right to bodily integrity and the means necessary for the proper development of life" (Mitchel, 1993, p. 31). These "rights" have a weak foundation because their only basis lies in a moral argument. They are not enshrined in the U.S. Constitution.

This line of reasoning provides an attractive argument for universal service, because it sidesteps complex cost–benefit questions. Once access to a service is accepted as a right, the cost–benefit equation becomes a secondary consideration. The society is then obligated to extend the service to everyone irrespective of cost. This is why, at times, champions of universal service have chosen to pursue a rights-oriented strategy. For example, Assemblywoman Gwen Moore adopted this strategy for her universal service campaign in California. She and her staff drafted legislation that asserted that access to a telephone was a basic right. The argument was grounded in a broad interpretation of the state constitution's "free speech" clause by the California Supreme Court (Jacobson, 1989). The rationale for this strategy was that "if the freedom to communicate is a fundamental right then access to the means of communication must also be a fundamental right. Without access, one cannot be a part of the telecommunicating community" (Jacobson, 1989, p. 59).

Universal Service and the Social System

The system benefit argument is more powerful than the one based on individual rights because it appeals to self-interest rather than altruism. The basic argument here is that the provision of a service on a universal basis makes it possible for the social system as a whole to function more efficiently. In the case of the telephone network, this logic takes the form of the network externalities argument that says that each additional subscriber increases the value of the entire network, because millions of other subscribers can now access the newest subscriber. The overall benefit to the system is likely to be more than the cost of subsidizing the service to those sections of the population that cannot afford telephone service. This benefit is seen as an economic justification for pouring resources into universal service.

The system benefit argument is not unique to telecommunications networks. It was articulated most clearly in the debate over universal education. The *American Monthly Magazine* contended that "viewed in a political light, education is imparted, not for the sake of the recipients, but for the state of which they are members" (Jackson, 1941, p. 63). The modern society is seen as a mechanism in which all the different parts have to work together. Therefore, it is in the interest of society that it educates all its members. "Where every individual thought and deed affected the social mechanism of the whole, it became the interest of the whole to provide the necessary education for its parts" (Ditzion, 1947, p. 65). The cold rational logic devoid of all sentiments is that the benefit that accrues to the individual is almost incidental. "Why does the state take money from your pocket to educate my child? Not on the ground that an education is a good thing for him, but on the ground that his ignorance would be dangerous to the state" (Strong, 1963, p. 99). Thereby the tax for universal service is not a levy on the rich for the benefit of the poor. It is the cost borne by the society for its own benefit (Carlton, 1966).

The only major problem with the system benefit argument is that it is difficult to measure the benefits, but the costs, which are easily measured, are very large. The universal medical coverage debate provides one of the few instances where the cost of not providing universal coverage is very visible. The failure to provide medical coverage on a universal basis leads to the excessive use of expensive emergency care facilities. A person without coverage does not seek medical help when a disease or sickness is in its earlier stages. The problem continues to fester until it becomes acute and requires emergency treatment. At this stage, it becomes a matter of life and death, and a hospital is unable to turn away a patient who does not have the ability to pay. The hospital bears this cost, but then passes it on to the rest of the population that has medical coverage. The end result of this chain of events is that the final costs are much higher than the amount the rest of the population has to pay to subsidize coverage for those who cannot afford it.

In addition to the economic benefits, other social benefits also have the potential for enhancing the overall system. The new service is often seen as "a fairy wand of social reform" (Ditzion, 1947, p. 24). Universal education was viewed as an antidote to poverty, intemperance, and social unrest. Rural electrification, which became "a symbol to save farmers," was expected to improve the quality of rural life and thereby stem the depopulation of rural areas (Nye, 1990, p. 304). Similarly, the telecommunications infrastructure is now seen as the vehicle for moving our society into the information age. The envisioned benefits range from the generation of new economic activity to a more cost-effective provision of education, medical, and other public services.

Universal Service and Humanity

The system benefit argument has mechanistic overtones. Universal service is seen as a means of enhancing the operation of a complicated mechanism—modern society. On the other hand, the humanity argument is organic in nature and points to the transcendental link between an individual and the rest of the humankind. The relationship between the individual and society is not functional—parts fitting into the system; it is ecclesiastical. The aspiration is to facilitate the union of humankind. Within this context, the communication technologies, both telecommunication and transportation, are viewed as the physical links that make the communion possible.

The large-scale networks are a very visible feature of modern society. They undergird its institutions. Dependence on these networks is so extreme that an industrialized society is not able to function without them. Yet, people have always had an uneasy relationship with them. This uneasiness can be traced back to the time of the emergence of the networks. People who once lived in isolated households increasingly found themselves placed as little nodes on a number of networks—sewage systems, water pipes, gas mains, hot-air ducts, electrical cables, telephone lines, and other systems. Although there was never any doubt that these networks enhanced the efficiency of day-to-day chores, they were disturbing, because "unlike a sewing machine, a stove, or a washbasin, an electrical cable, a gas main, or a sewage pipe was not finite. Each disappeared into the wall or the ground and then was connected to an invisible network that spread beyond the home" (Handlin, 1979, p. 452). These networks were objects of suspicion, because they tied us to strangers. They connected, in a strange way, isolated individuals in an increasingly fragmented society.

These networks also altered the private and public responsibilities in the life of the community. The public authority was cast over an expanding range of activities that were earlier left to the individuals. The affected areas included water supply, sanitation, street lighting, and education among others. "This substitution was full of significant implications for the common man in a society destined to become increasingly interdependent and insecure" (Curti, 1964, p. 288). The telephone bridged the gap between the private and public spheres of an individual's life on a more immediate level. "The telephone was the first electric medium to enter the home and unsettle customary ways of dividing the private person and family from the more public setting of the community" (Marvin, 1988, p. 26). This new link had a disconcerting impact on our lives. The sense of unease continues today. It is aggravated by telemarketers and computer hackers who have an opportunity to intrude into our private domains the moment we connect ourselves to the network.

The impact of the telephone has been particularly powerful, because the connections it creates have more to do with human interaction than the movement of materials or energy. The telecommunications networks are not mere conduits for transporting information. They are the symbolic threads that tie all of us together. Ironically, this sense of unity comes through physical connectivity with total strangers via technological systems. Jules Romains, the French writer, wrote about how the "anonymity of city life could be transcended as individuals used the experience of city life to build a common identity and to discover a common humanity" (Konvitz, 1985, p. 140). The life on the network is similar to Romains' description of urban existence. The network creates a possibility of communication but does not necessarily lead to it. The isolated individuals are tied together, although there is little interaction between them. The situation is analogous to that of an individual in an anonymous crowd.

The utilitarian and symbolic dimensions of telecommunication networks are fused together in a peculiar manner. The mechanistic aspect of the technology enables the transportation of messages, whereas the symbolic aspect fosters a sense of union. The roots of this dualism lie in the deep-seated cultural ideas about the nature of communication. The word *communication* has an interesting history. The words *communication* and *communion* share the same Latin root, *communis*. The original meaning of communication signified communion or shared participation in a common experience. The word had strong religious overtones (Czitrom, 1982). In the late 17th century, the concept was extended to include exchange and transportation of information and materials. Later, the term was also used to refer to technologies that made exchange possible—roads, canals, railroads, telegraph, and telephone (R. Williams, 1976). This meaning of communication has a clear mechanistic orientation, whereas the original meaning, communication as communion, conjures up the imagery of oneness of humanity and union with God.

Universal service is an embodiment of this dualism. It is the technological means that makes communion possible. The larger the number of humans reached, the greater the scope of the communion. Therefore, there is virtue in ubiquity. It is no wonder that each major advancement in communication technologies has created a sense of miracle (Czitrom, 1982). The birth of the telegraph stirred deep religious emotions. "The electric telegraph promised a unity of interest, men linked by a single mind, and the worldwide victory of Christianity" (Czitrom, 1982, p. 10). "Universal communication" became the catchword of the times. The religious discourse fused technology with the idea of communion, and the telegraph was seen as "the wonderful vehicle" for transforming the condition of man. For example, the establishment of telegraph connections with Europe in 1857 had a great impact on the American imagination (Pacey, 1990). The clerics saw God's guiding hand in the unfurling of this momentous development. The poets called it "a loving girdle round the earth" (Field, 1898, p. 203). The optimism was universal. Even a secular source such as *Scientific American* (1881) celebrated the telegraph for fostering a "kinship of humanity" (Fischer, 1992, p. 2). This euphoria can best be described as "organicist electric utopianism" (Thompson, 1990, p. 67). The new technologies were perceived as holding the promise of "the Universal Brotherhood of Universal Man" (Carey, 1989, p. 208).

This hope for oneness has lingered on and become a recurring theme in the discussions on new communication technologies. There is a cultural bias toward ubiquity that seems to be driving our thinking about communication technologies. It is no accident that grandiose claims were made for many communications technologies when they were still in their infancy. In 1838 Morse wrote that it would not be long before "the whole surface of this country would be channeled for those nerves which are to diffuse with the speed of thought, a knowledge of all that is occurring throughout the land; making in fact one neighborhood of the whole country" (Czitrom, 1982, pp. 11–12). In 1878 Bell predicted that "cables of telephone wire would be laid under ground, or suspended overhead, communicating by branch wires with private dwellings, country houses, shops, manufactories . . . and a man in one part of the country may communicate by word of mouth with another in a different place" (Kingsbury, 1915/1972, p. 90). This bias toward ubiquity is generic to all network technologies and not limited to telecommunications. However, it has not been limited to them. Walt Whitman, well-known poet, expressed the hope that the people will be united by a single railroad network that "covers all over with visible power and beauty" (Marx, 1964, p. 224). Calvin Colton, a Massachusetts clergyman, thought that the railroads had the potential of reuniting "the human race that had been dispersed at the time of the building of the Tower of Babel" (Kohn, 1957, p. 18). This discussion suggests that there is a natural connection between the craving for ubiquity and the desire for a communion.

POTPOURRI OF MOTIVES

On a number of occasions an industrialized society has managed to achieve a near universal provision of essential services. It is rather a remarkable achievement of modern civilization. Universal service has never been an issue in the centuries preceding the industrial revolution. This realization raises a number of questions. Why did egalitarianism suddenly become desirable? What purpose did it serve? And what were the motives for providing universal service?

When we examine the discourse about universal service, it first appears that there is a very enlightened perspective on the part of humankind. The publicly espoused motives include inalienable rights of every individual, enhancement of the democratic process, and the betterment of humanity. A closer analysis, however, reveals that the desire for universal service did not flow from the goodness of the human heart. Rather, the groups involved supported universal service because it furthered their private agendas.

The development of universal education provides a particularly revealing illustration of this phenomenon. The biggest impediment to the realization of the educational ideals embraced in the Declaration of Independence and the U.S. Constitution has been the dogged reluctance to "translate sentiments into appropriations" (Ditzion, 1947, p. 10). The founding fathers showered many eminently quotable praises on the importance of widespread education for the functioning of a democracy. However, limited action was taken to appropriate the resources for achieving this ideal. The net result was that education languished for over a century. This drama was played out most explicitly in Indiana. In 1816 the state constitution directed its lawmakers to establish an educational system that would be "free and open to all." At the same time, it asked the lawmakers to delay the undertaking "until circumstances will permit" (Meyer, 1965, p. 387). This qualifying clause provided a convenient loophole for less determined spirits, and the idealistic enterprise was put off for many years.

Eventually, universal education did strike roots. It was driven by machinations of the myriad players who were pursuing their own interests with regard to education. As Katz (1968) pointed out, "the extension and reform of education in the mid 19th-century were not a potpourri of democracy, relationalism and humanitarianism" (cited in Button & Provenzo, 1989, p. 94). The people agreed to pay taxes for a public school system because of other reasons. The 19th century ordinary Americans voted for universal education mainly because they wanted to neutralize the growing immigrant threat, most of whom were Catholics. A tax-supported public school was seen as the "principle digestive organ of the body politic" (Strong, 1963, p. 89). It was expected to Americanize the newcomers. This xenophobic fear elicited the greatest support from the vast majority of Americans. Others had more narrowly defined interests. The propertied elite were apprehensive of the consequences of universal suffrage, which had become a reality during the Jacksonian era in spite of their bitter opposition. They were scared of what they denigratingly called the "mob rule." Their more euphemistic position was that "education could play an important role in reconciling freedom and order" (Kaestle, 1983, p. 5). They wanted the masses to be educated so that they would exercise their newly won political rights in a responsible manner. Their greatest concern was that the masses should conform to the existing system and not destabilize it. Finally, there was the rising class of industrialists who wanted trained manpower for their ever expanding factories.

It would be incorrect to deny the contribution of individuals who were fired by the ideals of enlightened democracy. They were, however, very few in number. Furthermore, only those who were able to co-opt others' agendas into ever widening circles of "overlapping consensus"[2] were able to facilitate the emergence of universal service. They were coalition builders.

> The businessmen and property-owners generally were approached in the name of their own protection. The warring sects of Christendom were appeased with Bible-reading in "non-sectarian" schools. The workingmen and shopkeepers, who asked for a better living and for a realization of their theoretically enjoyed "equal rights," were given some concessions and the advice that education rather than political action would lead them to the Promised Land. (Jackson, 1941, p. 172)

Horace Mann, the great idealist and a practical coalition builder, best captured the essence of this struggle when he described universal education as the "balance wheel of the social machinery" (cited in Binder, 1974, p. 64). Although a very odd metaphor to attach to a rather hallowed undertaking, it served as a symbol that could subsume the various agendas of the different groups.[3]

The development of universal telephone service is also cloaked in a myth that projects its evolution as a product of an egalitarian ideal. The regulatory compact between the erstwhile Bell System and the regulators is credited with making this ideal a reality. Also, Theodore Vail, the chief architect of the Bell System, is cast as the great visionary behind the universal service concept. His often quoted words, "one policy, one system, and universal service," are accepted as a statement of an enlightened business philosophy. This storyline is, at best, full of half-truths (Mueller, 1993).

In 1907 Vail first used the term *universal service* to emphasize the need for a monolithic system that would allow a subscriber to talk to any other subscriber. The vision of universal service was presented as an alternative to the existing fragmented telephone networks that lacked interconnection. He championed universal service as an ideal because it furthered his "drive to achieve political support for the elimination of competition and the establishment of regulated monopoly" (Mueller, 1993, p. 365). The concept did not refer to geographic ubiquity or pro-

---

2. This term has been borrowed from Rawls (1985) who used it in a very different context. In his work on political conceptions of justice, he argued that for justice to be perceived as fair it must be supported by an "overlapping consensus," which he explains as "a consensus that includes all the opposing philosophical and religious doctrines likely to persist and to gain adherents in a more or less just constitutional democratic society" (pp. 225–226). In other words, the system of justice should be based on the common ground or "overlap" between the different schools of thought within a society. It is only then that justice will be accepted as reasonable by all sections of society. In this chapter *overlapping consensus* refers to the common ground or overlap in the agendas of the different players pursuing their own interests with regard to a new technology or service.

3. The other metaphors compared a schoolmaster to a constable, a school to a sentry box of liberty, and libraries to arsenals of democracy (Ditzion, 1947; Jackson, 1941).

vision of service to the entire population. The evolution was slow and gradual. Oettinger very succinctly described the process leading to the development of universal service:

> AT&T's Theodore Vail spoke about the idea of universal service around 1907. The idea was written into the preamble of the Communications Act of 1934, but there is very little legislative history on why and how it got in there. And, even though the words were grand, nothing really happened until around 1945. By then, penetration of telephones, which you described as being almost 95 percent these days, was only about 40–50 percent. So after 40 years of rhetoric, universal service as we now know it came into place somewhere between 1945 and the early 1960s, by which time we had begun to dismantle the apparatus which brought it to us in the first place. (Oettinger, quoted in National Governors' Association, 1988, p. 1)

The universal service concept took on an egalitarian color during the 1960s when competition and antitrust lawsuits threatened to unravel the Bell System. The potential danger to universal service now became a convenient defense for the preservation of the Bell System.

This analysis suggests that there is rarely any direct correspondence between rhetoric and the actual agendas of the different players whose actions influence the development of universal service. The discrepancy between publicly stated ideals and actual motives deserves a closer look.

The use of hyperboles has always been a consistent feature of political discourse. As Edelman (1964) pointed out, "If politics is concerned with who gets what, or with the authoritative allocation of values, one may be pardoned for wondering why it need involve so much talk" (p. 114). The reality is that "talk" greases the wheels of the resource allocation process and ensures acquiescence on the part of the general population. Verbose rhetoric is a necessary feature of political life, and "the employment of language to sanctify action is exactly what makes politics different from other methods of allocating values" (Edelman, 1964, p. 114). Therefore, the public policy formation process can be described as a "historically determined discursive practice, as a way of doing things with talk" (Streeter, 1987, p. 39). Furthermore, politics is a peculiar process that "begins in conflict and eventuates in a solution. But the solution is not the 'rationally best' solution, but the emotionally satisfactory one. The rational and dialectical phases of politics are subsidiary to the process of redefining an emotional consensus" (Lasswell, 1962, p. 185). In the case of universal service, the fusion of idealistic "talk" and private gain makes the emergence of an emotional consensus possible.

COMPETITION OVER A QUILTED TOPOGRAPHY

The fusion of "talk" and motive creates the will for action. There is another factor that solidifies this will and speeds up the process. It is the quintessential American ingredient—competition. It played an important role in the development of the telephone network. At first, Bell companies concentrated their energies on the urban markets. The neglect of rural areas and small towns led to the creation of a "reservoir of unsatisfied demand" (Brock, 1981, p. 107). As soon as the Bell patents expired, independent telephone companies mushroomed all across the rural landscape, thus forcing the Bell companies to enter the rural areas. The ensuing fight led to a reckless race to occupy virgin territory. The network investment decisions were not made on the basis of potential profitability, but on the burning desire to be the first to establish a telephone network in new territories. If it had not been for competition-induced "irrationality," the telephone network would not have grown at the pace it actually did. Thereby competition played a significant role in the development of universal telephone service even though, at that time, the telephone companies were only interested in capturing areas rather than serving everyone in the population. The later objective could only be achieved once Bell System became a regulated monopoly.

The fragmented nature of the peculiarly American sociopolitical terrain gives a spin to the interplay of competitive forces. The political landscape is divided over 90,000 governmental jurisdictions that include 50 states, over 3,000 counties, and a multitude of cities, villages, and townships that are further balkanized into special jurisdictions for administering water supply, sewage disposal, pollution control, highways, harbors, schools, hospitals, airports, police planning, zoning, parks, charities, and many other services (Zelinsky, 1973). Here, each political unit serves as a laboratory or a crucible for social experimentation where a unique set of circumstances often leads to a successful innovation. The competitive pressures then force other political units to follow suit. The development of universal suffrage provides an excellent illustration of this incremental process.

Universal suffrage was neither granted to U.S. citizens by benevolent founding fathers, nor was it bestowed in any single act of great magnanimity. Instead, it was extended in a very piecemeal manner as a result of a protracted struggle between competing political interests. A democracy is in many ways a game of numbers in which the side with the largest number of votes emerges as the winner. In a situation of limited democracy, where the suffrage is restricted, the numbers game can be influenced by expanding the population base of eligible voters. This possibility of manipulation has frequently led to fraudulent practices. When voting privileges were restricted to freeholders, "the practice of creating freeholds at the time of a crucial election was fairly widespread in the American colonies. When an office seeker and his friends thought an additional number of votes was necessary to win an election, it was not unusual for them to create small freeholds for the express purpose of manufacturing votes" (Williamson, 1960, p. 50). The other remedy was to ease the restrictions on suffrage so that a more favorably disposed population could be included in the voter pool. In Pennsylvania, the Quaker party used naturalization of aliens as the means for enlarging its voter base (Williamson, 1960). The Midwestern states extended suffrage to foreigners for economic reasons to "attract immigrants from other states to her unplowed fields" (Porter, 1918/1969, p. 18). As Porter (1918/1969) pointed out, "The philosophy of suffrage has always been more or less opportunistic, if the word is permissible. Suffrage qualifications are determined for decidedly materialistic considerations, and then a theory is evolved to suit the situation" (p. 4).

Once a new more liberal standard was set in a particular region, it created pressure for other regions to also loosen their franchise restrictions. The quilted nature of the U.S. political topography greatly accelerated the process. For example, in the 1770s, Vermont was the first state to do away with property or tax-paying considerations from the right to vote. However, "the dropping of the tax-paying qualification was perhaps not exclusively the result of democratic convictions" (Williamson, 1960, p. 98). The newly formed state did not have a tax-collecting machinery, and therefore a tax-paying qualification would have defranchised everybody. Later, as the state developed, a number of factors worked against the introduction of tax-paying qualification (Williamson, 1960). The establishment of this liberal principle in Vermont created pressure for change in New Hampshire, which soon discarded her tax-paying clause (Porter, 1918/1969). The other states also succumbed to the pressure with the passage of time. As Binder (1974) noted, "By 1850 twenty-seven states had adopted universal manhood suffrage, six had a small tax qualification, and only North Carolina continued to maintain a significant property tax" (p. 7). It was only a matter of time before North Carolina gave in to the pressures of the times.[4]

The universal suffrage spread in an incremental manner with developments in one state influencing those in others. "The gathering sentiment from year to year, modified and influenced in one state by the actions in others, would finally gain complete expression on the convention floor and be recorded in a new constitution which in turn would influence and modify the tendencies in the neighbor states" (Porter, 1918/1969, p. 18). The unrolling of universal suffrage was a series of compromises and political compacts that had very little to do with the espoused ideals. These negotiated interim compromises offered temporary advantages to different players and hence provided motivations for extending suffrage. Each step forward opened new doors and thereby had a snowballing effect. The entire experience has very aptly been described as the "strange phenomenon of suffrage being carried forward on a tide of fallacies and specious doctrine" (Porter, 1918/1969, p. 65).

The decentralized and incremental process described here is a unique feature of the U.S. federal structure, which has often been celebrated as "laboratories of democracy." The process allows for trial-and-error and seat-of-the-pants pragmatism and thereby much innovation (Osborne, 1990).

4. In other situations, the change often had roots in great bouts of hypocrisy. For example, Maine's decision to extend suffrage to its African-American population was not based on any sense of idealism. It resulted from a mixture of hypocrisy and calculated assessment of practical reality. The African-American population was a miniscule minority in Maine and therefore the "righteous men invoked high principles and lived up to them with punctilious consistency—when doing so could not harm the community in the slightest degree. They took pride in being magnanimous when there was no harm in it" (Porter, 1918/1969, p. 52). Although these moves by themselves were fairly hollow, they did set standards that pressured conformance from other states.

DISCUSSION

The current discourse on universal service is framed in terms of the choices that a society must make. According to OTA (1990), "defining universal service is, in effect, making choices about the nature of the society itself" (p. 254). Hadden (1991a) stressed the need for a "deliberate collective choice" (p. 74). The notion of a conscious choice has led to a fixation on a new definition for universal service. The assumption here is that once the benchmark is established, the decision-making process will be relatively easy. The analysis presented in this chapter beckons us to pause and reassess this assumption and explore alternate ways of framing the problem.

The previous discussions on the role of "talk," motives, and competition reveal that there is a great deal of consistency in the way U.S. society marshals its political will, emotions, and resources to attain the egalitarian ideal of universal service. There seems to be an archetypal pattern underlying the development of universal service in the United States that transcends the specifics of each individual service. The following discussion presents three generalizations that can be distilled from past experience and then examines their implications for the current debate.

Generalization 1. The publicly espoused ideals usually bear little correspondence with the actual motives of the different players. The public discourse is hypocritical, but it serves a useful purpose because the fusion of "talk" and motives creates the will for action. However, the momentum for extending a service to everyone builds only when there is an overlapping consensus or the convergence in the agendas of the different players.

Generalization 2. The quilted nature of the U.S. political topography provides varied arenas for the interplay of different actors and their agendas. The small and numerous political units facilitate innovation where each county and each state serves as a social laboratory. Once an innovation takes root in one county or state, it creates pressure for others to follow. The innovation spreads in a piecemeal manner, but the progress is rapid.

Generalization 3. The development of universal service is greatly accelerated by competition. Competitive impulses push the entities involved into making investments that they otherwise would hesitate to make. The county and state governments pour resources into infrastructure development so as to remain competitive with their regional rivals. The providers of commercial services go into areas that are not particularly profitable or cost effective in an effort to grab unoccupied territory before others do. This dynamic facilitates the creation of a ubiquitous infrastructure or service.

Past experience with other services suggests that expanded universal service for telecommunications will eventually emerge out of an intensely contested process. As Anderson (1985) pointed out, "the American political and problem-solving style is incremental, not synoptic—this country is wary of large-scale blueprints" (p. 280). This analysis leads one to believe that the current emphasis on the development of a new definition for universal service is a misdirected effort, because it is unlikely that it will ever be possible to formulate a definition that will be acceptable to everybody. Even if it is possible to forge a universally acceptable definition, it will have limited utility in the political arena, because the different players are likely to pursue their own private interests in spite of their public support for the new definition of universal service.

Contrary to the conventional wisdom, the development of universal service does not hinge on enlightened choices but on effective coalition building. The redefinition of universal service, therefore, is not as important as the development of an overlapping consensus that hitches the pursuit of private gain to the creation of a public good. The overlapping consensus has more to do with convergence of agendas and formation of coalitions than an explicit agreement on a course of action. Here the articulation of the ideal is of little relevance for the achievement of the ideal. The ambiguity and fuzziness of an ill-defined ideal is better suited for the emergence of an overlapping consensus, because the different players can then see their individual benefits in their own interpretations of the ideal. Furthermore, the process will be incremental and not be based on a grand plan—a new definition of universal service.

Within the context of this phenomenon, a localized strategy is more likely to succeed, because each state is, in effect, a social laboratory where a unique set of circumstances can lead to a conceptual breakthrough. The fragmented political landscape in the United States will serve as a natural test bed where different states will experiment with different strategies. The more successful innovations will be replicated by other states and the unsuccessful ones abandoned. Competition between different states will greatly accelerate the adoption of a new innovation. Every incremental expansion of the universal service package beyond the basic voice service in any one state will create competitive pressure on other states to follow suit. Therefore the proponents of universal service should attempt to accelerate this evolutionary process by directing their efforts to a few states that are most "vulnerable" to change. The competitive dynamic among the states will create a snowball effect, and the innovation will spread across the nation.

The emerging competition between different service providers is another issue. Although conventional wisdom states that competition will undermine the provision of universal service, past experience suggests that we take a second look at the relationship between competition and universal service. As discussed earlier, the present universal telephone service is to a large degree a product of competition. The increased competition will help lower prices and make advanced telecommunications services affordable for the general population. At the same time, it must be realized that the marginalized population will continue to need subsidized access.

Although these conclusions are congruous with the findings of the earlier research on interest group politics, there is a critical difference. Lowi (1969) very aptly described the policy-making process as a "parallelogram of forces" among groups where "the public interest is 'determined and established' through the free competition of interest groups" (p. 75). This phenomenon of interest groups influencing the development of communications policy has attracted much scholarly research. However, the focus of this literature has been on episodes where there appeared to be a collusion among the involved industry groups and political interests (Hazlett, 1990; Leiserson, 1942; Lowi, 1969; Mosco, 1982, 1989; Posner, 1971; D. Schiller, 1982; H.I. Schiller, 1970, 1981, 1986; Streeter, 1987; Wilson, 1980). This body of literature suggests that the influence exerted by different interest groups typically leads to politically optimal solutions where each party gets rewards in proportion to their political influence (Cohen, 1992; Hazlett, 1990). These studies provide a good backdrop for the ideas discussed in this paper. In fact two key conclusions of this study are corroborated by other studies dealing with interest group politics. First, the need for incremental and localized policies has been discussed in the literature on interest group politics and a diverse range of other fields (Anderson, 1985; Boorstin, 1965; Krasnow, Longley, & Terry, 1982; Martin, 1986; Zelinsky, 1973). Second, there has also been some discussion of how ambiguous concepts can play a unifying role. Krasnow et al. (1982) explained how the ambiguity of the "public interest" concept in broadcasting holds together the entire regulatory enterprise. The phenomenon described in this paper shares a similar dynamic but there is a key difference in terms of the linkage between the motivations of the different interest groups and the final outcome of the process. The different interest groups do not conspire to create universal service! The creation of universal service is an unanticipated outcome of the convergence in the agendas of different interest groups.

In terms of its philosophical orientation, this chapter perhaps comes closest to the pluralist interest group tradition as represented by Krasnow et al.'s (1982) *Politics of Broadcast Regulation.* What is shared is a rather sanguine view of a policy-making process that is shaped by the interplay of political influence exerted by the different interest groups. The somewhat noncritical stance is more a result of a fascination for the process rather than an absolution of the use of high ideals for self-interested motives. Lowi (1969) would perhaps find the tolerance for the shenanigans of interest groups rather unbearable. He believed that democratic ideals flourished only when there was a citizenry that exercised its judgments as individuals. He felt that any groupbased activity was a corruption of the democratic process. Marxist scholars would interpret the findings of this study in an entirely different way. They would argue that the processes described here are the very means by which the elites are able to impose their will over the rest of the population. Thus we see that interpretations will vary according to the ideological perspective through which the findings of this study are viewed.

The chapter explains the processes that will lead to the development of universal service. The analysis is not normative in nature as it only explains how the process works and not how it ought to work. The phenomenon of interest here is instances where major social goals are accomplished because of the emergence of an overlapping consensus among most unlikely groups. As the analysis in this study reveals, universal service usually emerges as an almost unanticipated by-product of the machinations of different groups whose interests at times converge in such a way that a much eulogized public good becomes a reality. The main contribution of this work is the explication of this process.

ACKNOWLEDGMENTS

I would like to thank David Waterman and M.B. Sarkar for their helpful comments and suggestions. I would also like to acknowledge the contributions of anonymous reviewers whose inputs helped me fill in the gaps and therby strengthen the overall argument.

REFERENCES

Anderson, O. W. (1985). *Health services in the United States: A growth enterprise since 1875*. Ann Arbor, MI: Health Administration Press.

Bellah, R. N., Madsen, R., Sullivan, W. M., Swidler, A., & Tipton, S. M. (1985). *Habits of the heart: Individualism and commitment in American life*. Berkeley: University of California Press.

Binder, F. M. (1974). *The age of the common school, 1830–1865*. New York: Wiley.

Boorstin, D. J. (1965). *The Americans: The national experience*. New York: Random House.

Bowie, N. (1990). Equity and access to information technology. In The Institute for Information Studies (Ed.), *The annual review, 1990* (pp. 131–177). Nashville, TN: The Institute for Information Studies.

Brock, G. W. (1981). *The telecommunications industry: The dynamics of market structure*. Cambridge, MA: Harvard University Press.

Button, H. W., & Provenzo, E. F., Jr. (1989). *History of education and culture in America*. Englewood Cliffs, NJ: Prentice-Hall.

Carey, J. W. (1989). *Communication as culture*. Boston, MA: Unwin Hyman.

Carlton, F. T. (1966). *Economic influences upon educational progress in the United States, 1820-1850*. New York: Columbia University Press.

Cohen, J. E. (1992). *The politics of telecommunications regulation: The states and the divestiture of AT&T.* Armonk, NY: M. E. Sharp.

Curti, M. (1964). *The growth of American thought*. New York: Harper & Row.

Czitrom, D. J. (1982). *Media and the American mind: From Morse to McLuhan*. Chapel Hill: The University of North Carolina Press.

Ditzion, S. (1947). *Arsenals of a democratic culture*. Chicago: American Library Association.

Dordick, H. (1990). The origins of universal service: History as a determinant of telecommunications policy. *Telecommunications Policy, 14*, 223–231.

Dordick, H. (1991). Toward a universal definition of universal service. In The Institute for Information Studies (Ed.), *Annual review, 1991: Universal telephone service: Ready for the 21st Century?* (pp. 109–139). Nashville, TN: The Institute for Information Studies.

Edelman, M. (1964). *The symbolic uses of politics*. Urbana: University of Illinois Press.

Field, H. M. (1898). *The story of the Atlantic telegraph*. New York: Scribner's.

Fischer, C. S. (1992). *America calling: A social history of the telephone to 1940*. Berkeley: University of California Press.

Gillan, J. (1986). Universal telephone service and competition: The rural scene. *Public Utilities Fortnightly, 117*(10), 22–26.

Hadden, S. (1991a). *Regulating content as universal service* (Working Paper, Policy Research Project: "Universal Service for the Twenty-First Century"). Austin: The University of Texas at Austin, Lyndon B. Johnson School of Public Affairs.

Hadden, S. (1991b). Technologies of universal service. In The Institute for Information Studies (Ed.), *Annual review, 1991: Universal telephone service: Ready for the 21st Century?* (pp. 53–92). Nashville, TN: The Institute for Information Studies.

Handlin, D. (1979). *The American home: Architecture and society, 1815–1915.* Boston: Little, Brown.

Hazlett, T. W. (1990). The rationality of U.S. regulation of the broadcast spectrum. *Journal of Law & Economics, 33,* 133–175.

Hills, J. (1989). Universal service: Liberalization and privatization of telecommunications. *Telecommunications Policy, 13,* 129–144.

Hudson, H. E., & Parker, E. B. (1990). Information gaps in rural America: Telecommunications policies for rural development. *Telecommunications Policy, 14,* 193–205.

Information Infrastructure Task Force. (1993). *The national information infrastructure: Agenda for action.* Washington, DC: National Telecommunications and Information Administration, National Information Infrastructure Office.

Jackson, S. L. (1941). *America's struggle for free schools: Social tension and education in New England and New York, 1827–42.* Washington, DC: American Council on Public Affairs.

Jacobson, R. (1989). *An open approach to information policy making: A case study of the Moore Universal Telephone Service Act.* Norwood, NJ: Ablex.

Kaestle, C. F. (1983). *Pillars of the republic: Common schools and American society, 1780–1860.* New York: Hill & Wang.

Katz, M. B. (1968). *The irony of early school reform: Educational innovation in mid-nineteenth century Massachusetts.* Cambridge, MA: Harvard University Press.

Kingsbury, J. E. (1972). *The telephone and telephone exchanges: Their invention and development.* New York: Arno. (Original work published 1915)

Kohn, H. (1957). *American nationalism: An interpretative essay.* New York: Macmillan.

Konvitz, J. W. (1985). *The urban millennium: The city building process from the early middle ages to present.* Carbondale: Southern Illinois University Press.

Krasnow, E. G., Longley, L. D., & Terry, H. A. (1982). *The politics of broadcast regulation* (3rd ed.). New York: St. Martin's.

Lasswell, H. (1962). *Psychopathology and politics.* New York: Viking.

Lavey, W. G. (1990). Universal telecommunications infrastructure for information services. *Federal Communications Law Journal, 42,* 151–190.

Leiserson, A. (1942). *Administrative regulation: A study in the representation of interests.* Chicago: University of Chicago Press.

Lowi, T. J. (1969). *The end of liberalism: Ideology, policy, and the crisis of public authority.* New York: W.W. Norton.

Martin, S. K. (1986). *Library networks, 1986–87: Libraries in partnership.* White Plains, NY: Knowledge Industries.

Marvin, C. (1988). *When old technologies were new: Thinking about electric communication in the late nineteenth century.* New York: Oxford University Press.

Marx, L. (1964). *The machine in the garden: Technology and the pastoral idea in America.* New York: Oxford University Press.

Meyer, A. E. (1965). *An educational history of the western world.* New York: McGraw-Hill.

Mitchel, C. (1993). Everyone has a right to medical care. In G. E. McCuen (Ed.), *Healthcare and human values: Ideas in conflict* (pp. 29–34). Hudson, WI: Gary E. McCuen.

Mosco, V. (1982). *Pushbutton fantasies: Critical perspectives on videotex and information technology.* Norwood, NJ: Ablex.

Mosco, V. (1989). *The pay-per society: Computers and communication in the information age.* Norwood, NJ: Ablex.

Mueller, M. (1993). Universal service in telephone history: A reconstruction. *Telecommunications Policy, 17,* 352–369.

National Governors' Association. (1988). *A hearing of the Subcommittee on Telecommunications: Universal service, summary transcript.* Washington, DC: Author.

National Telecommunications and Information Administration. (1988). *Telecom 2000: Charting the course for a new century* (NTIA Special Publication 88-21). Washington, DC: U.S. Government Printing Office.

National Telecommunications and Information Administration. (1991). *The NTIA infrastructure report: Telecommunications in the age of information* (NTIA Special Publication 91-26). Washington, DC: U.S. Government Printing Office.

Nye, D. (1990). *Electrifying America: Social meanings of a new technology, 1880–1940.* Cambridge, MA: MIT Press.

O'Connor, B. (1991). Universal service and NREN. In The Institute for Information Studies (Ed.), *Annual review, 1991: Universal telephone service: Ready for the 21st Century?* (pp. 93–140). Nashville, TN: The Institute for Information Studies.

Office of Technology Assessment. (1990). *Critical connections: Communications for the future* (Rep. No. OTA-CIT-407). Washington, DC: U.S. Government Printing Office.

Osborne, D. (1990). *Laboratories of democracy.* Cambridge, MA: Harvard University Press.

Pacey, A. (1990). *Technology in world civilization: A thousand-year history.* Cambridge, MA: MIT Press.

Pacific Bell. (1988). *Pacific Bell's response to the Intelligent Network Task Force Report.* San Francisco: Author.

Parker, E. B, Hudson, H. E., Dillman, D. A., & Roscoe, A. D. (1989). *Rural America in the information age: Telecommunications policy for rural development.* Lanham, MA: University Press of America.

Pool, I. S. (1984). Competition and universal service: Can we get there from here? In H. M. Shooshan, III (Ed.), *Disconnecting Bell: The impact of the AT&T divestiture* (pp. 112–131). New York: Pergamon.

Porter, K. H. (1969). *A history of suffrage in the United States.* New York: Greenwood. (Original work published 1918)

Posner, R. A. (1971). Taxation by regulation. *Bell Journal of Economics and Management Science, 2,* 22–50.

Pressler, L., & Schieffer, K. V. (1988). A proposal for universal telecommunications service. *Federal Communications Law Journal, 40,* 351–375.

Rawls, J. (1985). Justice as fairness: Political not metaphysical. *Philosophy and Public Affairs, 14,* 223–251.

Schiller, D. (1982). *Telematics and government.* Norwood, NJ: Ablex.

Schiller, H. I. (1970). The military-industrial complex and communications. In H. I. Schiller & J. D. Phillips (Eds.), *Super-state: Readings in the military-industrial complex* (pp. 161–172). Urbana: University of Illinois Press.

Schiller, H. I. (1981). *Who knows: Information in the age of the Fortune 500.* Norwood, NJ: Ablex.

Schiller, H. I. (1986). *Information and the crisis economy.* New York: Oxford University Press.

Streeter, T. (1987). The cable fable revisited: Discourse, policy, and the making of cable television. *Critical Studies in Mass Communication, 4,* 174–200.

Strong, J. (1963). *Our country.* Cambridge, MA: Harvard University Press.

Thompson, G. M. (1990). Sandford Fleming and the Pacific cable: The institutional politics of nineteenth-century imperial telecommunications. *Canadian Journal of Communication, 15*(2), 64–75.

Williams, F. (1990). The coming intelligent network: New options for the individual and community. In The Institute for Information Studies (Ed.), *The annual review, 1990* (pp. 179–210). Nashville, TN: The Institute for Information Studies.

Williams, F. (1991). *The new telecommunications: Infrastructure for the information age.* New York: The Free Press.

Williams, F., & Hadden, S. (1991). *On the prospects for redefining universal service: From connectivity to content* (Working Paper, Policy Research Project: "Universal Service for the Twenty-First Century"). Austin: The University of Texas at Austin, Lyndon B. Johnson School of Public Affairs.

Williams, R. (1976). *Keywords: A vocabulary of culture and society.* New York: Oxford University Press.

Williamson, C. (1960). *American suffrage: From property to democracy, 1760–1860.* Princeton, NJ: Princeton University Press.

Wilson, J. Q. (1980). The politics in regulation. In J. Q. Wilson (Ed.), *The politics in regulation* (pp. 357–394). New York: Basic Books.

Zelinsky, W. (1973). *The cultural geography of the United States.* Englewood Cliffs, NJ: Prentice-Hall.

11

Universal Service as an Appropriability Problem: A New Framework for Analysis

Milton Mueller
Rutgers University

A telecommunications network that enlarges its scope becomes more valuable to users. Indeed, a "universal" or ubiquitous communication infrastructure is recognized by nearly all societies as being of immense social and economic value. However, many believe that competition undermines the economic foundations of universal service.

The following paragraph summarizes the conceptual framework that now dominates the policy debate over universal service:

> The current system of affordable and widely adopted telephone service is based on rate averaging and cross-subsidies among routes, users, and locations. Competing networks undermine this support system because they are able to undercut the incumbent's prices in routes and services that are generating the surpluses, still benefiting from access to the areas controlled by the incumbent. This kind of "cream-skimming competition" will inexorably lead to unbundling of the routes and services of the public network and a deaveraging of the rates associated with each service component. As this happens, many end users, forced to bear the "full cost" of their particular network components, will not be able to afford it. So how will the subsidy system be maintained in the new, competitive environment?

In this framework, universal service is conceived as a problem in public finance. If society is to generate subsidies to finance universal telecommunications infrastructure, what is the most efficient and equitable mechanism for doing so? There may be some disagreement about the size of the subsidies needed, the efficiency and fairness of the mechanisms used to collect them, and about who should and should not pay them. But virtually everyone accepts the public finance paradigm as a given.

In this chapter I challenge this framework and attempt to make the case for a radically different approach to the problem. Using theories of property rights, I reframe universal service as a problem of appropriability. In economic theory, appropriability refers to the ability of a firm to control access to and pricing of its product in such a way as to capture its full economic value. A firm's ability to appropriate value from its products is dependent on the property structure prevailing in that industry. The concept of appropriability should be a crucial part of the current debate. In this writer's opinion, most of the alleged incompatibility between universal service and competition is a predictable result of a property structure that does not allow telecommunication networks to benefit from increasing their scope. With the right property structure, universality ceases to be a costly liability and instead becomes a valuable asset.

This chapter is an attempt to redefine the problem. It is not a rigorous and complete theoretical solution to it. Although airing these still-developing ideas may seem premature, the ideas contained here go so firmly against the grain of most current thinking that taking one step at a time seems optimal.

THE PREVAILING VIEW: UNIVERSAL SERVICE AS A PROBLEM IN PUBLIC FINANCE

I begin with an analysis and critique of the public finance approach to universal service support. Most previous analyses of the universal service pricing problem define universal service as the product of a subsidization scheme. Analyses of the problem cite "cost support systems" (OPASTCO, 1994) and "massive intra-industry transfers of funds" (Egan & Wildman, 1993) as the pillars of universal service. This approach to the problem has two elements. One is an essentially historical assertion that the current subsidy scheme was in fact responsible for creating the United States' ubiquitous telephone system. This issue is not directly related to the topic of this paper but is addressed in the Appendix. The other is that current competition is undermining this subsidy system and that without it, an equally universal telecommunications network is not economically viable.

The Tension Between Current Competition Policy and Universal Service Policy

For the past 15 years, advocates of competition have made interconnection to the PSTN the basis of competitive entry. Economists and lawyers have argued that interconnection to the PSTN should be made available on a nondiscriminatory basis to competitors as well as end users. Moreover, they have argued that the prices charged for the interconnection of competing networks should reflect only the incremental costs of supplying it. They have, in effect, equated the proper price of network access with the economic cost of supplying the facilities enabling access. Having won these arguments in policy circles, they have now discovered that their policies are undermining the economic foundations of universal service. Competing networks that are given special forms of low-cost access to the PSTN will serve only the most profitable markets and leave the supply of access to more costly or less high-volume markets to the regulated carrier. As this happens, the financial surpluses used to finance the higher cost routes and components of the PSTN will be competed away; each user will be required to bear the stand-alone cost of the particular set of facilities he or she uses.

Faced with this scenario, policy analysts now propose to revamp some kind of subsidy scheme to recreate the effects of the one they have spent the last 15 years tearing down. Indeed, an outright contradiction between the policy goal of introducing competition and the methods of universal service support has plagued common carrier telecommunications policy since the AT&T divestiture. The divestiture destroyed the old separations and settlements system, but recreated many of its economic effects with the NECA pool, weighted DEM charges, and the Universal Service Fund. In a partially competitive environment, these measures encourage uneconomic bypass and assorted inefficiencies, just as their predivestiture forebears did.

The current wave of policy concern proposes to fix this problem by systematically overhauling the universal service subsidy. Most economists are proposing some kind of value-added tax applied specifically to the telecommunications industry (Egan & Wildman, 1993; Einhorn, 1993; Noam, 1993; Teleport, 1994). An industry-specific tax and subsidy scheme is proposed because support from general tax revenues is not supposed to be politically feasible. The tax would be applied not just to long-distance carriers, but also to cellular, PCS, information service providers, and virtually everyone else in the industry. In other words, instead of eliminating the cross-subsidies and associated inefficiencies that developed during decades of regulated monopoly, we are now supposed to make them more extensive and generally applicable than ever before.

Critique of the Prevailing View

Is this progress? Aside from the fact that there is something a bit ridiculous about a policy regime that tries to undermine subsidies with one hand and save them with the other, there are four serious problems inherent in approaching universal service as a public finance problem.

First, under the proposed schemes carriers get no competitive advantage by enlarging their scope. In fact, all these approaches to universal service maintenance are based on the assumption that a universal telecommunications network is a costly liability rather than a valuable asset. Subsidies are supposed to compensate for this, but anyone familiar with the dynamics of the marketplace knows that competitive pressures and the profit motive are far more powerful than regulation and subsidies (Gillett, 1994).

Second, under a subsidy mechanism, there is no way to distinguish between "high costs" and obsolete or inefficient ways of doing things. Because the provision of service to high-cost areas will be removed from marketplace competition, costs must be taken as a given and "covered" by the subsidy. I do not doubt that, given current applications of technology, factors of density and loop size make it more expensive to provide traditional telephone service in rural areas. But as long as the rural telcos are collecting universal service subsidies, why should anyone bother to develop and deploy radically different, more efficient ways to serve these areas?

Third, universal service subsidies are not compatible with a free, open, and competitive marketplace. At best, they create an inherently unlevel playing field, in which there must always be a designated "carrier of last resort" with special obligations and protections. At worst, they require that telcos in certain areas be franchised monopolies. Panzar and Wildman (1993), for example, argued for leaving monopolies in place in rural areas, and showed convincingly how a mixture of competitive entry and universal service subsidies could have unintended and counterproductive effects.

Fourth, subsidization of universal service requires that the government rather than the market define the nature of basic service. Whoever administers the subsidy will have to decide what services are eligible for subsidies. This will give regulators a major role in determining the shape of the market for advanced information and communication services (Weinhaus, 1994).

UNIVERSAL SERVICE AS AN APPROPRIABILITY PROBLEM

At the heart of the current debate over universal service and competition lies an interesting contradiction. On the one hand, universal access is discussed as if it were a resource so valuable that no competitor could survive without it. Competing networks—competitive access providers (CAPs), interexchange carriers (IXCs), wireless alternatives, and so on, all claim that interconnection to the PSTN at reasonable rates and with equal technical conditions is essential to their success. Apparently, the ability to reach every telephone user, a capability that can only be supplied by the PSTN, is extremely valuable. At the same time that competing networks are claiming that they cannot survive without universal access, however, the local exchange companies (LECs) who supply it are claiming that it is a costly and unremunerative liability. How is it possible that something so valuable and so much in demand is also economically unsustainable? Something is wrong with this picture.

The answer to this question lies in the current property structure, or perhaps we should say the absence of property structure, in the public telecoms network.

A telecom network offers its customers an enormous bundle of services; that is, millions of connections to other users. Under the traditional industry structure consumers purchased access to all or most of these services at rates that sustained the system as a whole. When two or more competing networks are interconnected, both can offer users the same access scope, despite what may be major differences in their actual scope. This is the source of an appropriability problem. A competitor who buys only one unit of access into a universal network is technically able to resell access to all of the users connected by the incumbent, even though the competitor does not have to face the costs and risks of creating the entire network.

The competitor is thus able to appropriate some of the economic value of the other network's scope.

The appropriability problem in telecommunications is similar in a number of ways to the problem of intellectual property in information markets. A person who buys access to information, such as a book, a videocassette, or computer software, can resell that information by duplicating it. The consumer need purchase only one unit of the producer's output to acquire the ability (if not the right) to compete with the producer in the supply of that information. The purchaser of a car, by contrast, does not gain the ability to manufacture and resell multiple cars by virtue of that purchase. (Although individual cars can be resold, and this activity may have some impact on the market for new autos, "reproducing" cars requires investments in technology, labor, and materials thousands of times more expensive than the car itself.)

Likewise, a competing network provider who purchases access to the established telephone network for its customers is physically enabled to resell access to all of the users of the latter system. Just as the information reseller need not worry about the costs of producing information but only the cost of reproducing it, so the telephone access reseller need not worry about the cost of reconstructing the established network, but only about the cost of acquiring enough access into it to handle the traffic between the two systems.

There is another respect in which there is a significant economic analogy between the market for telecommunications access and intellectual property: Both markets are characterized by inherently imperfect substitution. The market for telephone access and for information products both have unique characteristics with respect to substitution by consumers. Any given information product is by definition unique. If it were not unique it would be a copy of some other information product. The degree of its uniqueness gives it a greater or lesser degree of monopoly power over those who wish to purchase access to it or the right to use it. The CD recording *Like A Prayer* by Madonna, for example, is an entity unto itself. No other recording is a perfect substitute for it. Although the presence of other female vocalists and other recordings by Madonna provide a significant check on the price commanded by Madonna's song, none are perfect substitutes for that specific recording. The argument regarding uniqueness is even clearer in the case of patents. The right of an inventor to appropriate the value of unique knowledge or technique is recognized by the law. Although it is possible to circumvent patent protection by developing technical substitutes, the closer one gets to perfect substitution the more likely one is to be prosecuted for violating the patent protection of the original inventor. Thus, all substitution choices in information markets are imperfect because the outputs of producers are not homogeneous. The market price of information products such as CDs, software, and so on, are thus determined not by the incremental cost of making a copy but by the value of the information.

The same is true of network access. Telephone access is not a homogeneous good. Telephone access in New Zealand is not a substitute for telephone access in New York. Each individual user and each group of subscribers is sui generis. One's decision about whether access to the network is worth the price depends on to whom one is obtaining access, just as the decision to buy a specific CD depends on who the artist is. The production of access by competing and non-interconnected networks results in inherently imperfect substitution choices for consumers, as each network provides access to only a subset of the users. By this I mean much more than that it is improbable that a new network will duplicate the scope of the established network in its entirety. Even if we assume that a new network could construct a physical infrastructure that paralleled completely that of the established network, the two networks would attract entirely different sets of subscribers. Consequently, the service provided by each network would differ radically because subscribers on one system would not be accessible through the other (remember, we are assuming that they are not interconnected). A consumer who switched from one network to the other would find that he or she cannot call some of the people on the other system. Only if the networks are interconnected can they offer access to identical subscriber sets. As before, interconnection in access markets is the equivalent of reproduction in information markets.

What are the implications of looking at the problem in this way? There are several, and they all run directly counter to the thrust of contemporary policy.

First, the primary resource that is sold by a telephone network is access to subscribers, not simply a set of physical facilities and functions. Moreover, it is the size of the access bundle offered by a network that increases its value.

Second, the price of access should be based on the market value of the access, not solely on the cost of the physical facilities required to provide it. To require telephone companies to make access available to competitors at a price that reflects only the incremental costs of the facilities used to provide access is as fallacious as requiring software producers to base the price of their product on the cost of producing and distributing the floppy discs that carry the software. In both types of markets, the marginal cost of extending access to an additional user is very low. But to make the product available at such a price is unsustainable.

Third, in both information and telecommunication markets, the problem of appropriability makes it necessary and legitimate for the owners of the information resource to discriminate in the prices charged for access depending on whether or not the user intends to resell it. This is true even though there is no difference in the incremental cost of supplying access to a reseller or an end user. Such discrimination is the only way a producer of access can cope with the appropriability problem; that is, benefit in some way from enlarging the scope of its network.

Appropriability based pricing is compatible with, and in fact presupposes, open entry into all telecommunications markets. If a network abuses its control of access and charges too much it simply creates an incentive for others to duplicate the access it provides. In this regard I reject the prevailing notion that LECs have an insurmountable "bottleneck" control over local access. The presence of CAPs, IXCs, cable television networks, and, soon, wireless PCS all offer alternative forms of access that act as a constraint on the LEC.

The next two sections apply this logic to specific cases in order to show that there is some empirical support for this approach to network competition.

A Historical Precedent: Telephone Competition Without Interconnection

Competition between the Bell system and the independents between 1894 and 1920 was structured as a system rivalry between unconnected local exchanges and toll networks. They were not interconnected because the prevailing interpretation of property rights gave competitors the option of deciding whether or not to interconnect with a competitor or another network. As I showed in another piece (Mueller, 1993), this type of property structure gave both competitors a powerful incentive to make their networks as universal as possible. The critical point, I hasten to add, is not just that "competition" helped to promote universal service, but that competition between networks that could appropriate the value of having a larger scope promoted universal service.

Even more surprising, during the early competitive period the competing telephone networks often sustained their large scope by averaging rates and costs. Most economists believe that intranetwork cross-subsidies or transfers of revenues would not exist in a competitive market. This belief is contradicted by the historical evidence from this period. Intrasystem averaging seemed to be sustainable as long as the competing systems were not interconnected. During the competitive era, the Bell system established many small exchanges in outlying areas and sustained them in part through toll usage revenues. That is, it paid the competing networks to add remote exchanges even if the small exchanges did not sustain themselves with local exchange access revenues alone, because the presence of many additional termination points increased toll usage and revenues in other exchanges. This was, to repeat, because the older concept of property rights in network access gave networks the right to appropriate the additional value derived from the scope of their networks.

Overnight Delivery Services

Overnight delivery services such as DHL, FedEx, and UPS are competitive network businesses. As networks supplying a bundle of package and letter delivery services, their economics are in some ways analogous to telecommunications networks. It is interesting to note, therefore, that these businesses, although competitive, still maintain uniform rates for services that incur different costs. It is also interesting to note that all of them offer universal service, or are constantly striving to improve and increase the universality of their distribution networks.

By uniform rates I mean that the price of sending an overnight DHL, FedEx, or UPS package is the same regardless of whether the package goes across town or across the country. Originally this was the case because (in the case of FedEx, the industry pioneer), all packages went to the Memphis hub regardless of their ultimate destination. Thus, all packages did have roughly the same costs. However, since that time regional sorting centers and package "bleed off" functions have been introduced, which undoubtedly lower the costs of regional or local distribution relative to national distribution. Yet the consumer prices have remained the same. Apparently, consumer preference dictates uniform national pricing, and this kind of pricing system is sustainable in a competitive market.

Courier express companies also must deal with the universal service issue. Each one recognizes that it is in their direct business interest to provide service to as many places as possible. Although service to points outside primary service areas may incur a surcharge, or result in slower service, competition forces each network to continually attempt to increase the number of places to which they can provide overnight service at the regular price.

In the overnight express industry, each network competes on a stand-alone basis. There is no price regulation and no common carrier or universal service obligation. It is interesting, but of course inconclusive, to speculate about what the economic impact of a different property structure might be. Assume, for example, that federal regulators required DHL and FedEx to exchange packages; FedEx could require DHL to deliver FedEx packages to places FedEx did not want to go, and vice versa. One suspects that uniform rates would end, as each courier sought to serve only those areas in which it had some cost advantage and left the rest to the others. One also suspects that the imputed costs each courier claimed to incur by serving the more remote regions would suddenly become very large, particularly if subsidies were made available to support them.

CONCLUSION: ACCESS PRICING, COMPETITION, AND UNIVERSAL SERVICE

In conclusion, market competition and network universality can be reconciled if the telecommunications industry adopts a property structure that allows networks to appropriate the increasing value created by enlarging a network's scope. Because current policies prevent networks from doing this, it is not surprising that the growth of competition seems to corrode the economic foundations of universal service.

APPENDIX
POPULAR MISCONCEPTIONS ABOUT UNIVERSAL SERVICE

The historical role of subsidies in creating a universally available telephone network is often misunderstood. A published research paper (Mueller, 1993) and a book manuscript (Mueller, 1995) went over these points in great detail, so I will merely summarize the major points here:

1. From 1907 until the 1950s, universal service meant interconnecting competing networks into an integrated monopoly, not putting a telephone into every home.

2. Universal service as a policy goal was not part of the 1934 Communications Act. The words do not appear in the Act, and in the extensive committee proceedings surrounding its passage there is no discussion about using regulation, rate av-

eraging, subsidies, or jurisdictional separations and settlement procedures to make telephone service more affordable or more universally available.

3. Under regulation, the separations and settlement procedures that developed between 1920 and 1950 were not originally conceived as a means of subsidizing telephone service in order to make it more affordable to larger numbers of people. Rather, it was a method of distinguishing between those portions of the rate base that fell under state and interstate regulatory authority. From the 1920s until the late 1960s, there is no historical evidence that regulators linked jurisdictional separations procedures to universal service goals in the modern sense.

4. At least 90% of the central office locations now in existence had already been established by 1915. Competition between unconnected networks made it incumbent on both the Bell system and the independents to extend their networks into as many places as possible.

5. Jurisdictional separations were used to subsidize local exchange access and rural areas after 1968, especially after the adoption of the Ozark Plan in 1970. But at that time, at least 85% of all U.S. households already had telephones. Thus, the contribution of subsidies to the popularization of the telephone was marginal.

REFERENCES

Egan, B., & Wildman, S. (1993, October). *Funding the public telecommunications infrastructure.* Paper presented at Universal Service in the New Electronic Environment Symposium, Benton Foundation and Columbia University CITI, New York.

Einhorn, M. A. (1993, October). *Recovering network subsidies without distortion.* Paper presented at Universal Service in the New Electronic Environment Symposium, Benton Foundation and Columbia University CITI, New York.

Gillett, S. E. (1994). *Technological change, market structure, and universal service.* Unpublished manuscript.

Mueller, M. L. (1993). Universal service in telephone history: A reconstruction. *Telecommunications Policy, 17*(5) 69.

Mueller, M. L. (1996). *Universal service: Competition, interconnection and monopoly in the making of American telecommunications.* Cambridge, MA: MIT Press/AEI Series on Telecommunications Deregulation.

Noam, E. M. (1993, October). *NetTrans accounts: Reforming the financial support system for universal service in telecommunications.* Paper presented at Universal Service in the New Electronic Environment Symposium, Benton Foundation and Columbia University CITI, New York.

OPASTCO. (1994). *Keeping rural America connected: Costs and rates in the competitive era.* Washington, DC: Organization for the Protection and Advancement of Small Telephone Companies.

Panzar, J. C., & Wildman, S. (1993). *Competition in the local exchange: Appropriate policies to maintain universal service in rural areas.* Unpublished reports.

Teleport Communications Group. (1994). *Universal service assurance: A concept for fair contribution and equal access to subsidies*. Staten Island, NY: Author.

Weinhaus, C. (1993). *What is the price of universal service? Impact of deaveraging nationwide urban/rural rates*. Tallahassee: University of Florida, Telecommunications Industry Analysis Project.

Weinhaus, C. (1994). *Redefining universal service: The cost of mandating the deployment of new technologies in rural america*. Tallahassee: University of Florida, Telecommunications Industry Analysis Project.

12

Access to Telecommunications in the Developing World: Ten Years After the Maitland Report

Heather E. Hudson
University of San Francisco

> *We believe that by the early part of the next century virtually the whole of mankind should be brought within easy reach of a telephone and, in due course, the other services telecommunications can provide.*
> Maitland Commission (1984, p. 4)

THE LINK IS STILL MISSING

In 1984, the Maitland Commission noted that telecommunications was a "missing link" in much of the developing world. More than ten years later, that statement is still true in many urban centers and throughout rural regions of developing countries. This chapter examines what progress has been made, what problems remain, and the effects of changing technologies and policies. It then proposes strategies to achieve the Commission's goal of bringing the "whole of mankind within easy reach of a telephone" by early in the next century.

In absolute terms, there are still significant gaps in access to telecommunications between industrialized and developing countries (see Table 12.1). Even where countries have invested in long-distance links through leasing satellite capacity, the "last mile" problem remains. With low teledensity, access to telecommunications remains limited (see Table 12.2).

TELECOMMUNICATIONS GROWTH DURING THE 1980S

To find out what progress has been made in meeting the Maitland Commission's goals, data were collected on telecommunications density (measured in lines per

TABLE 12.1
Information Gaps: Access to Telecommunications

OECD Member Countries	*Telephone Lines per 100 population*	*African Countries*	*Telephone Lines per 100 population*
Australia	47.1	Algeria	3.7
Austria	44.0	Benin	0.3
Belgium	42.6	Botswana	2.3
Canada	59.2	Central African Republic	0.2
Denmark	58.0	Gabon	1.9
Finland	54.2	Kenya	0.8
France	52.1	Liberia	0.2
Germany	44.0	Madagascar	0.3
Greece	43.7	Malawi	0.3
Iceland	53.9	Mauritius	7.4
Ireland	31.3	Morocco	2.5
Italy	41.0	Namibia	4.0
Japan	46.7	Nigeria	0.3
Luxembourg	52.9	Rwanda	0.2
Netherlands	48.9	Senegal	0.8
New Zealand	45.0	Seychelles	14.3
Norway	52.9	Sierra Leone	0.3
Portugal	32.1	South Africa	8.9
Spain	35.3	Sudan	0.2
Sweden	68.1	Swaziland	1.9
Switzerland	61.3	Tanzania	0.3
United Kingdom	45.3	Togo	0.4
United States	56.5	Tunisia	4.5
		Zambia	0.9
		Zimbabwe	1.2

Source: ITU, *World Telecommunication Development Report*, 1994.

TABLE 12.2
The Last Mile Problem: Developing Countries With Intelsat Domestic Leases

Countries	*Telephone Lines per 100 population*
Algeria	3.7
Argentina	11.1
Brazil	6.8
Chile	8.9
China	1.0
Colombia	8.4
Côte d'Ivoire	0.7
India	0.8
Indonesia	0.8
Libya	4.8
Malaysia	11.1
Mexico	7.5
Morocco	2.5
Mozambique	0.4
Thailand	3.1
Venezuela	8.7
Zaire	0.1

Source: ITU, *World Telecommunication Development Report*, 1994.

100 population) in 1980 and 1990. These data are presented in Table 12.3. Countries are grouped according to World Bank criteria, namely:

- Low-income economies
- Middle-income economies
    - lower middle income
    - upper middle income
- High-income economies

For low- and middle-income countries, telecommunications grew faster than the economy (percentage change in telephone lines/100 vs. percentage change in per capita GDP). Low-income economies appear to have made significant progress. Although their economies grew by only 4.2%, telephone density increased an average of 19.7%. However, the average telephone density was only .55 lines per 100 population. Lower middle-income economies grew by 12.3% on average during the decade, whereas telecommunications density improved by 53%. Yet there were only an average of 5.4 lines per 100 in these countries.

Upper middle-income countries showed growth of 23.9% in GDP, and 77.9% in telephone lines, with density increasing to 16.5 lines per 100. High-income economies averaged a percentage increase in GDP of 120.2%, and increase in telephone density of 20.9%. These data may indicate conditions closer to market saturation.

Although the data do indicate progress, the change is discouragingly modest in low-income and lower middle income economies. In lower income economies there is an average of only 1 telephone line per 200 people. As these countries tend to have at least 75 percent of their populations in rural areas, it is safe to assume that there are many rural regions in poor countries still without any access to telecommunications. A similar situation is likely to prevail in many lower middle income countries, where there is on average only 1 telephone line per 20 population. Again, since these countries also have a majority of their population living in rural areas, there are likely to be many rural communities without any telecommunications access.

TABLE 12. 3
Indicators: Percentage Change 1980–1990

Country Groups	*GNP/cap*	*Tel Lines/100*	*TV Sets/100*
Low income	4.2%	19.7%	172.2%
Lower middle income	12.3	53.0	220.9
Upper middle income	23.9	77.9	169.4
High income	120.2	20.9	17.7

Derived from: ITU, *World Telecommunication Development Report*, 1994; UN, *1993 Statistical Yearbook*; UN, *World Tables 1992*.

OTHER TELECOMMUNICATIONS INDICATORS

TVs Versus Telephone Lines

Data are also presented on TV sets per 100 population, with comparable figures for 1980 and 1990. A ratio of telephone lines to TV sets is computed, to provide an indicator of progress compared to broadcast communications access. In all groups of developing countries, television set access grew by more than 100%, with growth of 172% in low-income economies, 221% in lower middle income economies, and 169% in upper middle income economies. In high-income countries, TV sets increased only 17.7%, so that there were approximately as many TV sets as telephones.

In both low-income and lower middle income countries, television set access has grown at a rate 4 to 8 times that of telecommunications growth, despite relatively low growth rates in GDP. Part of this growth may be explained by the installation of television transmitters that provide a signal to large numbers of previously unreached people. If this is true, we must ask why there has been greater attention paid to extension of television transmission than telecommunications.

Table 12.4 shows the ratio of TV sets to telephone lines in Latin American and Caribbean countries. In the most economically developed countries, access to tele-

TABLE 12.4
Ratio of TV Sets to Telephone Lines in Selected Latin American and Caribbean Countries

Country	*Tel Lines per 100*	*TV Sets per 100*	*Ratio of TVs/Tel Lines*
Antigua and Barbuda	27.7	29.1	1.1
Argentina	11.1	22.2	2.0
Bahamas	30.6	22.5	0.7
Barbados	30.8	26.1	0.8
Bolivia	2.4	10.1	4.2
Brazil	6.8	21.1	3.1
Chile	8.9	20.5	2.3
Colombia	8.4	11.5	1.4
Costa Rica	10.5	15.1	1.4
Cuba	3.2	20.8	6.5
Dominican Republic	6.4	8.4	1.3
Ecuador	5.0	8.3	1.7
El Salvador	3.1	9.0	2.9
Honduras	2.1	7.8	3.7
Jamaica	6.8	13.2	1.9
Mexico	7.5	14.3	1.9
Peru	2.8	9.9	3.5
Trinidad and Tobago	14.3	31.5	2.2
Uruguay	15.7	23.2	1.5
Venezuela	8.7	17.1	2.0

Derived from: ITU, *World Telecommunication Development Report*, 1994.

TABLE 12.5
Ratio of TV Sets to Telephone Lines in Socialist and Former Communist Countries

Country	*Tel Lines per 100*	*TV Sets per 100*	*Ratio of TVs/Tel Lines*
Bulgaria	26.1	25.0	1.0
China	1.0	3.1	3.1
Cuba	3.2	20.7	6.5
Hungary	12.5	41.0	3.3
Korea PDR	4.8	10.9	2.3
Lao PDR	0.2	0.8	4.0
Mongolia	3.0	6.2	2.1
Poland	10.3	29.5	2.9
Romania	11.2	19.5	1.7
Poland	10.3	29.5	2.9
former USSR	13.8	38.0	2.8
Viet Nam	0.3	3.9	13.0

Derived from: ITU, *World Telecommunication Development Report*, 1994.

phones and TV sets appears to be about equal. However, in poorer countries, particularly where the telecommunications sector is government operated, there is far wider access to TV sets than to telephones. The contrast is even more striking if we examine socialist and former Communist countries, as shown in Table 12.5. Here, the ratio of TV sets to telephones may be more than 10 to 1.

Most countries do not have a specific policy that is intended to provide greater access to television than telephones; however, people can buy TV sets if they can afford them, whereas, if telephone lines are not available, there is no possibility of obtaining telephone service. Some socialist countries do intentionally foster access to mass media, and limit access to interactive communications that could be used for political as well as social and economic purposes. However, regardless of whether there is a deliberate policy, a country that provides television service without telephones not only deprives people of the means of getting help in emergencies, but of the means of having a voice in their own development.

Indicators of Entrepreneurship

Another approach to determining whether current strategies for telecommunications investment are somehow missing the mark is to examine indicators of communications entrepreneurship. Although comparative data are not available, the following activities in a country would indicate that there are entrepreneurs willing to offer communications services, and customers to support them:

- *Video shops:* Shops that rent video cassettes and/or video recorders and players. These are found in even relatively poor developing countries where there would appear to be very little disposable income in rural areas.
- *Cable TV systems:* Cable TV systems (government authorized or otherwise) that have been installed to provide access to TV channels (e.g., as from a

satellite) for a fee. The most striking current example is India, where cable TV systems have sprung up in urban neighborhoods to deliver programming from AsiaSat.

- *Kiosks and copy shops:* Entrepreneurs who offer communications facilities such as telephones and facsimile services. Some countries such as Indonesia have introduced this model for pay phone service, retaining government control over operation of the public switched network. Entrepreneurs typically retain a percentage of the toll revenues.

THE CHANGING TELECOMMUNICATIONS ENVIRONMENT

Telecommunications technology has changed dramatically since 1984. Perhaps the most telling evidence of change is the cover of the Maitland Commission report itself, which shows two rotary dial telephones. This is not to say that digital switching did not exist by 1984, but that it was not considered necessary or perhaps even appropriate for developing regions. A second indicator is that the Commission specifically identified only telephone service, and proposed access "in due course [to] the other services telecommunications can provide." Today, many of those services could be available as soon as telecommunications service is provided.

There are many recent technological innovations that can make telecommunications services more reliable and cheaper to provide. Among the new technologies:

- *Wireless technologies:* Advances in radio technology such as cellular radio and rural radio subscriber systems offer affordable means of reaching isolated rural customers. These technologies make it possible to serve rural communities without laying cable or stringing copper wire.
- *VSATs:* Small satellite earth stations are proliferating in developing regions, usually for distribution of television signals. However, VSATs can also be used for interactive voice and data, and for data broadcasting. Examples include bank networks in remote parts of Brazil and India's NICNET for government data services.
- *Digital compression:* Compression algorithms can be used to "compress" digital voice signals, so that 8 or more conversations can be carried on a 64-kbit voice channel, thus reducing transmission costs. Compressed digital video can be used to transmit motion video over as few as 2 telephone lines (128 kbps), offering the possibility of relatively low-cost video for distance education and training (Tkal, 1994).
- *Store-and-forward data:* Development organizations seeking cheap ways to communicate with field projects are using single satellite LEO systems for electronic messaging. For example, HealthNet provides store-and-forward data communications to small terminals in developing countries via a "microsatellite" (Clements, 1991).
- *Voice messaging:* Voice mail systems can be used to provide "virtual telephone service" to people who are still without individual telephones. Callers can leave messages in rented voice mail boxes, which the subscribers

can retrieve from a pay phone. For example, TeleBahia in northeastern Brazil is using voice messaging technology to serve small businesses owners. There certainly is not real-time interactivity, but voice mail provides a way for people to communicate in rural regions where postal services are slow or erratic and literacy levels are low. (A similar approach has been used in some U.S. homeless shelters to enable job seekers to have a way to be contacted by prospective employers.)

THE CHANGING POLICY ENVIRONMENT

In addition to proliferating technologies, we are witnessing proliferating models of restructuring the telecommunications sector in various countries. The major models are shown in the following.

Ownership

- *Autonomous Public Sector Corporations:* The first strategy for creating incentives to improve efficiency and innovation in the telecommunications sector is to create an autonomous organization operated on business principles. This often seen as an intermediate step between a PTT structure and some form of privatization.
- *Privatized Corporations:* Privatization models range from minor investments by private companies to joint ventures between private carriers and governments to full privatization without any government stake or with a small government "golden share."

Structure

- *Monopoly:* Most countries began with a national monopoly model that is being eroded. Most maintain some level of monopoly, for example in the local loop, but alternative providers using wireless and fiber are also beginning to challenge the assumption of natural monopoly in the local loop.
- *Open Entry:* An intermediate step between national monopoly and competition is a policy of open entry for unserved areas. For example, the United States, Finland, and the Philippines have small companies or cooperatives that were formed to provide services in areas ignored by the national monopoly carrier.
- *Competition:* Competition can range from terminal equipment (now commonly competitive in most countries, including developing countries) to new services such as cellular telephony, to value-added services such as packet data networks to full competition in the network.

IMPLICATIONS FOR PLANNING

Changing Assumptions

Changing rural economies and needs are likely to result in new and changing demands for telecommunications services, as the introduction of new technologies is changing the economic viability of rural telecommunications.

- *Voice and Data:* Although basic voice communication is still the first priority, many users now have requirements for data communications as well, particularly facsimile and relatively low-speed data communications. Thus transmission channels must be reliable enough to handle data as well as voice traffic.
- *Urban and Rural:* The availability of relatively low-cost radio and satellite technologies for serving rural areas makes it possible to reach even the most remote locations, and to base priorities for service on need rather than proximity to the terrestrial network.

The combination of increased demand and lower cost technologies makes rural areas more attractive for investment. As a result, other institutional structures besides public sector monopolies may also be suitable for rural service provision.

Setting Goals and Targets

Before taking major steps to encourage investment or restructure the telecommunications sector, planners should set national telecommunications goals. Nations in general seek to improve educational standards, to provide health care to all, to create jobs, and to reduce disparities between haves and have nots, both urban and rural. Telecommunications can contribute to many of these goals.

These general development goals must be translated into specific telecommunications goals, which might include:

- *Universal access to basic communications.* Access may be defined using a variety of criteria such as:
  - population: for example, a telephone for every permanent settlement with a minimum population.
  - distance: for example, a telephone within x kilometers of all rural residents;
  - time: for example, a telephone within an hour's walk or bicycle ride of all rural residents.
- *Reliability.* Standards for reliable operation and availability; quality sufficient for voice, facsimile, and data communications.
- *Emergency Services.* A simple way to reach help immediately, so that anyone, including children and illiterate adults, would be able to call a hospital, police, and so on.
- *Pricing.* Pricing based on communities of interest; for example, to regional centers where stores and government offices are located; to other locations where most relatives are located (surrounding villages, regional towns, etc.).

In North America, we have advocated that, in order to ensure that telecommunications technologies and services can be put to optimal use for rural development, the basic goal should be to provide in rural and remote areas affordable access to telecommunications and information services comparable to those available in urban areas (Parker and Hudson, 1995). The underlying rationale is that universal access to information is critical to the development process.

Although planners may want to modify this goal for lower income countries, there is no longer a compelling technological or financial reason to limit rural services. The same technologies that are used to transmit voice can also transmit facsimile and data, and, through digital compression, video as well. As noted earlier, access criteria may differ in rural areas, but they may be actually be comparable to access criteria in high-density urban areas, where the goal is not to provide a line for every dwelling, but access for everyone through public phones in kiosks, shops, common areas, and so on.

It is important to note that this goal is in effect a "moving target": It does not specify a particular technology, but assumes that as facilities and services become widely available in urban areas, they should also be extended to rural areas. Information can be accessed and shared through a range of technologies such as satellite earth stations, microwave and cellular radio links, optical fiber, and copper wire. Indeed, the technologies used to deliver the services in rural areas may differ from those installed in urban areas; for example, satellite links and radio networks may be less costly for rural communications than optical fiber or even copper wire.

Next it is necessary to devise a set of strategies to achieve these goals. Strategies are needed to create incentives to increase telecommunications investment, and to drive the investment toward achieving these goals.

INDUSTRY STRUCTURE

The Maitland Commission (1984) paid little attention to the structure of the telecommunications sector, beyond advocating that telecommunications be set up "as a separate, self-sustaining enterprise, run along business lines" (p. 38). At the time, many developing countries were still running telecommunications through a government department with revenues subsidizing the postal services, and often turning foreign exchange earnings over to the national treasury. Today, a majority of developing countries are running their telecommunications administrations as autonomous government-owned enterprises, and many are in the process of privatizing these operations.

Yet, as the data show, a more entrepreneurial national monopoly may not have adequate incentives to invest in facilities to accomplish the goals outlined here, given the unmet demands of business and upper middle class residential customers in the cities. The following are some strategies that can create incentives to invest in rural and less profitable areas:

- *New Services: Franchise or Competition.* The introduction of a new service may be accelerated by issuing licenses for franchises. This approach has been used for cellular radio in Argentina and Mexico, for example. It allows foreign investors with the necessary capital and expertise to provide the service more quickly than it could be offered through the PTT. Satellite services such as data communications may also be offered through one or more private licensed carriers. For example, private banking networks using VSATs have now been authorized in Brazil.

- *Local Companies.* Although in most countries there is a single carrier that provides both local and long-distance services, it may make sense to delineate territories that can be served by local entities. In the U.S., the model of rural cooperatives fostered through the Rural Electrification Administration (REA) has been used to bring telephone service to areas ignored by the large carriers. Local enterprises are likely to be more responsive to local needs, whether they be urban or rural. An example of this approach in urban areas is India's Metropolitan Telephone Corporation established to serve Bombay and Delhi. Local companies also provide telephone service in Colombia. Cooperatives have been introduced in Hungary. A disadvantage of this approach is the need for local expertise to operate the system, which is likely to be in particularly short supply in many developing countries.
- *Franchises for Unserved Areas.* Another approach to serving presently unserved areas is to open them up to private franchises. Large carriers may determine that some rural areas are too unprofitable to serve in the near term. However, this conclusion may be based on assumptions about the cost of technologies and implementation that could be inappropriate.

  It should be noted that wireless technologies could change the economics of providing rural services, making rural franchises much more attractive to investors. Whereas companies such as GTE and US West are selling rural franchises, other companies with a more optimistic assessment of rural profitability are buying them. For example, Rochester Telephone has bought properties in the rural east and midwest. Citizens Communications spent $1.1 billion to buy 500,000 access lines, primarily in the rural western US. And Pacific Telecom, the parent of Alascom (recently sold to AT&T), has also recently bought rural properties.
- *Resale.* Third parties may be permitted to lease capacity in bulk and resell it in units of bandwidth and/or time appropriate for business customers and other major users. This approach may be suitable where some excess network capacity exists (e.g., between major cities or on domestic or regional satellites).

INCENTIVES

Another strategy that may be used with a variety of institutional structures is to introduce incentives that are designed to achieve policy goals such as extension of telecommunications services into rural areas. These may include:

- *Incentive Regulation.* Some countries and U.S. states have introduced changes in regulation that allow the carriers considerable pricing flexibility in return for meeting certain conditions (e.g., price caps). An alternative to financial incentives would be a management by objectives approach where policymakers and/or regulators would set objectives and carriers would be rewarded for achieving them. These objectives could include service upgrades such as extension of service to rural areas or meeting quality of service targets. For example, the Philippines is requiring new franchisees to install a specified number of lines in a currently unserved rural area.

- *Investment Incentives.* Several countries including Indonesia and Thailand have encouraged investors to build new facilities through schemes known as Build Operate Transfer (BOT) where the investors build the system, operate it, and receive a percentage of the revenues for a specified period, and then turn it over to the government. Joint ventures may also include incentives for investment in rural areas.
- *Service Incentives.* Some countries have encouraged private entrepreneurs to offer telecommunications services. For example, in Rwanda, entrepreneurs may install telephones in kiosks that also sell soft drinks and newspapers. The entrepreneurs receive a percentage of the revenue, and typically stay open much longer hours than post offices, and provide a secure location for the telephone. A similar approach is used in Indonesia.
- *Limiting Exclusivity.* Although investors may require a predictable industry environment to commit capital, countries must resist pressure to issue indefinite or very long-term licenses. The technology and the industry are changing too fast for countries to assume that what seems adequate investment and performance today will be adequate five years—let alone ten years—from now. Thus, franchise awards should be for five years or less; and exclusivity agreements should not exceed five years.

OTHER SOURCES OF REVENUE

Internal cross-subsidies have been the primary means of sustaining rural services in the past. However, with the restructuring of the sector, if competition is introduced or even contemplated for some services, subsidies must be separately accounted for, so that monopoly services are not used to subsidize competitive services. If subsidies are required, they should be targeted for specific services or classes of customers. Funds may be generated in several ways:

- *Pooled revenues.* Carriers may pool a percentage of their revenues, which would then be allocated to provide services in high-cost areas. In the U.S., a high-cost fund set up after the divestiture of AT&T is administered by the National Exchange Carriers Association (NECA).
- *Taxes on usage.* A tax on usage may be imposed, with the revenue used to provide services that would otherwise not be economical. This is one of the models used to support universal service in the U.S.
- *Taxes on revenues.* Carriers or service providers may be taxed on revenues generated, with these taxes allocated for service upgrades or rate subsidies.

- *License or franchise fees.* Carriers may be charged fees for a franchise, use of spectrum, or other resource. These fees could be allocated to a fund for upgrading or extending services.
- *Aggregating demand.* Rural areas often lack economies of scale that would make provision of new services attractive. However, government is a usually a major telecommunications user, but government traffic is often carried on dedicated networks. One approach to aggregating demand would be to require that all government traffic be carried on the public switched network. Government expenditures would then generate revenue that could be used to upgrade and extend the public network.

MONITORING PROGRESS

No matter what approach or combination of approaches countries choose to adopt, they must have some way of monitoring progress toward their goals. Incentives have been stressed because most countries do not have the legal history or regulation, or sufficient available expertise to staff regulatory bodies. However, these countries can establish a small oversight group with the legal authority to require licensed carriers to provide data on the number of lines available, quality of service, sample period traffic data, and so on.

A second strategy is for this oversight group to schedule regular opportunities for users to present their needs and problems to carriers. Formal hearings may not always be appropriate, but there needs to be some mechanism for carriers and users to share information, and for regulators to be made aware of user issues and perspectives.

BRINGING TELECOMMUNICATIONS WITHIN REACH OF ALL

These strategies are designed to reflect the changing technological, policy, and financial environments of the 1990s. In particular, they are designed to reflect three themes:

- An awareness that telecommunications goals will be moving targets because of changes in technology and user needs.
- A broadening of the definition of "public interest" beyond the simple assessment of price to customers, which is the indicator most often used in industrialized countries.
- An assumption that incentives are likely to be more successful than regulations in encouraging development-oriented investment, but that sanctions must be available if agreed-on targets are not met.

Innovation in policy must match innovation in technology if the Maitland Commission's goal of bringing all of mankind within reach of telecommunications is to be reached by the turn of the century.

REFERENCES

Al-Ghunaim, A. R. K. (1988). *The missing link and after.* Kuwait: Kuwait Government Printing Press.

Clements, C. (1991, May) *HealthNet.* Cambridge, MA: SatelLife.

Cooperman, W., Mukaida, L., and Topping, D.M. (1991, January). The return of PEACESAT. Proceedings of the Pacific Telecommunications Conference, Honolulu.

Gallagher, L. and Hatfield, D. (1989). *Distance learning: Opportunities in telecommunications policy and technology.* Washington, DC: Annenberg Washington Program of Northwestern University.

Gore, A., Jr. (1991). Infrastructure for the global village: Computers, networks and public policy. *Scientific American, 265*(3), p. 150.

Hardy, A. P. (1980, December). The role of the telephone in economic development. *Telecommunications Policy.*

Hudson, H. E. (1990). *Communication satellites: Their development and impact.* New York: The Free Press.

Hudson, H. E. (1992). Developing countries' communications: Overcoming the barriers of distance. *Froehlich/Kent Encyclopedia of Telecommunications* (Vol. 4, pp. 351–368).

Hudson, H. E. (1993). Maximizing benefits from new telecommunications technologies: Policy challenges for developing countries. In M. Jussawalla (ed.), *Global telecommunications policies: The challenge of change.* Westport, CT: Greenwood.

International Commission for Worldwide Telecommunications Development (The Maitland Commission). (1984, December). *The missing link.* Geneva: International Telecommunication Union.

International Development Research Centre. (1989). *Sharing knowledge for development: IDRC's information strategy for Africa.* Ottawa: Author.

International Telecommunication Union. (1994). *World telecommunication development report.* Geneva: Author.

Johnson, T. (1991, April 26). Microsatellite that turns information into medical power. *The (Manchester) Guardian.*

Lawton, R. A. (1990). Telecommunications modernization: Issues and approaches for regulators. Columbus OH: National Regulatory Research Institute.

Mayo, J. K., Heald, G.R., Klees, S.J. (1992). Commercial satellite telecommunications and national development: Lessons from Peru. *Telecommunications Policy, 16*(1), 67–79.

National Research Council, Board on Science and Technology for International Development. (1990). *Science and technology information services and systems in Africa.* Washington, DC, National Academy Press.

Parker, E. B., Hudson, H.E. (1995). *Electronic byways: State policies for rural development through telecommunications.* Second edition. Washington, DC: Aspen Institute.

Parker, E. B., Hudson, H.E., Dillman, D.A., & Roscoe, A.D. (1989). *Rural America in the information age: Telecommunications policy for rural development.* Lanham, MD: University Press of America.

Saunders, R., Warford, J., and Wellenius, B. (1994). *Telecommunications and economic development*, (2nd ed.). Baltimore: Johns Hopkins.

Tietjen, K. (1989). *AID rural satellite program: An overview.* Washington, DC: Academy for Educational Development.

Tkal, L. (ed.). (1994). *Technology survey report: Educational technologies 1994.* Redfern, Australia: Open Training and Education Network.

United Nations. (1993a). *1993 statistical yearbook.* New York: Author.

United Nations. (1993b). *World tables 1992.* New York: Author.

V

THE INTERNET AND THE NII

13

Internet Cost Structures and Interconnection Agreements

Padmanabhan Srinagesh
Bellcore

WHAT IS THE INTERNET?

The Internet is a network of networks connecting a large and rapidly growing community of users spread across the globe. Individuals use the Internet to exchange e-mail, to obtain and make available information on file servers, and to log on to computers at remote locations. In July 1994, there were 3.2 million host computers on the Internet. The number of hosts on the Internet has been doubling every year since 1989. The number of users on the Internet is not known with any certainty, but estimates range from 2 million to 30 million users. The revenue generated by Internet service providers (ISPs) is not known with certainty, either. *The Wall Street Journal* of June 22, 1994 reported that NEARNet, an ISP serving New England, had annual revenues of $5 million in 1994. AlterNet, a national ISP, reported an annual revenue of $11 million for 1994, and a growth rate of 50%. Maloff (1994) estimated that 1994 revenues for all ISPs were $118 million, more than double the revenues in 1993.

The Internet is a loose federation of networks, each of which is autonomous. There is no central point of control and no overarching regulatory framework. More detailed descriptions of the Internet's history and organizational structure can be found in numerous sources, including Comer (1991) and Mackie-Mason & Varian (1994).

INTERNET SERVICES AND THEIR COSTS

ISPs offer their customers a bundle of services that typically includes hardware and software, customer support, Internet protocol (IP) transport, information content and provision, and access to individuals and information sources on the Internet.

The service mix varies across providers and over time. Customers usually obtain an access link from their location to the ISP's nearest node. Although many ISPs will arrange for this connection and pass the cost on to the customer, the access link is not usually considered a service offered by the ISP. Access through an 800 number or other options where the called party pays for the telephone call (such as Feature Group B access) are the major exceptions.

Access to the Internet is a minor miracle that is often taken for granted. There are dozens of commercial ISPs offering a variety of service options. At the low end are dial-up accounts with limited electronic mail capability suitable for some individuals. At the high end is 45 Mbs connectivity that is suitable for institutions with sophisticated campus LANs, such as large universities. At the level of basic e-mail connectivity, all customers can reach, and be reached by, the same set of people and machines. In this sense, the Internet is like the public switched telephone Network (PSTN). A difference between the Internet and the PSTN is that broad connectivity on the Internet resulted without explicit regulations or government mandates on interconnection. A major purpose of this chapter is to describe the economic environment that resulted in this connectivity, and to analyze how fundamental changes in the economic environment will impact the connectivity of the Internet in the future.

As a prelude to this analysis, I describe the cost structure of Internet service provision. Two caveats are in order. First, I focus on the incremental costs of Internet service provision, with occasional references, where relevant, to examples of Internet costs that are directly borne by end users. The Internet has been built incrementally on a very expensive infrastructure that includes the facilities of the telephone companies, the computing environment of end users (including department LANs and systems administration) and the campus LANs around which the original regional networks were built. ISPs pay for a part of this infrastructure through their purchase of leased lines and, in some cases, payments to universities for rent and local administration. For the most part, the joint costs of the infrastructure are picked up directly by end users and are not included in the prices charged by ISPs. Second, in the absence of results generated by a more methodical approach, I rely on anecdotal evidence. In the following, I list the major categories of cost and describe which elements of cost are sunk, fixed, and variable. The analysis is a useful first step toward understanding Internet competition and interconnection arrangements.

Costs of Hardware and Software

Customers have a choice between dial-up and leased line access to the Internet. Dial-up access is of two types: shell accounts and Serial Line Internet Protocol (SLIP) or Point to Point Protocol (PPP) accounts. With a shell account, a customer uses his computer, a modem, and communications software to log onto a terminal server provided by the ISP. The terminal server is connected to the Internet, and the customer can use the Internet services that the ISP has enabled for the shell account. As most potential purchasers of shell accounts already own a computer and a modem, the incremental hardware and software costs on the user's end are negligible.

An ISP offering dial-up service must purchase a host computer, a terminal server, a modem pool, and several dial-up lines to the telephone network. In March 1993, the World (an ISP selling shell accounts in Boston) reported using a Solbourne SPARC-server as its host computer. At that time, the host computer was reported to have 256 MB of memory and 7 GB of disk space. The World had 65 incoming lines serving approximately 5,000 customers. By August 1994, the number of subscribers had doubled to over 10,000. The World's SPARCserver was reported to have 384 MB of memory and 16 GB of disk space. In 1994, the World did not report on the number of incoming lines, but there are reliable estimates that this number has doubled.

The costs of supporting shell accounts are partly fixed and partly variable. When the number of customers and the amount of usage increase, increases in computer memory, disk space, modems and the number of incoming lines may be necessary. The link from the ISP to the Internet (typically through one of the major backbone providers such as AlterNet, PSI, or the NSFNET) may also need to be upgraded. These upgrades are usually lumpy.

SLIP and PPP accounts require software in the customer's host computer to packetize data according to the IP protocol suite and format the packet for transmission over a telephone line. A suitable modem is required. The hardware and software costs of SLIP connectivity are comparable to similar costs for shell accounts. However, SLIP software has been difficult to configure in the past, and a SLIP account has often been priced higher than dial-up shell accounts. At the ISP end, costs are incurred in purchasing dial-up routers and inbound telephone lines. No terminal server is required. Additional customers and additional use will eventually result in additional costs for ports on the router and upgrades in the link from the router to the rest of the network.

Leased line customers typically have multiple users who are already connected to an enterprise network consisting of one or more LANs. Internet connectivity for these customers often requires the purchase of a router and Channel Service Unit/Data Service Unit (CSU/DSU). AlterNet's prices for equipment suitable for 56 Kbs and T1 connections are about $2,500 and $5,700, respectively. At the ISP end, additional hardware costs may include the purchase of matching CSU/DSU and a port on a router, or an additional router if existing routers are loaded fully. Hardware costs increase in a lumpy manner with the number of customers. Upgrades of the ISP's internal links may also be necessary. Prices for leased line service vary considerably among providers.

The hardware and software costs already described are part of the costs of obtaining Internet service, as is the cost of the users' computers, and the LAN infrastructure in which large customers have invested. This is an important feature of Internet economics: Substantial elements of cost are borne by the user and not the ISP. Consequently, user costs are considerably higher than the charges set by the ISP. The incremental costs of Internet connectivity are small in comparison to the larger investments that potential customers have already made.

Costs of Customer Support

ISPs incur support costs when a customer is acquired, on an ongoing basis during the business relationship, and when the business relationship is terminated. Service establishment may require a credit check, consultation with the customer on the appropriate choice of service options, a billing record that accurately reflects the customer's selected options, facilities assignment, configuration of the ISP's network to recognize the new customer, analysis of the network infrastructure for possible upgrades to support the added load, and other activities necessary to maintain service at the level expected by customers. In addition, some initial debugging may be required to ensure that the hardware and software at both ends of the connection interoperate. Ongoing customer support is required during the business relationship. Large corporate customers may upgrade their LAN hardware and/or software, and this may require reassignment of IP addresses and reconfiguration of their Internet link. Individual dial-up customers may upgrade their operating system software, or install new Internet search tools, and they may require help with configuration. ISPs must also undertake network management and maintenance activities to assure an acceptable quality of service. Costs at service termination include a final settling of accounts, and reconfiguring routers and domain name servers to ensure that the records accurately reflect the termination of the relationship. Although all customers require some support, the level and cost of supporting customers varies widely across individual customers. A technical description of support activities can be found in Lynch & Rose (1993).

BARRNet's service description in January 1993 (obtained via anonymous ftp) provides some information on the nature and cost of service activation. When it was founded in 1986, BARRNet did not view customer support as a component of its service mix:

> BARRNet was conceived and implemented as a network of networks. It connects "sites" or "campuses" rather than individual computers. Our assumption has been generally that our member sites operate their own networks, and support their own users. BARRNet is then more a provider of "wholesale" network service than "retail" service."

This view had changed substantially by 1993, when a wide range of support services were offered. For example, by 1993, T1 connectivity was offered in two flavors. Full Service had a nonrecurring fee of $17,000. With this option, BARRNet owned, operated, and maintained the hardware at the customer's end, provided spares and upgraded the software as necessary. The Port-only option required the customer to provide the router at its location, and to assume responsibility for configuration, management, and maintenance at their end. The nonrecurring fee for Port-only service was $13,000 (24% less than for Full Service). It can be inferred that the cost of configuring, managing, and maintaining a router added up to approximately $4,000 over the expected lifetime of the contract.

Other elements of customer support were offered as unbundled options. The Basic Internet Connectivity Package, priced at $1,500, included assistance in acquiring an Internet number and domain name, specification of a hardware platform for domain name service and e-mail, configuration of the platform, and training for one person in the maintenance of the platform. The Deluxe Internet Connectivity Package, priced at $3,000, offered the following additional services: specification of equipment to secure internal networks, configuration of packet filters, configuration of secure mail servers on the internal network, and configuration of a Network News server. Additional consulting services such as specification of technical platforms, remote monitoring of internal links, and training, were available at $125 per hour.

Currently, BARRNet's equipment and installation fee for a Full Service T1 customer is $13,750, and a 2-year prepaid contract for service is $22,800. For high-speed connections (T1 and 56 Kbs), the nonrecurring charge for equipment and service activation exceeds the ongoing charge for a year's service. For low-speed service, the installation fee is about half the annual service fee. BARRNet's current service activation fee varies from $13,750 for Full Service T1 to $1,300 for 14.4 Kbs leased line or dial-up SLIP or PPP access. Port-only T1 service costs $2,000 less at installation than Full Service. Whereas the additional charge of the Full Service option has fallen by 50% between 1992 and 1994, the hourly charge for consulting services has risen to $175 per hour, an increase of $50 per hour. This suggests that the costs of standard support tasks have fallen, as the cost of customized advice has risen sharply.

AlterNet provides another data point. They charge a nonrecurring fee of $5,000 for T1 service; the fee does not include necessary hardware and software, but does include help in configuring the customer's router. Although these charges vary across ISPs, similar patterns emerge: The charge associated with account activation is a very significant component of a customer's cost. To the extent that there is active competition among ISPs the structure of prices reveals, at least partly, the structure of ISP costs; and customer support appears to be significant. In sum, there appears to be considerable expense and effort involved in connecting a new customer to the Internet, and considerable variation in cost across support services and customer types.

IP Transport

In this section, I discuss some relevant aspects of IP transport: the method by which an IP packet is transported from one location to another. Within a campus LAN, the entire IP packet (data plus header) is treated as a data unit by the LAN, encapsulated in a LAN packet, and transported in accordance with the LAN's protocols. An important consequence is that LAN protocols will determine access to the LAN's bus. From an economic point of view, the significance is that the IP header cannot be used to allocate a potentially scarce resource (access to the bus) without accommodation by the underlying LAN protocol. The identification of bottlenecks and the design of resource allocation mechanisms in a layered architecture is complex and beyond the scope of this chapter.

On wide area networks built on private lines contention for the scarce resource takes on a different form. A router receives packets from various interfaces, consults its routing table and forwards each packet on the appropriate outward link. When the incoming rate exceeds the outgoing rate, packets can be temporarily queued. The IP header has a Type of Service (TOS) field that can, in principle, be used to manage queues in routers, but this function has not been implemented widely. Bohn, Braun, Claffy, & Wolff (1993) described an episode during the mid-1980s when router queues were managed to offer priority to some delay-sensitive uses of the network. They propose that the TOS field be implemented and used to manage access to congested links. When IP runs over private lines (or any time-division multiplexed service) the IP protocol can be used to implement smart markets or other congestion management schemes without any change in protocols at lower layers in the stack.

Wide area networks that use so-called "cloud technologies" (also called fast packet services) such as Frame Relay (FR), Switched Multi-Megabit Data Service (SMDS) and Asynchronous Transfer Mode (ATM) raise a different set of economic issues. Fast packet services statistically multiplex packets over time slots carried on underlying physical facilities. The fast packet technology treats the IP packet as a Protocol Data Unit, just as LANs do. As was the case with LAN technologies, the IP header, by itself, cannot deal adequately with resource allocation issues. In the following, I briefly describe the role of IP transport in the provision of Internet services.

Costs of Transporting IP Packets over Private Lines

In the late 1980s most ISPs had internal backbones consisting of routers connected redundantly to one another by private lines ranging in speed from 56 Kbs to 1.5 Mbs. Most lines were leased from telephone companies. There were a few exceptions. NEARNET, in Boston, had its own wireless Ethernet (10 Mbs) backbone connecting five nodes, and BARRNet had its own wireless link between the University of California at San Francisco and Berkeley. Statistical multiplexing of IP packets on the underlying leased lines led to significant cost savings. The cost of transporting IP packets was determined by leased line tariffs, the costs of the routing hardware and software at the nodes, and the ongoing costs of monitoring the network and remedying supply disruptions in a timely manner. These costs were fairly substantial. Mackie-Mason and Varian (1993) estimated that the costs of leased lines and routers amounted to 80% of total NSFNET costs. The cost of the Network Operations Center amounted to another 7%. With high transport costs, the ability to use bandwidth efficiently through statistical multiplexing is a major benefit. However, it should be noted that the NSFNET service provided by ANS (a nonprofit joint venture of IBM, MCI, and the State of Michigan) was part of a research experiment in high-speed networking funded by the NSF. Almost all NSFNET customers were large regional networks, not end users and ANS' cost structure differed significantly from that of other ISPs. An estimate based on an analysis of several midlevels suggests that IP transport accounts for 25% to 40%

of a typical ISPs total costs. Although efficient use of bandwidth was important for these ISPs, it was not as important an issue.

For many ISPs, transport costs are sunk over the business planning horizon. A brief digression on leased line tariffs may be useful, both for an understanding of IP transport costs and because the evolution of leased line prices may foretell similar developments on the Internet. Currently, most long-haul transmission links are provided over optical fiber. The major cost of constructing fiber optic links is in the trenching and labor cost of installation. The cost of the fiber is a relatively small proportion of the total cost of construction and installation. It is therefore common practice to install "excess" fiber. According to the FCC's (1994) *Fiber Deployment Update*, between 40% and 50% of the fiber installed by the typical interexchange carriers is "dark"; the lasers and electronics required for transmission are not in place. The comparable number for the major local operating companies is between 50% and 80%. Private lines are provided out of surplus (lit and unlit) capacity available in the networks constructed by telephone companies. The incremental cost of providing private line service is determined by the costs of lighting up fiber if necessary (lasers plus electronics at the ends), the costs of customer acquisition (sales effort and service order activation), and ongoing costs of maintaining a customer account. Private line tariffs must recover these incremental costs and contribute to the very substantial sunk costs of the underlying facilities. Furthermore, these tariffs are set in a very competitive environment. The effect of this competition has been to drive down the price of leased capacity. According to *Business Week*, private line prices have fallen by 80% since 1989 ("Dangerous Living in Telecom's Top Tier," 1994).

The cost structure described, together with competitive forces, has resulted in two increasingly common features of the price structure: volume discounts and term commitments. The standard interLATA private line tariff consists of a nonrecurring charge and a monthly charge based on the airline mileage between the two locations to be connected. Customers can select optional features at an extra charge. The standard charges vary with the bandwidth of the private line, but there are no usage-sensitive charges. Private line tariffs offer discounts based on volume and term commitments. AT&T's Accunet 1.5 (T1) tariff offers a discount of 57% to customers whose monthly bill for a specified bundle of services, including T1 lines (at standard rates) exceeds $1 million, and who commit to maintaining that level of expenditure for five years. The volume discount may reflect the fact that large customers are more desirable: It may be less costly selling to one customer with a $1 million bill than to 1,000 customers each of whom has a $1,000 bill. The term commitment may be the telephone company's response to the long-run cost structure of building physical networks and the high cost of churn: As there are fixed costs of service activation and termination, companies seek to provide their customers with incentives to be loyal.

In a competitive environment with excess capacity, there is a tension between the large sunk costs of physical networks and very low incremental costs of usage. On the one hand, the need to recover sunk costs suggests using price structures with high up-front charges and low (or zero) usage rates. On the other hand, with significant excess capacity present, short-run profits can be increased by selling at any price above incremental cost. Economic theory would suggest that the pricing outcome in this situation might be unstable, unless regulatory forces or other influences inhibiting competition were present.

The consequence of the leased line tariff structure already described for the cost of IP transport is straightforward. Given a high nonrecurring service order charge, ISPs with leased line backbones have an incentive to size their needs over a 3- to 5-year period, and commit to a level of purchase determined by projected demand. In a rapidly growing Internet, this can result in substantial excess capacity among ISPs in the short run. The incremental cost of carrying IP packets will be close to zero. (If private lines charged for usage, this would not be true.) However, the sunk costs of IP transport can be substantial. An examination of ISP network maps in mid-1993 suggested that none of the national providers had backbones large enough to qualify for AT&T's largest discount. However, many ISPs were large enough to qualify for smaller, but nevertheless substantial, discounts, on 3- to 5-year contracts. Competition among these ISPs may be subject to the economic tension present in the private line market. Indeed, the use of volume discounts and term commitments is emerging in the ISP market. ISPs typically charge their T1 customers twice the rate they charge their 56 Kbs customers, even though the T1 customers have 24 times the bandwidth. Term commitments can be seen in BARRNet's price structure. 56 Kbs customers are offered a 17% discount over monthly rates if they take a 2-year prepaid contract.

There are at least three types of ISPs whose cost structures do not fit the model described here. One type is represented by Sprint, which offers a national IP service, SprintLink. As Sprint owns a large national fiber-optic network with substantial excess capacity (45% dark in 1992), it faces a lower incremental cost for transport provision than other ISPs who lease lines. On the other hand, it has far higher sunk costs. The second type of ISP is represented by small midlevels or regional networks. These ISPs obtain access to the global Internet by connecting to a larger ISP. Very often, this larger ISP is ANS, which historically provided interregional connectivity to the regional networks sponsored by NSF. The third type of ISP is the small reseller, which appears to have grown rapidly in the last year or so. To the larger ISP, the reseller often appears to be a customer with 56 Kbs or T1 access. The midlevel or reseller has a very small (perhaps nonexistent) backbone: Customers are responsible for the connections to the reseller's node (and there may be just one node), and the reseller purchases the connections to the larger ISP (and there may be just one connection). The small ISP/reseller has relatively small sunk costs and little excess capacity. According to an article in *Forbes* (1993), small providers working out of their basements require an initial investment of $30,000 for electronic gear and about $1,000 per month for telephone connections. For these providers, incremental costs for transport are relatively high as significant volume discounts are not applicable on the links to their customers and to the larger ISP.

Not surprisingly, the range of prices charged by service providers varies widely. In a competitive environment where the cost structures of different providers are radically different, where average costs are very different from incremental costs, and where there is substantial excess capacity in one key input (raw bandwidth), the equilibrium outcome is not obvious. The prices for shell and SLIP accounts vary dramatically among providers, and serve to make the point. Alternet, which leases an extensive international backbone, charges $20 per month for the basic account, another $10 for e-mail service, another $10 for USENET news, and $3 per hour for direct dial to an Alternet POP. The one-time fee is $99. At the other extreme, the Connection, a local provider serving the 201 area code in New Jersey, charges $10 per month for a shell account, has no fees for e-mail or USENET news, no usage charge, and will waive the sign-up fee of $20 for customers who sign a 12-month contract. Scruz-Net, a small network in Santa Cruz, offers single-host SLIP/PPP connectivity at 28.8 kbps for $25 per month, with an allowance of 100 free hours. JVNCnet, a commercial provider with a backbone spanning many states, offers SLIP access at $59 per month and $4.95 per hour. Whether these large price variations are accompanied by large quality variations is not known.

In sum, the cost structure of IP transport provision varies considerably among ISPs. The four broad classes include providers who own a physical network, national backbones based on leased facilities, small regional networks, and resellers. Sunk costs of transport are highest for the first type and lowest for the last. Variable (with number of customers and usage) costs are lowest for the first type and highest for the last. Prices across providers are highly variable.

The Impact of Fast Packet Technologies on IP Transport Cost

The introduction of new fast packet services such as Frame Relay, SMDS, and ATM may have a significant impact on an ISP's cost structure and its role as a provider of low-cost transport. Fast packet services (also referred to as cloud technologies) statistically multiplex variable size packets or fixed size cells onto time slots carried on an underlying physical facility. IP packets ride on top of this statistically multiplexed service. As was true of LANs, a fast packet service will treat an IP packet (header plus data) as a data unit, add its own header, and transport it over the underlying network in accordance with its own rules. Currently, NEARnet and PSI run IP over Frame Relay, Cerfnet runs IP over SMDS, and AlterNet runs IP over Ethernet over ATM. The additional statistical multiplexing gains of IP transport over those obtained by the underlying "cloud" service will be less than the gains obtained when ISPs used private lines. The extent to which an ISP can offer additional multiplexing gains will be determined in part by the proportion of the traffic over the physical link that is generated by the ISP; the higher this proportion, the greater the potential gain generated by the ISP.

Early tariffs for fast packet services had monthly rates that varied with bandwidth. There were no charges for usage or for distance. One example is a Frame Relay tariff filed by US West in September 1992. This tariff offers a small discount (about 10%) to customers who commit to a 5-year contract. MCI's SMDS price structure (announced in August 1994) is considerably different. There are usage and distance charges, but the usage charges are capped at relatively low levels

(*Business Wire*, 1994). For large users (with an access speed of 34 Mbps), MCI's SMDS price can vary from $13,000 to $20,000 per port per month, depending on the customer's usage. There is no mention of a discount for term commitment. This may be because MCI is the only interexchange carrier (IXC) to offer SMDS, and is not concerned with customer churn.

Both tariffs show a movement away from the deep term discounts that characterize private line charges. ISPs who lease fast packet services from others will have a weak incentive to sign long-term contracts for their backbones. A smaller proportion of their transport costs will be sunk. In addition, connectivity among multiple ISPs can be established at significantly lower costs than was possible with private lines. Once an ISP pays a flat rate to connect to a fast-packet cloud, the incremental costs of virtual connections to multiple ISPs are very small. With MCI's SMDS service, for example, there are no additional costs involved in communicating with multiple SMDS customers (although an ISP will have to pay for each port it connects to on the SMDS cloud). Small ISPs with a few nodes can reach out to anyone else on the national SMDS cloud without investing in a national backbone consisting of multiple private lines. With Frame Relay and ATM service, there are incremental costs of reaching additional sites on the cloud, as Permanent Virtual Connections (PVCs) must be configured and managed. However these costs are relatively small (as little as $1.19 per month in US West's tariff for Frame Relay).

Even if IP transport provides minimal multiplexing gains when it is run over a cloud technology, IP service will perform important functions. These include uniform global addressing (e.g. Frame Relay has reusable addresses with only local significance) and wide connectivity, protocol conversion across varying LAN, MAN, and WAN technologies, and an important "bearer service" role that helps insulate embedded investments from ongoing technical change in network hardware. Most of these functions can be performed at the edge of the network, and IP may migrate to the network border over time. This may accelerate the evolution of some ISPs into systems integrators, or "market-makers," in the terminology of R. Mandelbaum & P. Mandelbaum (1991).

The Impact of Multimedia Traffic on IP Transport Costs

The share of transport in ISP costs is likely to change as new multimedia applications grow in popularity. Voice, video, data, and images differ in the requirements they place on the network, and raise difficulties for the use of IP transport in its current form.

Efficient coding schemes vary greatly for the different media; an excellent discussion of coding schemes can be found in Lucky (1989). Voice can now be digitized and compressed by a factor of 16 to 1 on commercially available chips, such as Qualcomm's Q4400 vocoder, which claims to achieve near-toll voice quality at less than 10 Kbs. Video compression using the MPEG standard allows for VCR picture quality at a bandwidth of 1.5 Mbs. Video applications over the Internet (Mbone and CU-SeeMe) use a different compression scheme. The Mbone uses the

JPEG standard to digitize each frame and transmits data using UDP and IP tunneling. These techniques offer low picture quality at 100 to 300 Kbs (2 to 10 frames per second). Transmission of data files requires no specific bandwidth; there is a trade-off between bandwidth and delay. For many current applications (e-mail, file transfer, and even fax) 9.6 Kbs is adequate, and store-and-forward techniques are acceptable. High-quality image transfer (such as that needed in medical applications) requires considerably more bandwidth; schemes using lossy JPEG allow for the transfer at 56 Kbs in reasonable time.

If bandwidth were essentially infinite, variations in bandwidth use would not be a problem for mechanisms like IP that treat all packets equally. However, most ISPs pay for added bandwidth, and must treat it as a scarce resource. A simple computation highlights the problem. At 300 Kbs for a video session on the Internet, it takes only 150 simultaneous sessions to congest a link on the NSFNET, the Internet's major backbone, with the highest speed links. The congestion created by video use is pernicious; it destroys some valuable mechanisms that are part of the Internet's discipline and efficiency. Transmission control protocol (TCP) is used by host computers to provide a reliable byte stream to the applications that are run by an end user. TCP selects a widow size, which determines the number of packets it can send to the other side before stopping for an acknowledgment. Large window sizes allow for faster throughput. With the implementation of the slow start mechanism, TCP monitors round trip times, and if it detects congestion, reduces the window size and contributes to better system behavior. Video sessions use User Datagram Protocol (UDP). Unlike TCP, UDP does not reduce its transmission rate during periods of congestion. Other users running data applications over TCP pay disproportionately in delay when video sessions congest any link. If congestion should get severe, TCP users may have an incentive to stop using the slow start mechanism.

ISPs recognize that there is a problem with high bandwidth users and uses. In the absence of pricing solutions that can be implemented with the IP header structure, they have resorted to blunt instruments. For example, anonymous ftp sessions on NEARnet's server begin with a welcome message announcing that only paying subscribers may access the Internet Talk Radio files. In view of the fact that some of these files are greater than 30 MB, the prohibition appears to be motivated by a desire to preserve bandwidth. If the Internet supported a more sophisticated billing mechanism, or if the cloud technology underlying IP service supported multiple quality of service types, blanket prohibitions such as this might not be necessary. Current work on IPng and real-time protocols may solve the bandwidth allocation problem at some point in the future. In the meantime, IP remains a very cost-efficient transport mechanism for applications (like e-mail and image transfer) that are not affected by delay.

Summary of Transport Costs

The costs of providing IP transport represent a substantial fraction (25–40%) of an ISP's cost. This proportion will fall as less costly fast packet services are more widely deployed. However, the increase in the use of multimedia applications may result in a proportionally greater increase in the need for bandwidth. The tension between satisfying customers with bandwidth-intensive needs and satisfying customers with low-bandwidth applications cannot be efficiently resolved with current technology. Mackie-Mason & Varian (1993) suggested a smart-market mechanism that allows bandwidth to be efficiently priced. An attractive feature of their pricing scheme is that it generates the correct signals for investing in new capacity. More work needs to be done in this area, taking into account the more complex layered structure that is now emerging in the Internet. In addition, the issue of bundling or unbundling transport, support, and information content needs to be addressed by new pricing approaches.

Information Content and Provision

There is usually a fixed cost associated with the production, formatting, and organization of information suitable for database applications and a low cost of duplication (Perritt, 1991). Some examples suggest that the revenue generated by information provision will greatly exceed the revenue generated by the underlying transport. Audiotex services (900 and 976 numbers) usually set per minute charges that are many multiples of standard toll charges. Further supporting evidence can be found in the price list published by Dialog. The connect time charge for transport has traditionally been in the $3 to $12 per hour range, depending on the network used to access Dialog. Once the user has logged on, the charge for database access ranges from $15 to $300 per hour. Huber, in the *Geodesic Network* (U.S. Department of Justice, 1987), reported that online services spend only 8% to 10% of their expenses on local and long-distance transport. Approximately 40% to 45% of their expenses are spent in acquiring information content, and approximately 45% is spent on sales, marketing, and administration.

For much of its life, the Internet has offered “free” information. This reflects the Internet’s roots in the academic community, which encourages the free dissemination of scholarly research. In addition, various government agencies have begun to use the Internet as a means of making public information available in electronic form. For example, the FCC posts its Daily Digest on an ftp server. Archives of Usenet groups and the ongoing contributions to newsgroups provide another source of free information. The use of Web-servers to establish company “presence” on the Internet represents yet another source of “free” information; the supplier of the information pays to display advertisements in a nonintrusive way.

There is, however, a rapid increase in the number of subscription-based information services on the Internet. All the major information services (Dialog, Orbit, etc.) offer telnet access to subscribers. Dialog provides an itemized bill that separately lists charges for transport and database access. For-pay service appears to be taking root in the university community too. The ORION system at UCLA charges for access and usage, with a minimum charge of $25 per month (CERFnet News, 1991). The clari.* hierarchy in Usenet is available only to paying customers. The charge is $75 per site plus $1 per user at the site.

It appears likely that the Internet will see a variety of free and for-pay information services develop. Whereas for-pay information services are accustomed to paying for all transport charges incurred by their clients, and billing back for network use, providers of free information may resist this arrangement for the recovery of transport cost, as they have no paying client to bill back. Currently, the Internet satisfies the academic community's needs. However, if budgetary pressures on academia should increase, universities may feel the need to charge for the use of information they produce, and the needs of this community may align with those of the online industry.

There is a considerable research effort under way on a variety of security, privacy, and billing mechanisms that will support commercial information provision over the Internet. The experience of the online industry suggests that the commercial potential of information content and provision will be significantly greater than the cost of the underlying transport, and the needs of information providers may have a significant influence on payment mechanisms for transport.

The Bottom Line on Costs

The early Internet was developed to meet a specific goal: the interconnection of academic sites for the purpose of open scholarly research and education. In meeting this goal, the Internet was not just successful, it was too successful. The rapid growth of the Internet into sectors of the economy that it was never designed to serve (such as banks and online information services) has revealed some gaps in capability that were not important to early users, but are very important to the new users. These include higher levels of customer service, greater reliability, assured security, and privacy and billing mechanisms. In response to these changing market needs, the nature of Internet service and the cost structures of service provision are being transformed. At the same time, the spread of fast packet services is reducing the Internet's value as a provider of cheap transport. As competitors continue to emphasize service quality as a differentiating factor, the share of transport in total cost will fall, and the share of existing and new support services and sales expenses will rise. As Noll (1994) pointed out, AT&T's expenses on sales and advertising for voice services grew rapidly in absolute and relative terms after equal access was implemented and competition grew more intense.

ECONOMICS OF INTERCONNECTION

This section develops a brief economic history of interconnection agreements on the Internet, with a view to understanding future developments.

The Early Years: 1986–1991

A logical starting point for a discussion of Internet interconnection agreements is the first NSFNET backbone, which was constructed in 1986. The NSFNET was the top tier of a growing network of networks organized as a three-layer hierarchy.

The second tier in the hierarchy was made up of midlevel networks, each consisting of 10 to 30 universities. Many midlevels were formed with some funding from the NSF. Each midlevel attached to a nearby NSFNET node. The bottom layer consisted of campus LANs, which attached to a midlevel network.

The Internet's hierarchical structure, crowned by a single backbone, resulted from early decisions by Internet architects. In discussing their work, Comer (1991) stated: "They came to think of the ARPANET as a dependable wide-area backbone around which the Internet could be built. The influence of a single, central wide area backbone is still painfully obvious in some of the Internet protocols that we will discuss later, and has prevented the Internet from accommodating additional backbone networks gracefully" (pp 33–34).

Early routing protocols such as the gateway-to-gateway protocol (GGP) and the exterior gateway protocol (EGP), bear out Comer's observation. Internet routers typically use hop-by-hop destination routing. The router reads the destination address in the packet header, looks up its routing table, and forwards the packet on the next hop. Routing protocols specify the rules by which routers obtain and update their routing tables. Exterior or border gateways that connect a network to the rest of the Internet use only the network portion (the first 8 to 24 bits of the 32-bit IP address) of the destination address for routing purposes.

GGP partitioned Internet gateways into two groups: core gateways controlled by the Internet Network Operations Center and noncore gateways controlled by others. The core gateways contained full routes. All other gateways were permitted to maintain partial routes to destinations that they were directly connected to, and point a default route up the hierarchy toward a core gateway. This arrangement simplified routing at the core and noncore gateways, and reduced the amount of information that routers had to exchange over the network. The set of core gateways and the links connecting them together formed the backbone at the top of the hierarchy. It is worth noting that a major justification for a single backbone was simplicity in routing, and this was unrelated to economies of scale in transport. GGP does not easily accommodate multiple backbones connected to one another at multiple points.

I provide a brief example (based on Comer's discussion) that highlights a technical problem and a pricing puzzle that arises when multiple backbones are multiply interconnected. Suppose two coast-to-coast backbones are interconnected on the East and West coasts. Suppose Host 1 located on the East coast on Network 1 wishes to send a packet to Host 2 located on the East coast on Network 2. It would make sense for the packet to go through the East coast interconnect point. Suppose Host 1 wished to send a packet to Host 3 on Network 2 on the West coast. If Network 1 had a better (less congested) backbone, it may make sense for Network 1 to transport the traffic across the continent, and deliver it to Network 2 at the West coast interconnect. Routing will depend on the host address, not just on the network portion of the address. Routing schemes that require this level of detail are not scalable: The size of routing tables would increase too rapidly with the growth of the Internet.

Apart from such technical problems, there is an economic problem. For transcontinental communication between users on different backbones, which backbone should carry the traffic? In the absence of settlements, it is not clear that either network has any incentive to volunteer for the job. As long as backbones are not congested, the issue of transport will not be a weighty one. But in a world of rapidly growing traffic, when frequent upgrades are needed, there are good reasons for an ISP to advocate routing arrangements that use another ISP's backbone instead of its own. This is not merely a theoretical possibility. A relatively large proportion of traffic between sites in Mexico transits the U.S. over the NSFNET.

Of course, all ISPs cannot pursue the strategy of shifting traffic onto others' backbones successfully. An alternative is for some form of settlement among interconnected networks. The most important criteria for an efficient settlements mechanism are that it should not impose high administrative costs, that it should provide the correct incentives for routing, and that the net flow of funds should allow all suppliers to recover their costs. As the Internet goes through periods of substantial excess capacity, followed by periods of congestion and capacity expansion, different settlement mechanisms will be required. The smart market mechanism described by Mackie-Mason & Varian (1993) can adapt well to these changing circumstances. More work may be needed to accommodate routing protocols to the smart market mechanism.

The alternative to competing peer backbones is a single backbone. The drawback with this alternative is that there will be no competition for the provision of a key service: routing and long-haul transport. Although economists have developed sophisticated regulatory schemes to deal with the lack of competition, the practical difficulties in implementing these schemes can be enormous. The early architecture of the NSFNET apparently avoided these issues by selecting a simple routing scheme and the architecture it implied. As the net was not commercial, there was little danger of monopoly pricing, and much to be gained by centralizing the routing function.

The next generation routing protocol, EGP, addressed several weaknesses of GGP by introducing the notion of an *autonomous system*. Interested readers are referred to the discussion of EGP in chapter 14 of Comer (1991). Despite its advances, EGP shared the key drawback of GGP. As Comer pointed out: "EGP is inadequate for optimal routing in an architecture that has multiple backbones interconnected at multiple points. For example, the NSFNET and DDN backbone interconnection described in Chapter 13 cannot use EGP alone to exchange routing information if routes are to be optimal. Instead, managers manually divide the set of NSFNET networks and advertise some of them to one exterior gateway and others to a different gateway."

The Current Framework: 1991–1994

From the very beginning the business plans of key ISPs appeared to be inconsistent with one another. ANS provided a bundle of services that included full routing and long-haul transport. The new backbones had constructed national networks of their own, and had no need to purchase transport or routing from ANS. However, they did need to offer their customers full access to all Internet sites. As most of these sites were connected to ANS, an interconnection agreement with ANS would have met their customers' needs. The commercial ISPs argued that such connectivity (without routing and transport) should be settlement free. The rationale advanced by the commercial providers was as follows. When a customer of one ISP communicates with a customer of another ISP, both customers benefit. Each customer pays his ISP for his use of the ISP's network. Both ISPs are paid by their customers, and there should be no further need for the ISPs to settle. Proponents of this view recognized that their argument did not apply to transit traffic. But when there are only two networks involved, there is no transit traffic, and no settlements are required. It was this philosophy, together with the inability of the new entrants to obtain interconnection agreements with ANS on terms acceptable to them, that led to the formation of the Commercial Internet Exchange (CIX) in August 1991. The three founding members were CERFNet, PSI, and AlterNet. The CIX members agreed to exchange traffic without regard to type (commercial or R&E) and without settlements. The CIX router was installed in Santa Clara and managed by PSI, and other founding members leased private lines from their networks to the CIX router.

Initially, ANS did not join the CIX. It formed a for-profit subsidiary, ANS CO+RE, and proposed a gateway agreement that would lead to full connectivity. At its core, the gateway agreement consisted of three parts. First, determine separate attachment fees for commercial and R&E customers. Second, use statistical samples of each attached network's traffic to estimate the proportion of R&E traffic. Third, charge each attached network a weighted combination of the commercial and R&E attachment fees, with weights obtained from the sample. A portion of the revenue generated by the gateway agreements was to be put in a pool that would be used to upgrade the NSFNET infrastructure. The proposals from ANS and the CIX members had little in common. Nevertheless, after some negotiation it was agreed that ANS and the CIX members would interconnect without settlements. ANS did not pay the CIX membership fee, and CIX members did not pay ANS for NSFNET services.

In October 1993, the CIX, apparently without warning, blocked ANS traffic from transiting the CIX router. At this point, ANS (through its subsidiary CO+RE) joined the CIX and full connectivity was restored. In the 10 months following this episode, CIX membership rose from about 20 to about 70 ISPs.

The right of resellers of IP service to transit the CIX router has been, and continues to be, debated. The Membership Agreement has, for at least 2 years, contained some rules suggesting that assured connectivity was limited to the "direct" customers of member ISPs. The term was not defined in the Agreement. It now appears that, beginning in November 1994, resellers of IP services would have their packets blocked at the CIX router if they did not join the CIX and pay the $7,500 annual membership fee. The cost of resale has gone up. AlterNet requires resellers to purchase a special wholesale connection that costs about three times as much as a retail connection. In addition, AlterNet requires resellers to use a complex addressing scheme and routing protocol (BGP4) rather than the simpler PPP protocol used by end users. PSI does not sell wholesale connections. Sprint apparently treated resellers just like its other customers, but new CIX rules together with the ability of the CIX to filter routes may affect Sprint's policy on resale. The growth of resellers and the change in the way established ISPs treat them is an interesting and unsettled phenomenon.

In order to simplify the exposition, only the roles of the NSFNET and the CIX in Internet connectivity have been discussed so far. There are several additional arrangements that are important for assuring connectivity. The more complex arrangements described in the following were made possible by developments in routers and routing protocols. Routers can now accommodate large routing tables, reducing the need for default routes. The current routing protocol of choice, Border Gateway Protocol 4 (BGP4), has automated the maintenance of routing tables and is flexible enough to accommodate routing policy based on configuration information. BGP4 also permits the use of more flexible addressing and aggregation of routes. Routing technology is not an absolute constraint to the development of a multiply connected multiple backbone architecture.

The Metropolitan Area Ethernet-East

The Metropolitan Area Ethernet-East (MAE-East) began as an experimental interconnect arrangement developed by AlterNet, PSI, and Sprintlink. Currently there are 14 members. MAE-East differs from the CIX in two important ways. First, MAE-East is a distributed Ethernet service (provided by Metropolitan Fiber Systems) spanning a wide geographic area. An attraction of a cloud service like MAE-East is its cost, compared to that of a physical connection to a single router. Second, there are no multilateral agreements in place at MAE-East. ISPs need to work out a set of bilateral agreements that meet their needs for connectivity. Currently, none of the bilateral agreements are in written form. It has been unofficially reported that there are no settlements at MAE-East. Every provider accepts all traffic from and delivers all traffic to any ISP with whom a bilateral agreement exists. The transaction costs of multiple bilateral negotiations can be high.

The CIX has announced that by September they will place the CIX router on Pacific Bell's SMDS cloud. The CIX will retain its multilateral connection agreement, which reduces the transaction cost of establishing connectivity. These developments suggest that there is broad agreement on the benefits of using cloud technologies as interconnection mechanisms. But the large ISPs have not yet settled on a common business agreement for interconnection. The simultaneous existence of two very different interconnection models (multilateral and bilateral) naturally raises the question: How many types of interconnection agreements can we expect to see in equilibrium? If a standard agreement emerges, will it look like the CIX agreement (multilateral), the MAE-East agreement (bilateral), or something else?

Analysis of Interconnection Agreements

Consider first an architecture based on the CIX model: multiple backbones interconnected at a single point (an XIX), exchanging traffic among the direct customers of all members without settlements. Members pay a fixed membership fee. Assume that ISP networks are based on leased lines and not on fast packet services. Advantages of this architecture are: (a) membership in the XIX is necessary and sufficient for connectivity to the Internet, (b) members whose routers cannot carry the full complement of routes can keep local routes and point a default to the XIX; and (c) competition among backbone providers is feasible (supported by the routing technology).

Disadvantages of this architecture are: (a) if too many networks use the XIX as the only interconnect point, the XIX router could experience congestion and provide poor service; (b) end users who are geographically close to one another, far away from the XIX, and on different networks will experience needless delay as their packets make the round trip journey to the XIX; (c) small regional providers who join the XIX will have the same reach as large ISPs who have invested in national backbones. However, small regional networks will have smaller sunk costs, and can offer lower prices than the national providers. Cost recovery may become a significant issue for the larger providers. However, if the national backbones cannot recover costs and go out of business, the small regionals lose the connectivity that their customers want; (d) given their cost structures, large ISPs with national backbones will set prices that are not proportional to bandwidth. Small resellers can arbitrage the difference profitably for customers who do not require much support, as they do not have large sunk costs to recover. There will be little incentive for anyone to invest in a national backbone (facilities-based or leased).

The first two disadvantages can be avoided by setting up multiple points of interconnection between ISPs with national backbones. This will reduce the load on the XIX router, and reduce latency. The third difficulty can be removed by restricting membership to ISPs with national backbones, or requiring small regional ISPs who join the XIX to pay settlements to the larger ISPs. Both these possibilities are hinted at in the CIX's membership agreement. Arbitrage by resellers can be handled by prohibiting resale, or raising the price to resellers. Large ISPs have already taken this step. The current situation may be stable.

The network topology of an ISP using fast packet services may be quite complex. An ISP's customers may be scattered all over the globe, and different cloud technologies may be in use at the various customer locations. The ISP can use the IP protocol to integrate its network over these disparate clouds. The ISP will provide customer support, some network management and possibly information content. The ISP will not provide the multiplexing function that reduces the cost of underlying transport; this function will be performed by the firms producing the underlying clouds. In this environment, the original CIX philosophy may be an attractive model for interconnection, and the distinction between large ISPs with national backbones and small regionals disappears. Every provider purchases access to the designated underlying clouds, and shares the costs of the underlying transport by paying the price charged by the supplier of the cloud. The costs of interconnection become symmetric. As some (large) ISPs do not have sunk costs that other (small) ISPs can leverage off, the incentives to interconnect will not be hampered by gamesmanship.

As transport charges faced by ISPs fall, the prices they charge their customers will not be proportional to their access speed, and resellers may continue to find the Internet a profitable business. The treatment of resellers in this environment is difficult to predict. The economic forces governing resale in the Internet are not very different than those at work in the market for interLATA voice traffic. An excellent discussion of resale in the long-distance market can be found in Briere (1990). The rapid growth of resellers in both the voice market and the Internet may raise difficult issues regarding the stability of competition.

ARCHITECTURES FOR INTERCONNECTION

A fundamental question that is not often asked is why networks should interconnect. This question does not have an obvious answer when the networks in question are virtual and not physical networks, and when "network interconnection" is used to refer to a business relationship between network service providers. At a purely physical level, it is true that a continuous path is needed between the equipment used by the communicating parties. However, three examples presented in the following suggest that full connectivity among users does not require all intervening networks to establish business relationships with one another. The optimal degree of interconnection is part of a larger architectural decision.

Consider first the case of an end user who wishes to purchase a private line to connect two points, one on the West coast and one on the East coast. One option currently available to the customer is to lease (from the New York LEC) a private line to an IXC POP, lease from the IXC a private line from the New York POP to the California POP, and lease a private line from the California LEC linking the California POP to the customer's location on the West coast. The end user pays a separate charge to each of the three networks for their segments of the private line. Whereas the three networks must agree on technical matters (timing and format of the signal), there need be no explicit business arrangements linking the network service providers that provide the private lines. Interconnection on the physical level does not require the network providers to maintain explicit business relationships. The customer may choose to designate one of the three networks to be its agent, and this would result in business agreements among the three networks; but such a designation is not required. The choice between the two arrangements would seem to hinge on transaction costs (who bills whom, who reports and coordinates repairs of outages, etc.).

For the second example, consider a community of electronic mail users who belong to networks that are not connected to one another. Suppose this community wants to establish a bulletin board or newsgroup like service to which they can post, and on which they can read each other's contributions. One solution is for the networks to interconnect and arrange for the transfer of e-mail. Another possibility would be for one of the users in the set (or even a third party) to establish e-mail accounts with all the networks to which the community is connected, and to operate a mail relay that passes e-mail transparently across networks. Which of these options is a lower cost solution? It is not clear that there are economies of scale in the customer support required for small operations of this sort. If all e-mail prices were based only on usage, with no monthly fees, then there would appear to be no cost differences between the two alternatives. Indeed, with pure usage-based pricing, each member of the list could join *all* the networks at no additional cost. If e-mail prices consist of flat monthly fees, with no usage charge, then the solution based on a mail relay may be more expensive. This solution requires that at least one user belong to all networks, and that user faces higher e-mail charges than he would if the networks were interconnected. This may (or may not) be offset by lower costs for customized support. In this example, the connectivity of the e-mail networks is not necessary to support full e-mail connectivity among users, and other alternatives can conceivably be more efficient than full network interconnection. The support costs associated with different arrangements are one important determinant of efficient interconnection arrangements.

Finally, consider a hypothetical situation where ISPs enter into bilateral interconnection agreements that result in a fragmented Internet. Suppose that there are two sets of ISPs, each of which is fully interconnected, but there are no interconnection agreements between the two sets. If all ISPs sit on a fully interconnected SMDS cloud, end users can be fully connected to one another by joining one ISP in each set. With a cloud technology like SMDS, a customer who joins two ISPs does not have to purchase two private lines or access ports, and there will be no additional costs associated with the need to reach two ISP nodes rather than one. If ISP costs are based on usage, and not on pipe size, the end user may see no additional costs to joining a fragmented Internet.

On the other hand, if (as is true today) many SMDS and Frame Relay clouds are not interconnected, ISPs straddling these clouds can provide full interconnectivity among users. Thus, a customer in Boston who is connected to NEARnet over a Frame-Relay link can communicate seamlessly with a customer in San Diego who is connected to Cerfnet over an SMDS cloud.

The relative costs of different interconnection modes discussed here appear to depend on a variety of prices, support, and transaction costs and not on the costs of raw bandwidth alone. The earlier discussion of Internet costs suggested that support activities are becoming a larger component of overall costs. Owners of physical networks who provide a full range of services to end users may spend very little on the underlying physical facilities. According to Pitsch (1993), transmission costs account for only 3% of AT&T's annual expenses. The modeling of support and other service-related costs appears to be important not just because they are a significant component of services end users pay for, but also because they have an important bearing on interconnection arrangements.

The economics of interconnection agreements is complex when the there are multiple layers of virtual networks built one over the other. Any layer in this chain has its costs determined by the prices charged by the virtual network below it, and its prices, in turn, determine the cost structure of the layer above.

What is the economic rationale for pricing in this layered structure? For illustrative purposes, consider a common set of services underlying the Internet today. At the very bottom of the hierarchy, real resources are used to construct the links and switches that constitute the first virtual network. In the emerging digital environment, time division multiplexing (TDM) in the digital telephone system creates the most evanescent of outputs (SONET time slots lasting 125 microseconds) out of very long-lived investments (including conduits and fiber optic cables). The pricing of these time slots determines the cost structure of the first layer of virtual networks (currently, ATM services) created on top of the TDM fabric. When multiple providers with sunk costs attempt to sell a very perishable good (time slots), unit costs will not be constant, but will decline with volume. Perfect competition will not be viable. If there is considerable excess capacity, no equilibrium may exist, and some providers may exit the market. If providers at this level do reach an oligopolistic equilibrium, will their price structure involve volume discounts and term commitments? If they do (as is the case with private line tariffs), then providers of ATM services (the next layer in the hierarchy) may be faced with relatively large sunk costs and their unit costs will not be constant. Again, perfect competition will not be viable. If providers at this level do reach an oligopolistic equilibrium, will their price structure involve volume discounts and term commitments? If so, the next level of service (SMDS and Frame Relay) will not be characterized by constant unit costs, and a perfectly competitive equilibrium will not be possible. The same questions of the existence of equilibrium and the use of volume discounts and term commitments arise, and will keep arising as we move upstream.

The fundamental economic problem arises from the large sunk costs required to build physical networks, and the technological reality that optical fiber has so much capacity that it is prudent for a network to lay large amounts of excess capacity during construction. A possible conclusion is that there are economies of scale, and the industry should be treated as a natural monopoly. It is too late for this solution. There are four large nationwide fiber optic networks (and some smaller ones) and 95% of all households are passed by both telephone and cable TV wires. In addition, alternative access providers have built fiber rings in every major business district, and most business customers in these areas have a choice of alternative service providers for voice and data communications. How might competition evolve in these circumstances? Huber (1993) suggested one interpretation of the apparent stability of some prices and market shares in the long-distance market. He concluded that competition is apparent and not real. A regulatory umbrella prevents instability. Noam (1994) discussed the stability of open interconnection (more specifically, common carriage) in these circumstances, and concludes that non-discriminatory contracts cannot survive in a competitive environment similar to the one described above. The success of resellers of Internet access, and the strong reaction to them by incumbents with national backbones, are consistent with Noam's view.

The Internet is one component of a very complex environment, and shares costs with other services (such as long-distance calling) that run on the same underlying transmission links. Of more immediate interest is the relationship of the Internet to some new services (SONET, ATM, SMDS, and Frame Relay) over which IP can be run. There are clearly many alternative interconnection arrangements in this layered structure that can result in full connectivity among end users. What is the socially optimal set of interconnection arrangements? Is full interconnectivity of virtual networks at each layer of the hierarchy necessary for optimality? If not, what is the minimal acceptable set of arrangements? How will this be impacted by vertical integration and vertical partnerships? Will an unregulated market provide this level of interconnectivity at acceptable cost? If not, what forms of regulation are optimal? Is the bearer service proposal of the Open Data Network (in *Realizing the Information Future: The Internet and Beyond,* by the Computer Science and Telecommunications Board) an optimal form of regulation? Clearly, economic theory can contribute to the analysis of this problem, but the work has barely begun.

CONCLUSION

The market for Internet services is highly competitive, and cost structures are driving pricing decisions. As transport costs fall and firms seek to differentiate their services, support costs will tend to rise as a fraction of total ISP cost. Support costs are not proportional to the bandwidth used by customers, and so prices will not be proportional to bandwidth. Small resellers may be able to arbitrage across the tariffs offered by large ISPs, and leverage off their own lack of sunk costs. Possible responses to arbitrage include a flat prohibition on resale, and special wholesale prices. Both these strategies are currently in use in the Internet. Nevertheless, ISP resellers represent the fastest growing segment of the ISP market (Maloff, 1994).

It is not clear that the current price structures and interconnection arrangements represent an industry equilibrium. The wide variation in prices for essentially comparable services, the growth of resale, and the current dissatisfaction with the CIX (as expressed on the mailing list *compriv*) are symptomatic of an evolving market where prices have not lined up neatly with costs. Part of the problem stems from the sunk costs of creating physical networks, and the resulting trend toward long-term contracts for services that resemble transmission links (i.e., private lines). This pricing strategy results in relatively large sunk costs for ISPs who create a national backbone using private lines. Competition among firms with sunk costs can be problematical, especially when there is excess capacity. At the physical level of fiber-optic links, there is a good deal of excess capacity; as owners of this fiber enter the market as ISPs (as Sprint has done and MCI is about to do), the distinction between high average embedded costs and low (or zero) short-run incremental costs may lead to repeated and unstable price cuts. Owners of physical networks may decide to avoid potentially ruinous price competition by integrating vertically and differentiating their service. Customer support, information content, and reliability are three elements of product differentiation that are in common use in the Internet. The announcement by the CIX that it will filter resellers' traffic suggest that another differentiating factor may be assured connectivity: CIX members can guarantee greater connectivity than a reseller who may be blocked at the CIX router. This may be a cause for future concern if Internet connectivity comes to be viewed as another means of differentiating an ISP's service.

An important problem that remains to be solved is the determination of economically efficient interconnection agreements. It is argued in this chapter that a careful economic analysis of this problem needs to focus on the layered structure of services and the support activities that are required to transform raw bandwidth into communications services that customers will pay for.

ACKNOWLEDGMENTS

I would like to thank Jeff Mackie-Mason, Stewart Personick, Thomas Spacek, and Hal Varian for comments on an earlier version. All remaining errors are mine alone.

REFERENCES

Briere, D. D. (1990). *Long distance service. A buyer's guide.* Boston: Artech House.

Comer, D. (1991). *Internetworking with TCP/IP* (Vol. 1). Englewood Cliffs, NJ: Prentice-Hall.

Huber, P. (1993) Telephones, competition and the Candice-coated monopoly. *Regulation, 2.*

Lucky, R. W. (1989). *Silicon dreams: Information, man and machine.* New York: St. Martin's.

Lynch, D. C., & Rose, M.T. (1993). *Internet system handbook* Reading, MA: Addison Wesley.

Mackie-Mason, J. K., & Varian, H.R. (1993). *Some economics of the Internet,* (Working paper). Department of Economics, University of Michigan, Ann Arbor.

Mackie-Mason, J. K., & Varian, H.R. (1994). Economic FAQs about the Internet. *Journal of Economic Perspectives*, 75–96.

Maloff, J. (1994). *1993–1994 Internet service provider marketplace analysis.*

Mandelbaum, R. & Mandelbaum, P. (1992). "The strategic future of the mid-level networks." In B. Kahin (Ed.), *Building information infrastructure* New York: McGraw Hill.

Noam, E. (1994). Beyond liberalization II: The impending doom of common carriage. *Telecommunications Policy, 18*(6), 435–52.

Noll, A. M. (1994). A study of long distance rates: Divestiture revisited. *Telecommunications Policy,* 18(5), 335–362.

Perritt, H. H. (1992). "Market structures for electronic publishing and electronic contracting on a national research and education network: Defining added value. In B. Kahin (Ed.), In *Building information infrastructure. New York:* McGraw-Hill.

Pitsch, P. (1993). Earth to Huber. *Regulation,* (3).

14

Realizing the Information Future: Technology, Economics, and the Open Data Network

Marjory S. Blumenthal
National Research Council

The concept of a national information infrastructure (NII) that is more advanced and better integrated than the networks and other information delivery elements in use today has captured the imagination of many. The administration; Congress; trade, professional, and public interest organizations; companies; and individuals have all expressed NII visions in relatively general terms. Those visions link improvements in the NII to business development, productivity, and job growth as well as benefits in education, health care delivery, the conduct of government, and the quality of life. The widespread expectations for progress toward a better NII reflect several forces, such as the cumulation of corporate and public sector experiences with networking; improvements in the sophistication, ease of use, and affordability of enabling computing and communications technologies; and expectations for lucrative markets. Although advancing the NII may be timely, how it will advance, when, where, at what cost, and to whose benefit are all uncertain. Economic aspects, notably costs, cost incidence, and pricing possibilities, will vary with alternative technical, business, and public policy paths.

The Computer Science and Telecommunications Board of the National Research Council examined critical issues in advancing the NII in a report entitled *Realizing the Information Future: The Internet and Beyond* (Computer Science and Telecommunications Board [CSTB], 1994), which serves as the basis for this chapter.[1] *Realizing* focuses on technology-related issues, characterizing features central to the NII ideal and providing a technical framework or architecture called the Open Data Network. It relates technical objectives, as well as the practical

1. The author of this chapter was a principal author and editor of *Realizing*, working in conjunction with an expert study committee. This chapter represents her interpretation and packaging of related material, and should not necessarily be viewed as a statement of the National Research Council.

problems of implementing some of the social objectives typically posed for an enhanced NII (e.g., equitable access, protections for privacy and First Amendment rights), to the experiences of the research and education communities with the Internet. The Internet illustrates how network value can increase with the size and diversity of the user community, the power and sophistication of applications, and the capability provided by the infrastructure.

The NII is a complex of facilities, goods, and services that is evolving, as opposed to something that will be created from whole cloth. Its evolution is driven by myriad private investments, many already under way or planned. The ongoing commercialization of the Internet—epitomized by changes in the concept, scope, and scale of the National Science Foundation's NSFNET, which has provided backbone facilities for the Internet—presents in microcosm many public policy challenges associated with fostering an advanced NII. Fundamental is the transition in emphasis from a relatively focused program predicated on government support to targeted education and research communities to a much larger, more diffuse construct emerging from private investment decisions.

The new NSFNET will be a much smaller, higher speed backbone network than what the original had grown to be since the mid-1980s. It nevertheless may provide a new laboratory for assessing NII-related questions, including those arising from the Internet evolution generally in the areas of physical construction, operation, access, pricing, and, in particular, interconnection (Srinagesh, 1994). For example, the new NSFNET provides for a set of network access points (NAPs) for traffic exchange among network service providers, to be managed initially under the NSFNET program but intended to become privatized (see Fig. 14.1). Although

FIGURE 14.1
Elements of the next-generation NSFNET

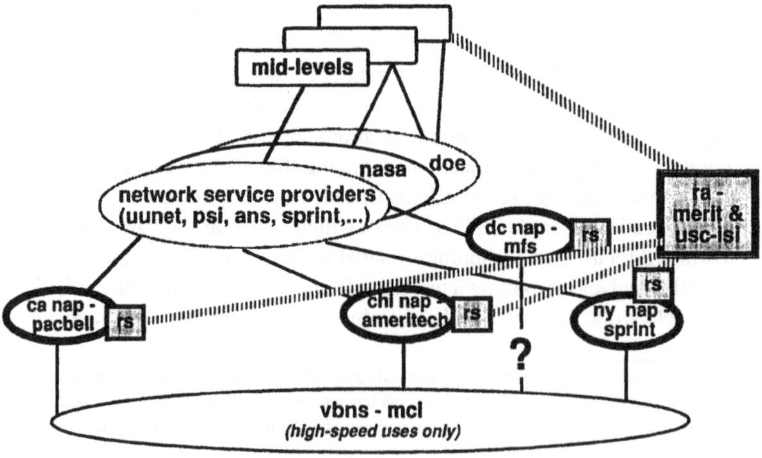

Figure provided by the National Science Foundation.

functionally modeled after the existing Commercial and Federal Internet Exchanges (FIXes), the NAPs raise new questions about how interconnection is established, priced, and governed. These issues are of growing importance in the commercialized Internet and for the advanced NII.

Although the Internet presents many lessons and many benefits, it is not now nor may it ever grow up into the advanced NII contemplated by many; it may not even survive as an identifiable construct. Accordingly, a full examination of NII interconnection-related issues should also address other internetworking contexts and models, such as networks supporting automated teller machine and credit card operations or wireless (cellular telephony, paging, and personal communication system) communication systems, which provide evidence of what does and does not work where disparate private parties interconnect facilities owned by different parties and exchange traffic and compensation. Figure 14.2 provides a conceptual picture of the internetworking/interconnection problem presented by the evolving NII.

This chapter briefly characterizes the Open Data Network concepts advanced in *Realizing* (CSTB, 1994), discusses Internet cost and financing issues, and outlines some considerations relating to pricing associated with commercialization of the Internet and its further integration into a larger information infrastructure. The chapter's goal is to relate key technological possibilities to economic choices, drawing on insights and issues raised by the Internet experience. It provides information and clarifies some persistent misimpressions about the Internet in the interests of informing and motivating further, more detailed economic analyses. It emphasizes the perspective of the infrastructure user, characterizing issues and implications for providers of infrastructure facilities and services.

FIGURE 14.2

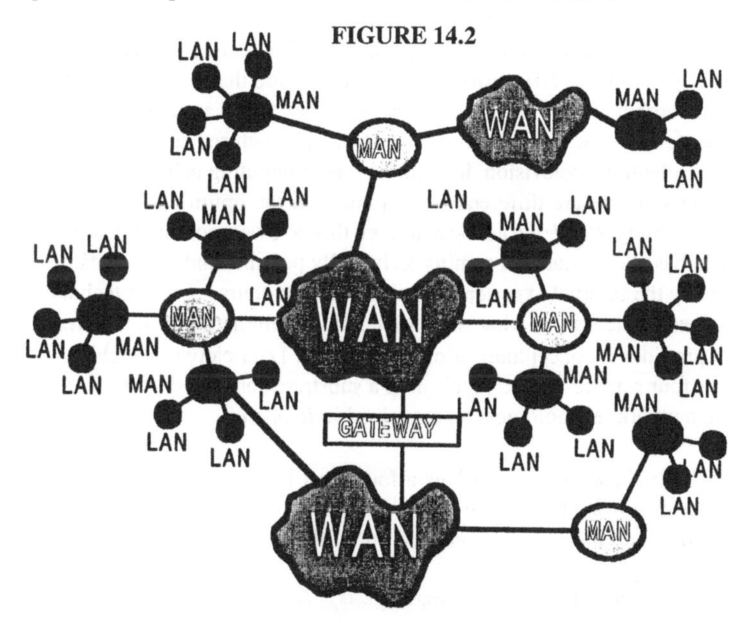

Reprinted with permission from *Realizing the Information Future: The Internet and Beyond.* Copyright 1994 by the National Academy of Sciences. Courtesy of the National Academy Press, Washington, DC.

WHAT THE NII CAN BE: THE OPEN DATA NETWORK

Realizing the Information Future articulated a specific vision for the NII that is characterized by openness, independence from specific technologies, large-scale and decentralized organization, heterogeneity (e.g., of component systems and services), trustworthiness (providing protections for privacy, security, intellectual property rights), accounting mechanisms (supporting various pricing and billing arrangements), and flexible or extensible basic service (to be generally available but likely complemented by other, more expensive services). Those qualities are supported by a technical framework or architecture called the Open Data Network, with *data* used as a generic term to cover all kinds of information that may be transmitted in all kinds of formats.

Openness has many guises. It applies to access by different kinds of users, network service delivery by different kinds of providers, and delivery of different kinds of higher level services and applications, as well as openness to change and evolution. The Internet illustrates this kind of multifaceted openness, having accommodated developments of many kinds that were not available or anticipated when it was conceived: new technology (local area networks [LANs] and later such underlying technologies as Asynchronous Transfer Mode and frame relay communications), applications (electronic mail and later such recent innovations as the World-Wide Web with its popular Mosaic interface), and providers (now a growing variety of commercial and nonprofit network and information service providers). Perhaps most importantly, the Internet demonstrates openness to individual users as providers of information content and services, because of its support for two-way or symmetrical communication.

Because the market cannot explore a space that the infrastructure precludes,[2] it is essential in developing the NII to strive for an architecture that is most open to possibilities. An architecture that can support interoperability among different kinds of services (e.g., voice, video, and comparatively unstructured data) becomes more feasible as telephony, television distribution, and computing achieve greater technical similarity. There are differences of opinion on the optimal degree and scope of interoperability. *Realizing* advocated more than a lowest common denominator approach, based on what is becoming technically possible and the proofs of concept offered by the Internet experience, in particular. Of course, what is technically possible does not determine what is economically or commercially attractive.

Figure 14.3 provides a conceptualization of the Open Data Network architecture. It includes four service-related layers atop a substrate composed of the various underlying network technologies. As explained in *Realizing*,

> The Open Data Network . . . involves a four-level layered architecture configured as follows: (1) at the lowest level is an abstract bit-level service, the *bearer service*, which is realized out of the lines, switches, and

2. The author owes this phrase to one of the principal authors of *Realizing*, Dr. David D. Clark of the Massachusetts Institute of Technology.

FIGURE 14.3
A four-layer model for the Open Data Network

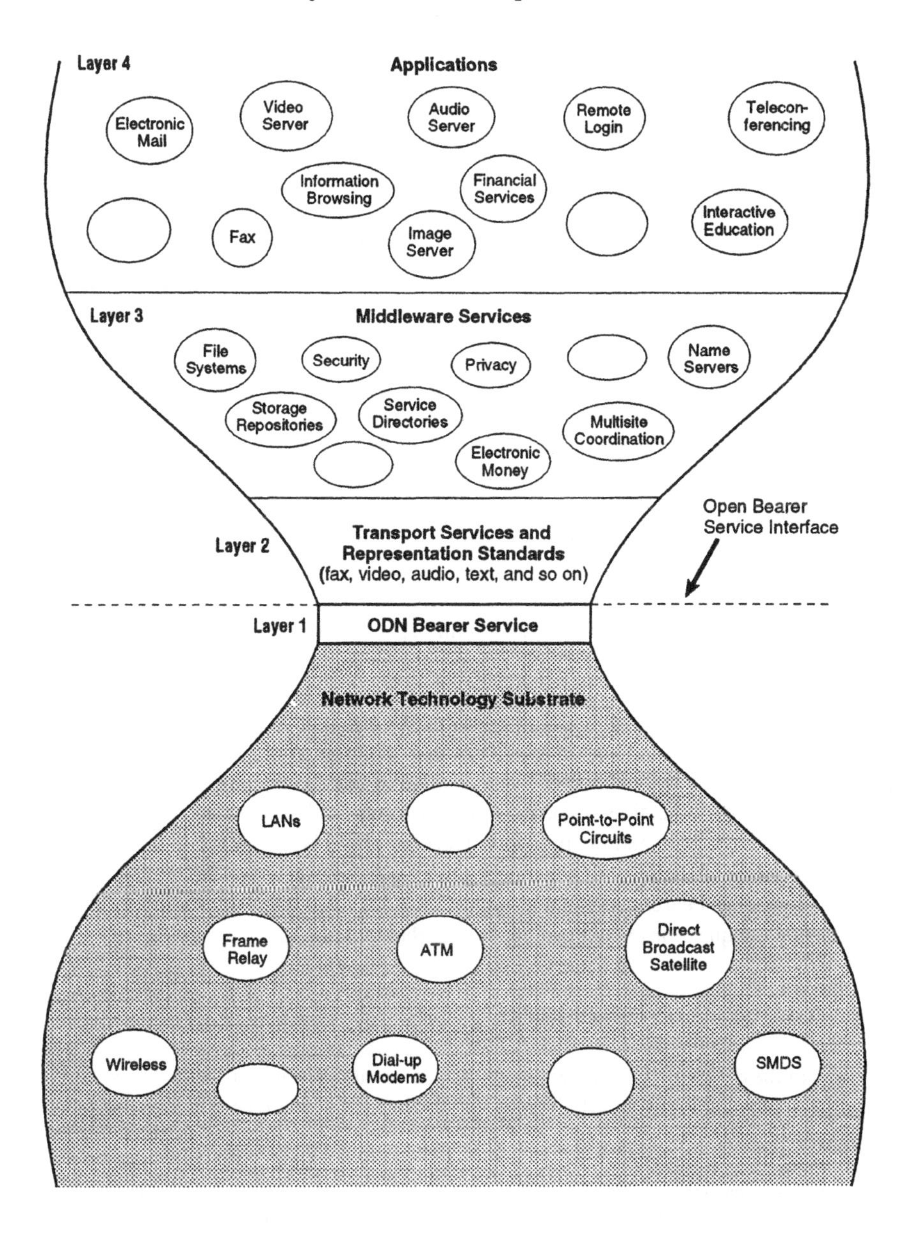

> other elements of networking technology; (2) above this level is the *transport* level, with functionality that transforms the basic bearer service into the proper infrastructure for higher-level applications (as is done in today's Internet by the TCP protocol) and with coding formats to support various kinds of traffic (e.g., voice, video, fax); (3) above the transport level is the *middleware*, with commonly used functions (e.g., file system support, privacy assurance, billing and collection, and network directory services); and (4) at the upper level are the *applications* with which users interact directly. (p. 5)

The layers highlight the concept of interfaces. Interfaces serve as both boundaries and service definitions.

The Open Data Network architecture is designed to allow and encourage competition among providers of services at each layer. It is also designed to permit (although not require) the separation of providers of facilities or network services within the technology substrate from providers of information and other services at the upper (transport, middleware, and applications) layers. This separation or unbundling potential is created by calling for a minimal set of defined functionality at the bearer service level. That minimality of specification is what is depicted by the narrowing in at the hourglass waist in Fig. 14.3. Its concept is a response to the varying technical requirements of applications and the varying performance characteristics of technologies. The bearer service, by hiding the detailed differences among the network technologies, minimizes technology-imposed constraints on higher level service and application design. Consequently, the bearer service is referred to as technology-independent. In the Internet environment, the Internet protocol or IP layer in the Internet protocol suite is such a bearer service, implemented as software in Internet "host" devices.[3] Another, more limited example of a bearer service is the 3.5-kHz voice channel in telephony, which not only supports the original objective of voice communications but has also been able to carry traffic between data and fax modems. The Open Data Network architecture implies at least the possibility of separately priced network services (technology substrate elements), bearer service, and higher level services. That possibility, in turn, raises many questions about pricing and market structures.

INTERNET FINANCING: PERCEPTIONS VERSUS REALITY

Financing of the evolving NII involves more than corporate investments and returns needed to deploy facilities and deliver services. There is another perspective on

3. As the Internet experience shows, a bearer service is something that can be implemented by multiple entities playing multiple roles: IP is what different kinds of providers of Internet access—carriers, value-added networks, information service providers, regional networks, and so on—and different users—campus networks, laboratories with LANs, computers of varying sizes supporting shared databases or individuals, and so on—all employ to enable them to participate together in the internetworking environment known as the Internet.

cost, that of the end user or customer. *Realizing* examined this perspective, drawing on the experiences of research, education, and library users of the Internet and other networks. The impacts of changes in federal funding and the changing exposure to prices experienced in the research and education communities provide insight into the larger problems of expanding information infrastructure access nationally.

The commercialization of the Internet arouses uncertainty and anxiety among its historic constituents because of the seeming clash of cultures between, on the one hand, the nonprofit, largely publicly supported research and education domains and, on the other, the largely for-profit entertainment, telephony, and cable television domains that dominate business and national press coverage of NII issues. The anxiety also reflects widespread misunderstanding about actual Internet financing and costs. It is important to understand actual sources of costs and their trends to achieve more realistic discussion and intelligent choices about NII financing, investment, and use.

One of the most prevalent and persistent misperceptions about the Internet is that it is free—many wrongly believe that all costs are paid for by the federal government, which provides Internet access at no charge to users in the research and education domains. The reality is that federal Internet support is relatively limited and highly leveraged. Total Fiscal Year 1994 funding for the National Research and Education Network (NREN) program, which included payments related to the operation of federally sponsored Internet components (NSFNET, such mission agency networks as Energy Sciences Network and the NASA Science Internet, and the FIXes that interconnect them) as well as support for related research and application development, was under $150 million, and the federal investment in NSFNET was on the order of $30 million (CSTB, 1994). Federal investment on the order of tens of millions of dollars contrasts with private investments of tens to hundreds of billions of dollars in telephony, cable, and other fundamental infrastructure (some of which is used by so-called federal networks). Note also that the federal investment in NSFNET was complemented by in-kind and other investments by IBM and MCI through Advanced Networks and Services Inc., the company that operated NSFNET under a cooperative agreement with NSF, and also by other parties (including state governments) that have contributed to the regional (feeder) networks stimulated by NSF, which could not afford to expand NSFNET directly beyond a certain scale.

A related source of misunderstanding concerns the overall level and importance of the long-haul communications component of Internet costs. The cost of network service includes the cost of local access and the cost of long-haul transmission. Overall, the long-haul communication cost for U.S. Internet users has amounted to a few dollars per person per year. It is relatively inexpensive because it relies heavily on facilities sharing: long-distance communication within the Internet uses packet switching to achieve transmission efficiencies that are not possible in the networks used for telephony. Thus, the fact that the federal government is paying for only a fraction of the Internet cost implies that reduction of that support will not be the catastrophe feared in many sectors of the research and education communities.

Most of the costs associated with the Internet relate to local access (which is generally priced as a function of distance to the nearest network node[4]) and to local infrastructure (computer systems, local area networks, software, associated peripheral equipment and software, associated staff and training). These costs are borne by users. Additional cost is borne by users for the information services they access; the cost of access to the network is just the beginning. Part of the confusion among individual users is that most of these costs are paid by the institutions that employ or educate them—few individual users actually see a bill for their use, except for those who procure individual accounts directly with Internet access providers (e.g., for some kinds of home access).[5] A more ubiquitous, advanced NII does suggest the possibility that in the future, individual accounts may become more common.

Local access costs can be high, reflecting various conditions in the "last mile" and the local drop. Relatively limited actual and potential sharing in conventional telephony transmission and in residential access generally is one reason that the costs of access are higher than long-haul network backbone costs. Averaged across all connections to commercial Internet services in early 1994, the average charge for access at 56 Kb/sec was $15,000 per year and the average charge for 1.5 Mb/sec access was $36,000 per year (CSTB, 1994).

Within user organizations—homes, offices, laboratories, schools, and so on—local infrastructure includes computer systems, modems or digital connection units, software, local area networks, and training and support services. For example, taking into account installation and learning-related costs, a typical corporation's average 5-year cost of owning a personal computer may run to $40,000 (Fisher, 1994). Local infrastructure costs loom large, with increasing activity taking place outside of the network in end-user facilities to support such activities as multimedia e-mail and distributed database retrieval. If Internet patterns of use are a harbinger of advanced-NII patterns, they emphasize activity and system intelligence at the periphery of the network—that is, in users' facilities—rather than within the network, as has been the emphasis in telephony.[6]

User costs, including both local access and local infrastructure, are sufficiently high that they can pose a key problem for those just starting up—an entry barrier of sorts. Despite mass media reports of the growing number of Internet users, within the research and education communities connectivity remains uneven and

4. There is some speculation as to whether costs may abate with broader deployment of wireless technologies, but currently wireless technologies may not support the bandwidth needed for some applications, such as those emphasizing real-time graphics.

5. Many in the research and education communities are funded by grants, which typically cover costs for such infrastructure as voice telephony and photocopying but not computer networking. Grant reimbursement could facilitate budgeting and choices, but it raises serious administrative problems (ranging from the need for corresponding accounting systems, which have costs, to issues about how much reimbursement would be channeled to cover overhead costs), and access to grants varies widely, raising questions about equity.

6. Of course, this is not to suggest that equipment within the network is not sophisticated today or becoming more so; both are true.

far from universal (both across and within institutions). The NREN program, of which the Internet is the heart, has focused on schools, laboratories, and libraries because it is cheaper to provide Internet capabilities to a set of institutions than deploy them universally. One of the attractions of the NII concept is the prospect of considerable economies of scale in serving education and libraries through a larger information infrastructure whose economic justification comes from other commercial users. The general "population" of businesses and residences provide a large base for cost recovery and profit generation. By contrast, if only a few sites in a community are connected to a network, it is necessary to allocate to them not only the cost of the local access link from each site, but also the costs of the link from the community to a more distant network access site. This situation illustrates the tension between system power or capability and universality or breadth of access to that capability. That tension suggests that for economic reasons there will always be differentials in who has access to what, although the floor should continue to rise as the cost of delivering better capability falls. Who gets what, when, and at what cost depends, of course, on what proves profitable (as well as on requirements imposed through public policy, including regulation, which are not examined here).

PRICING PROSPECTS

The Internet illustrates how a network that would not exist under perfect competition can grow under the aegis of a "welfare-maximizing planner" (Economides & Himmelberg, 1994). Under artificially low effective prices and with a separation of activity from prices, use and innovation have flourished in the research and education communities. Because an information infrastructure that is truly national involves investment and expense far beyond any level contemplated for the federal government, it implies a need for a broad base for cost recovery, which in turn drives consideration of pricing.

With commercialization, growth, and integration with other kinds of information infrastructure, end user charging of some kind for the Internet seems almost inevitable (although it poses significant administrative or transaction costs and other practical problems). It would, of course, bring into focus questions about what portions of budgets for research, education, and libraries should go for networking versus other activities or resources. It is axiomatic among economists that prices can help achieve efficient trade-offs among inputs—those that complement use of networks (e.g., desktop systems, other resources for gaining access) and those that substitute for networking (e.g., other means of communication and gaining access to information). Pricing is also valuable for inhibiting excessive use of shared information infrastructure. On the other hand, a price that is too high too soon will inhibit reasonable levels of use and therefore the development of the network markets that will make up the NII.

Overly high prices would prematurely inhibit experimentation, affecting the rate and possibly the direction of information infrastructure growth. The Internet experience and experiences with various corporate and for-profit networks demonstrate the value of experience and experimentation. Trade-offs and choices about spending on information infrastructure and associated goods and services may be uninformed absent networking experience. Further, experimentation yields insights into potential benefits and costs of information infrastructure. It may thereby help people to learn how to lower costs. On the other hand, competition among Internet access providers to serve a clientele that is in many instances budget constrained suggests that there may be negligible margins to support service donations, cross-subsidies, or hypothetical new taxes or other mechanisms to support "needy" users. Over time, it can be assumed that growth in the volume of users will result in lower average prices, given that there are economies of scale; that an increasing number of service providers creates competition that will also place a check on prices; and that pricing schemes may motivate new protocol developments that can result in lower costs. All of these developments would take time to unfold, and none is inconsistent with some degree of continued government support for research and education users over the near term, within the broader context of government support for research and education.

Among the questions that the commercialization of the Internet raises for pricing are the following: Should individual users face charges based on their level of use? Is there a preferred structure of pricing for infrastructure access and use?

To address those questions it is necessary to consider how pricing should relate to cost. The relationship between cost and price seems relatively well established in telephony, given the long history and influence of regulation—but that relationship is poorly understood by many, and often seemingly indirect. The price–cost relationship is even murkier in the Internet, which is free of some of the extraordinary requirements (e.g., cross-subsidies, as for support of universal access) found in telephony, and which lacks the accounting mechanisms that could support relevant data gathering and analysis as well as billing. The price–cost relationship is at best a subject of speculation in the emerging NII, given its composition from many systems and industries with different cost structures. The situation grows even harder to analyze as technologies change in the underlying networks, thus altering cost structures. Several examples and their implications are outlined in Srinagesh (1994).

At this time, there appears to be little variation in cost with use for typical uses of the Internet (electronic mail and file transfer). Most of the Internet's costs are fixed—circuit and hardware costs—and in the absence of congestion, the incremental cost of additional traffic is essentially zero. Consistent with this cost structure is an Internet payment model based primarily on a fixed price for access, with so-called best-effort service—delivery of service by sharing the available resources equally among those who at one moment want to use the network. Technical means are used to control congestion at times of higher traffic, but delay may occur. By contrast, the telephone network charges either according to the amount of use that is made of the network, or at rates independent of usage (at least up to some volume); regardless of the number of users, it offers a constant quality of service. If Internet costs were to continue to be relatively invariant with use, then a high cost of adding in the accounting mechanisms needed to support usage-based charging could be difficult to justify. The level of that cost is currently a subject of speculation.

The prospect of costs varying with use grows with the rise of applications and systems supporting more graphics- and video-intensive traffic plus integrated audio and other kinds of traffic (e.g., multimedia e-mail), which require much more bandwidth and more predictable quality of service than conventional electronic mail and nontime-sensitive file transfers. This trend has raised concerns about increasing congestion and a need for pricing to inhibit congestion. A related concern is whether and how to relate pricing to type of traffic, a concern that affects the economics of unbundling of basic transmission facilities from basic and more enhanced services. As *Realizing* points out, "Technically, a bit is a bit, but voice and video bits are priced very differently today. Voice-level charges discourage video traffic (which requires far more bits), while video-level pricing would allow voice to be carried essentially for free" (p. 192).

Another factor that can affect cost, at least in the short term, is the mechanisms for achieving openness, the integration of services and symmetrical communications properties implied by the Open Data Network architecture. The costs of delivering NTSC-standard video to televisions has been optimized by cable system operators, for example, but added costs might be incurred to achieve delivery of the same video programming to personal computers.[7] The general phenomenon is a familiar one in engineering: a general and flexible solution may be more expensive, in the short term (i.e., with today's technologies), than more specialized or proprietary approaches. In the long term, the more general approach may provide for easier shifts among end products, support for a broader range of uses and users by a given investment, and so on; in particular, it can provide a form of insurance against unanticipated changes in market demand. The degree to which providing more general and flexible technology is actually more expensive is hard to judge, although some degree of greater cost is consistent with anecdotal evidence supplied by interested parties (i.e., network and information service providers who benefit today from more closed, proprietary delivery systems). To the extent it is a problem, however, in the near term market solutions are not likely to promote an Open Data Network; it may be hard to build the business case. For this reason, *Re-*

7. *Realizing* provided the following discussion:

As the delivery systems migrate from NTSC encoding of video to digital encoding, such as MPEG-II, the video becomes more amenable to processing by both specialized and general digital processors. While the digital video may be delivered over the network in a highly specialized and cost-effective manner, the encoding can be defined (as MPEG-II is) so that it has at least one form that separates the actual video information from the details of the delivery method. A capability for converting the video into this format at customers' premises allows it to be transferred to and processed by other end-node elements such as general-purpose computers. This arrangement is a very powerful one: on the one hand, it does not affect the coding or delivery of the presumed high-volume information, which can be delivered in whatever cost-reduced manner the industry prefers; on the other hand, it simultaneously supports a general bearer service and provides a way to move the video into that more general format as needed at the end node. (p. 88) *alizing* called for exploration of incentives to foster engineering of access circuits in a manner consistent with an Open Data Network (see p. 91).

The research and education communities are very budget constrained. They therefore prefer a flat-fee or subscription pricing scheme because it makes expenditures predictable. Although these communities relate their preference to need, other user communities, from residential to commercial, also have indicated a sufficient preference for flat-fee pricing that it has become increasingly available in the marketplace. Today one can find flat rates that vary with such factors as the speed of access lines, time spent using the system, and volume of packets transmitted. Nevertheless, flat-fee pricing is resisted by many in the vendor community and by some economists. Although it makes both expenditures and revenues predictable, it places more risk on the provider, depending on how costs vary with use. Flat-fee pricing can also give rise to congestion, by disconnecting user costs from use, and with a single flat fee can result in the subsidizing of high-volume users by low-volume users.

The most likely adverse consequences of flat-fee pricing should be avoidable by having a set of service packages with different flat fees corresponding to different levels of service. A set of differently priced packages, instead of a one-size-fits-all package, would both best accommodate different kinds of users and promote the kind of experimentation needed for the information infrastructure to continue to grow. Multiple service tiers would allow for giving higher access priority to higher value uses of scarce bandwidth resources (e.g., real-time transfer of medical imagery for time-sensitive health care applications; MacKie-Mason & Varian, 1993). It would also allow for a choice between variable and predictable qualities of service in the market, and it would support the use of pricing and other means to shift usage to off-peak periods. Another goal for the pricing scheme may be to balance long-term typical demand with occasional, possibly spontaneous demand for higher grades of service. *Realizing* assumed that there would be a variety of pricing schemes and that the optimal pricing scheme would vary over time, as technologies and markets changed.

An issue raised by the Internet experience is whether the payment for information infrastructure access and use must be fully monetized. Internet users often distribute information and services without charge or at cost, suggesting the possibility of some kind of quid pro quo arrangement exchanging in-kind resources for access. Such an arrangement has been explicitly factored into a Pacific Bell program, for example (Pacific Bell, 1994). Such a creative approach would be consistent with the apparent cost structures of information resources. Because the production of information is subject to economies of scale, varying prices charged to different classes of users—including charging lower prices for some who otherwise could not purchase the information—may help to optimize the production of information products (Noll, 1993). Although the price charged reflects many considerations (ranging from ownership of intellectual property to the costs of assembling and packaging the information for customers), the actual costs of getting into and extracting information from a database are very, very low. Most of the marginal costs for an individual user are associated with physical access (including, in addition to communications links, the software that adds a user's address to the network for billing and connecting purposes).

CONCLUSION

If the NII evolves in a way that is consistent with an Open Data Network, a high degree of technological, product, and market innovation could be anticipated, and a variety of users and uses for the infrastructure could be supported. This flexibility and generality is particularly valuable in light of difficulties predicting which services will be profitable and, overall, what mix of services will prevail. How the NII does evolve, at a given point in time, will depend substantially on the apparent costs and profitability of alternative options. The commercialization of the Internet, which started out as a government experiment, provides a laboratory for assessing patterns of innovation and use, and also for assessing issues relating to cost, cost incidence, and pricing. Inferences should be tempered by the recognition of real differences between the research and education context and the business and consumer communications markets that argue for caution in generalizing from the Internet experience; those differences are also relevant to considerations of pricing and targeted subsidy issues.

REFERENCES

Computer Science and Telecommunications Board, National Research Council. (1994). *Realizing the Information Future: The Internet and Beyond.* Washington, DC: National Academy Press.

Economides, N., & Himmelberg, C. (1994, October). *Critical mass and network size.* Paper presented at the Telecommunications Policy Research Conference, Solomons Island, MD.

Fisher, L. M. (1994, March 27). Reining in the rising hidden costs of PC ownership. *New York Times*, p. F10.

MacKie-Mason, J., & Varian, H. (1993, May). *Pricing the Internet.* Paper prepared for a conference, Public Access to the Internet, held at the John F. Kennedy School of Government.

Noll, R. G. (1993). The economics of information: A user's guide. In *The knowledge economy: The nature of information in the 21st century* (pp. 25–52). Nashville, TN: Institute for Information Studies.

Pacific Bell. (1994, February). *Pacific Bell to link public schools and libraries to communications superhighway* [press release].

Srinagesh, P. (1994, October). *Internet cost structures and interconnection agreements.* Paper presented at the Telecommunications Policy Research Conference, Solomons Island, MD.

15

Roles for Electronic Brokers

Paul Resnick
Massachusetts Institute of Technology

Richard Zeckhauser
Harvard Kennedy School of Government

Chris Avery
Harvard Kennedy School of Government

The information superhighway promises to facilitate the exchange of products. Broadly, we can think of all such exchanges as electronic commerce, even though some products, such as messages on computer bulletin boards, may be free, and physical transactions must be consummated elsewhere. (For example, a seller and buyer would exchange the product and payment by mail.) Most predictions about commercial opportunities on the information superhighway focus on the provision of information products, such as video on demand, and on new sales outlets for physical products, as with home shopping. We believe that electronic brokers, as intermediaries between buyers and sellers, can help to create more efficient markets, both for information products and physical products. Most simply, they can reduce buyer search costs and arrange to pay for information that would not be provided without payment.

Such services raise two important policy questions. First, how do we weigh privacy and censorship concerns against the provision of information in a manageable form? Whenever information products are brokered, privacy and censorship issues come to the fore. When the broker is a computer rather than a person, the stakes are raised; a computer can more easily perform potentially troubling operations involving large amounts of data processing. Second, how should the provision of brokering services be organized? Should they be integrated with information provision, or separated? Is brokering a natural monopoly?

We first identify a number of vital activities that electronic brokers could perform. Next, we describe one prototype service, a Better Bit Bureau, in more detail and discuss its policy implications. Finally, we argue that brokering services should be vertically separated from information provision, and we discuss the tensions between the advantages of competition and monopoly in the provision of brokering services.

THE VALUE OF ELECTRONIC BROKERS

Producers and consumers interact directly in a marketplace: Producers provide information to customers, who select from among the available products. In general, producers set prices, but sometimes they are negotiated. However, direct negotiations are sometimes undesirable or infeasible. Fortunately, intermediaries, whether human or electronic, can redress five important limitations of privately negotiated transactions.

Search Costs

It may be expensive for providers and consumers to find each other. In the bazaar of the information superhighway, for example, thousands of products are exchanged among millions of people. Brokers can maintain databases of customer preferences and reduce search costs by selectively routing information from providers to consumers. Furthermore, producers may have trouble accurately gauging consumer demand for new products; many desirable items may never be produced simply because no one recognizes the demand for them. Brokers with access to customer preference data can predict demand.

Lack of Privacy

Either the buyer or seller may wish to remain anonymous, or at least to protect some information relevant to an exchange. Brokers can relay messages without revealing the identity of one or both parties. A broker can also make pricing and allocation decisions based on information provided by two or more parties without revealing the information of any individual party.

Incomplete Information

The buyer may need more information than the seller is able or willing to provide, such as information about product quality or customer satisfaction. A broker can gather product information from sources other than the product provider, including independent evaluators and other customers.

Contracting Risk

A consumer may refuse to pay after receiving a product, or a producer may give inadequate post-purchase service. Brokers have a number of tools to reduce risk.

First, the broker can disseminate information about the behavior of providers and consumers. The threat of publicizing bad behavior or removing some seal of approval may encourage both producers and consumers to meet the broker's standard for fair dealing. Second, if publicity is insufficient, the broker may accept responsibility for the behavior of parties in transactions it arranges, and act as a policeman on its own. Third, the broker can provide insurance against bad behavior. The credit card industry uses all three tools to reduce providers' and consumers' exposure to risk.

Pricing Inefficiencies

By jockeying to secure a desirable price for a product, providers and consumers may miss opportunities for mutually desirable exchanges (Myerson & Satterthwaite, 1983). This is particularly likely in negotiations over unique or custom products, such as houses, and markets for information products and other public goods, where free-riding is a problem. Brokers can use pricing mechanisms that induce just the appropriate exchanges. One intriguing class of mechanisms requires a broker because the budget balances only on average: The amount the producer receives in any single transaction may be more or less than the amount paid by the customer, and the broker pays or receives the difference.

The information superhighway offers new opportunities for brokering services. First, brokers are especially valuable when the number of participants is enormous, as with the stock market, or when information products are exchanged. Second, many brokering services require information processing; electronic versions of these services can offer more sophisticated features at a lower cost than is possible with human labor. Finally, for delicate negotiations, a computer mediator may be more predictable, and hence more trustworthy, than a human. For example, suppose a mediator's role is to inform a buyer and a seller whether a deal should go through, without revealing either's reservation price to the other, because such a revelation would influence subsequent price negotiations. An independent auditor can verify that a software mediator will reveal only the information it is supposed to; a human mediator's fairness is less easily verified.

BETTER BIT BUREAUS

Information overload plagues computer network users (Malone, Grant, Turbak, Brobst, & Cohen, 1987); there is simply not enough time to sift through all the available information or even a significant fraction of it. In response, several subscription services provide selective filtering: A consumer specifies a profile consisting of a few words or topics; periodically the service sends the consumer news articles that match the profile.

In this analysis, we consider the use of other people's subjective evaluations to help route messages. Subjective evaluations are valuable to consumers who are deciding which products to buy or how to spend their time. For example, we read magazines devoted to product evaluation before purchasing cars and appliances. We ask our friends and read reviews by professional critics when deciding which movies to see or where to eat. Before committing to a new job or school, we ask current employees what the firm is like and inquire of college students what it would be like to go to their school.

Of course, there are drawbacks to subjective evaluations. Not all evaluations are equally trustworthy; there may be differences in effort or expertise, or conflicts of interest due to the financial stakes of the evaluators. Moreover, consumers' tastes may differ, so that an evaluation that is trustworthy to one person may not be to another. It is costly to gather and distribute evaluations, and it takes time for an individual to process them. The high transaction costs of sharing evaluations reduces their use except when they may influence an expensive purchase, such as a car, or when the evaluations are entertaining, as with movie reviews.

Computers can reduce the cost of gathering, distributing, and processing evaluations, especially evaluations of information products such as computer bulletin board messages. By analogy to the Better Business Bureau, we call a broker that shares evaluations of information products a Better Bit Bureau. A Better Bit Bureau that helps people choose information products is called a *collaborative information filtering service*. We describe a research prototype, called GroupLens (Resnick, Iacovou, Suchak, Bergstrom, & Riedl, 1994), that implements collaborative information filtering. Other researchers are also exploring related services (Goldberg, Nichols, Oki, & Terry, 1992; Hill, Stead, & Rosenstein, in press; Maltz, 1994; Shardanand & Maes, in press). Commercial brokering services based on shared evaluations are just beginning to appear (Nichols, 1994).

GroupLens collects evaluations of computer bulletin board messages. After reading each bulletin board message, a user enters a number from 1 to 5. The user's computer forwards the numeric rating to a Better Bit Bureau, which may distribute it to other Better Bit Bureaus. The GroupLens Better Bit Bureaus implement a rating aggregation scheme that takes account of differences in individual tastes. It employs the heuristic that "people who agreed in the past will likely agree again." Thus, in predicting whether a particular person will enjoy a particular message, it weights more heavily ratings from people who agreed substantially with the person in the past.

CENSORSHIP AND GROUPTHINK

To what extent is collaborative filtering censorship, and therefore objectionable? Here we define censorship as any activity whereby a third party prevents or inhibits one party from communicating with another. Although collaborative filtering fits this definition of censorship, we argue that it is a benign form. Instead, the danger may be a splintering of society, where each individual listens only to others with like views.

One filtering mechanism already built into some software for browsing computer bulletin boards is the ability to create a "kill" file that suppresses all messages containing a certain string of characters. For example, a user might put a subject line in a kill file to avoid all follow-up messages, or a name to avoid all messages from that person. This is not censorship, because no third party is involved: One person's kill files do not affect what anyone else receives.

A second filtering mechanism is the "moderated" newsgroup. A moderator, acting much like the editor of a periodical, receives all messages and decides which to post to the newsgroup. In most cases, computer bulletin board moderators screen messages for conciseness and relevance to a particular topic, rather than the positions they argue for, but some abuses of power are inevitable. A moderated newsgroup often coexists with an unmoderated one that addresses the same topic, so that an author has an alternative place to post a message rejected by the moderator.

Collaborative filtering fits somewhere between kill files and moderated newsgroups. It fits the previous broad definition of censorship—some people's evaluations can cause other people not to read a particular message. This form of censorship seems benign, however, for two reasons. First, the power to censor is distributed among many evaluators, so the damage from any one person's abuse of power is limited. Second, each person relies on ratings from a different set of evaluators. In GroupLens, the computer identifies which evaluators' opinions to weight most heavily. Alternatively, users might specify that only evaluations from particular friends or celebrities be used. In either case, no one is forced to take suggestions from an incompatible evaluator.

Collaborative filtering may be a viable alternative to the "free speech versus censorship" debate about television programming (Brynjolfsson & Resnick, 1993). It may provide a way to supersede regulations concerning the broadcast of nudity and to quiet agitation for restrictions on the broadcast of violent material. One proposed technology, nicknamed the V-chip (Andrews, 1993), would allow individuals to automatically block out all violent material from appearing on their own TV sets. Broadcasters would send a "V" signal along with any violent shows they broadcast and new TVs equipped with the chip would detect the signal. Thus, a parent could program the TV not to show anything accompanied by a "V" signal between 3 p.m. and 6 p.m., the unsupervised after-school hours. The drawback is that technology can not obviate the old question of who decides what is violent enough to receive the "V" rating; proposals range from industry self-labeling to a government-appointed independent board.

With collaborative filtering, it would not be necessary to designate any official rating board. Anyone could publish a set of evaluations (on many dimensions, not just violence). If a wide range of evaluations were available, a parent could program a TV to pay attention to any censor, from Action for Children's Television to Howard Stern, or a group of neighbors. Today, people practice collaborative filtering on an informal basis, but there are too many programs and too many evaluators to keep track of. Technological support, in the form of broadcasting many independent evaluations, could make collaborative filtering an effective alternative to government regulation.

Yet there may be a danger more subtle than censorship in giving people greater control over what television programs they watch or what messages they read. No individual rights are abridged if Rush Limbaugh's evaluations were to screen out politically liberal messages; liberals could still reach people (including open-minded conservatives) who want to hear what they have to say. It may be dangerous for society as a whole, however, to fragment into like-minded groups. Today's inadequate technology for filtering information may force people to be exposed to contrary viewpoints. This may have a positive value for society despite its negative value for individuals. On the other hand, ineffective information filtering is hardly an efficient way to ensure cross-fertilization. With collaborative filtering, boundary-crossing individuals may naturally introduce interesting new ideas into self-selected interest groups.[1]

Some people may mistrust collaborative filtering, either out of misunderstanding or fear that the system will be manipulated to perform a more pernicious form of censorship. One difficulty is that in a very large network, it may become infeasible for a user to decide how to weight each evaluator. Each user must either lose some control, by not understanding whose opinions the filter is relying on, or bypass the potential benefits of strangers' evaluations. In GroupLens, users must trust the computer's mechanism for picking compatible evaluators.[2]

PRIVACY

Protecting the privacy of evaluators and their information is another important policy concern. Contemporary standards of fairness require that many documents, ranging from letters to the editor to personnel evaluations, be signed, and that one's accuser be identified in court. Signed evaluations are less likely to be unfair and, over time, people can identify trustworthy evaluators. Perhaps requiring signed evaluations could help people develop friendships; matching people based on their tastes in movies might be useful as part of a dating service. On the other hand, evaluators may prefer not to sign their evaluations. For example, most professional journals employ blind reviews so that reviewers will not fear retribution from authors. Evaluators on computer networks may not even want anyone to know what bulletin boards they read, much less their opinions of particular messages.

GroupLens offers a compromise between signed and anonymous evaluations; each user signs all evaluations with one pseudonym. There is no simple way to identify the person behind the pseudonym, but the Better Bit Bureau can still match evaluators with similar tastes.

The use of pseudonyms does not completely resolve the privacy issue. First, by preserving privacy we sacrifice some information. With signed evaluations, it is possible to draw on external information when choosing which evaluators to pay attention to. Knowing that Evaluator A is C. Everett Koop is a much faster way to

1. More sophisticated possibilities might also be created. Though each individual might prefer to be insular, he might agree to receive contrary views if others would do so as well, leading to a general agreement on a system that provides some contrary exposure to all.

2. A mistrustful user might subscribe to several Better Bit Bureaus that follow different filtering rules, much as medical patients often ask for a second opinion from another doctor.

learn of his expertise than to examine all his past evaluations. One solution to this problem may be to provide some descriptive attributes of each evaluator.

Second, some information about the evaluator is leaked even without the name and a knowledgeable observer may be able to guess an evaluator's identity from past evaluations. Publishing descriptive attributes along with pseudonyms, such as the fact that Evaluator B is an economist, would exacerbate this problem. Even without guessing the evaluator's identity, an observer may be able to infer information that the evaluator would prefer to hide. For example, suppose the evaluator wishes to hide the very well-kept secret that most mainstream economists have socialist leanings. An analysis of many evaluations could support this claim even if it could not pinpoint the identity of any individual socialist.

A potential compromise is to pay people for the loss of privacy. A broker could pay one amount for signed evaluations, somewhat less for pseudonymous evaluations, and very little for anonymous evaluations. An interesting policy question is whether people should be allowed to sell their privacy in this way, because it is nearly impossible to anticipate all the potential uses of large databases of personal information. Moreover, selling one's privacy may create externalities; revealing one's identity might reveal information about similar people.[3]

PAYING FOR EVALUATIONS

Even if we can effectively and appropriately aggregate all evaluations, a second major problem will arise: Too few evaluations will be produced. In GroupLens, all evaluations are generated as a result of self-interested activities. When someone evaluates a document, possibly helping vast numbers of others, he is not compensated. What can we do to assure a sufficient supply of evaluations? Elsewhere we present a series of mechanisms that can be helpful (Avery, Resnick, & Zeckhauser, in preparation-a); here we review the issues and our findings, using very simple examples. We analyze three types of social inefficiencies that are likely to occur if evaluators are not compensated, and then consider the difficulties of creating a compensation scheme that is fair and that can induce the socially optimal set of evaluations.

Suppose that two people are deciding whether to read a message. Initially, they both think it is equally likely that the message is "good" or "bad." The two players' evaluations of messages are perfectly informative; when a message is "good" to one player it is "good" to the other as well. There are two critical parameters for each potential reader; the values of reading and evaluating a good message and a bad message. Call them r_i and s_i respectively. Either or both can be greater or less than zero, but typically we would expect $r_i > 0 > s_i$. A player who does not read

3. Vote selling offers an interesting analogy. I am hurt when others who are like me sell their votes, but whether they do or not, it is still worthwhile for me to sell my own. Similarly, my privacy is lost if others like me sell theirs, yet I might sell mine. In either case, a like-minded group might agree to preserve their votes or their privacy, even though each would sell without an agreement.

the message receives a payoff of zero, regardless of its content. We assume that each player knows the other's payoffs.

If evaluators receive no compensation, three types of inefficiency are likely to arise. We illustrate these with three scenarios that differ only in the payoffs to each player. First, too few evaluations may be produced. It will sometimes be socially beneficial for individuals to read messages and evaluate them even if they expect a negative value on average. In Scenario (a), Player 1 benefits by 10 units ($r_1 = +10$) if she reads the message and likes it, but loses 12 units ($s_1 = -12$) if she reads and dislikes it, reflecting the value of the time she spent reading the message. Second, people may evaluate voluntarily, but in the wrong order. In Scenario (b), Player 1 gains more from a good message, but has less time to spare, so it is better for Player 2 to read first, even though her expected value is negative and Player 1's is positive. Third, as shown in Scenario (c), people who expect a positive utility from reading may wait anyway, in the hope that others' evaluations will enable a better informed decision.

There are two initial strategies for each player: read immediately (R) and wait (W). A player who waits can make a fully informed decision if the other player reads immediately. If both players wait, they decide simultaneously whether or not to read in the second round, without the benefit of any additional information, and then the game ends.

We assume that reading a good message provides the same benefit in the first and second periods, which in most practical circumstances will only be separated briefly. Without any discounting, however, waiting will be a weakly dominant strategy. We therefore assume a sliver of discounting when required for tie-breaking: Given a choice between the same expected payoff now or at some future time, a player will choose the earlier payoff.

We assume the players care only about maximizing their own payoffs, without regard to the payoffs of the other players. The initial probability that the message is good is assumed to be $p = \frac{1}{2}$. Figure 15.2 shows the games in the 2 x 2 normal form after converting individual outcomes to expected payoffs.

In (a), the social optimum is for either person to read immediately and the other to wait. The reader has an expected loss of 1; $r_i p + s_i(1 - p) = \frac{10}{2} - \frac{12}{2} = -1$. The expected value for the player who waits is $r_j * 1 + s_j * 0 = \frac{10}{2} = 5$. Unfortunately, waiting is a strictly dominant strategy for both players: No matter what Player i does, Player j gains by not reading the message initially. The only Nash Equilibrium, the outcome that results if each player takes the other's action as fixed, is (W, W), giving each a payoff of 0. Thus, Case (a) demonstrates the natural tendency

FIGURE 15.1
Payoffs from reading in three scenarios

	Good	Bad
Player 1	+10	-12
Player 2	+10	-12

(a) underprovision

	Good	Bad
Player 1	+40	-20
Player 2	+10	-12

(b) improper ordering

	Good	Bad
Player 1	+12	-10
Player 2	+12	-10

(c) wasteful value claiming

FIGURE 15.2
Normal forms of the games

(a) underprovision

	Read	Wait
Read	−1, −1	−1, 5
Wait	5, −1	0, 0

(b) improper ordering

	Read	Wait
Read	10, −1	10, 5
Wait	20, −1	10, 0

(c) wasteful value claiming

	Read	Wait
Read	1, 1	1, 6
Wait	6, 1	1, 1

to underprovide information in equilibrium, because neither player takes account of the value that her first-round evaluation provides to the second-round decision of the other player.

In (b), Player 1's stakes are higher: She can gain more if the message is good, but the cost of her potential mistake is also higher. The social optimum is for Player 2 to read the message first and then recommend it to Player 1 if it is good (*W*, *R*, in the lower left of Table b.) In effect, Player 2 should act as the king's taster: If the taster does not get sick, it is safe for the king to eat. Left to himself, however, Player 2 will refuse the role of human guinea pig (a -1 payoff), preferring to wait (at least a 0 payoff). Player 1 realizes that Player 2 will not provide an evaluation. Either a bit of altruism or a sliver of discounting is sufficient to make Player 1 prefer to read in the first round rather than the second, for an expected payoff of 10. Thus, (*R*,*W*) is the unique Nash Equilibrium, producing a payoff of (10, 5) compared to the larger social payoff of (20, 1) when Player 2 reads first. Case (b) demonstrates the failure of the Nash Equilibrium to generate the optimal order of message reading.

In (c), both players are willing to read immediately, and their payoffs are symmetric. The social optimum is for one to read and the other to wait. If the players could coordinate their actions, they might flip a coin to decide which should read, thereby giving the other the benefits of waiting. Without any coordinating mechanism, however, they are likely to engage in a costly game of waiting, not unlike a game of chicken, hoping to be the last to decide in order to gain information provided by the other's action. The inclusion of a discount factor δ illustrates these incentive problems.

The unique symmetric equilibrium calls for each player to read immediately with probability $\frac{1-\delta}{5\delta}$. As δ approaches 1, the probability that either player will immediately read the message vanishes. For both players, it is worse to wait a period, yet that is the most likely outcome. The costs of the waiting game are represented by equilibrium payoffs of 1 for each, the same payoff as derived from reading immediately. In equilibrium, attempts to claim the surplus dissipate it.

TRADE-OFFS IN COMPENSATION SCHEMES

Evaluations as a potential public good: Distribution is essentially costless and the use of an evaluation by one person does not reduce its value to anyone else. Provisioning of public goods is problematic for several reasons, most notably because of the free rider problem: People do not pay their "fair share" because they hope that others will purchase the good and share it. An ideal compensation system would generate the socially optimal number of evaluations and satisfy these three criteria: same action same price (SASP), budget balance, and voluntary participation. Budget balance and voluntary participation are widely accepted goals in the design of allocation mechanisms. *Budget balance* means that the amount collected from some players equals the amount paid out. *Voluntary participation* requires that no individual would rather drop out of the game than participate.

The remaining condition, we believe, is new to theory, although it is widely invoked in practice and is a natural consequence of market-allocation processes. We label it *same action same price*, or SASP. It requires that individuals engaging in the same action receive the same compensation, and that any offer made to one individual be offered to all. Normal markets, where pricing does not depend on the identity of the customer, follow the SASP constraint. A wealthy person who would willingly pay $10 for a gallon of fresh milk if prices rose that high, still pays only $2 at the store, the same price as an impoverished mother of four. In fact, we often feel vaguely cheated when a producer manages to price discriminate. For example, air travelers who pay the full fare for seats equivalent to those occupied by supersaver passengers may be resentful. In essence, SASP is a no-envy condition far removed from its usual fair division context (Crawford & Sobel, 1982; Glazer & Ma, 1989; Varian, 1974). SASP is highly relevant to traditional public goods financed through taxation; similarly situated citizens are taxed the same amount, independent of their preferences.

Unfortunately, it is not possible to achieve all three conditions as well as efficient allocation, even if everyone's value of waiting for evaluations is public knowledge (Avery, Resnick, & Zeckhauser, in preparation-b). The proof is by counterexample. Two individuals, A and B, value a public good by the amounts 3 and 5, respectively. The cost of provision is 7. It is efficient to buy the good, because the total payoff, 8, is more than the cost. To secure voluntary participation, A can be charged no more than 3. SASP requires B to pay the same amount as A. Thus the most that can be raised is 6. To purchase the good at a cost of 7, the budget must be unbalanced.

Our difficulty is neither the free-rider problem nor asymmetric information; the problem arises because beneficiaries have different valuations. Even if all agents were willing to contribute their true values, thereby obviating the free-rider problem, the combination of SASP and voluntary participation limits them to paying the minimum value of any participant. Some degree of price discrimination among them is required to raise sufficient funds to pay for the public good.

If everyone's preferences are public knowledge, it is possible to achieve efficient allocation along with any two of the three desired properties. Which of the three conditions can be most readily sacrificed?

Consider first relaxing the voluntary participation constraint. Suppose an allocation scheme could force some people to read and provide evaluations for less compensation than their actual cost of doing the work. If everyone who uses the evaluations pays an equal share of the total compensation payments, budget balance and SASP would be maintained. Even if forced participation were palatable from a policy perspective, it would not work without a mechanism to ensure that the people assigned to evaluate give their best effort. As a form of resistance to such coercion, people might provide random or even deliberately misleading evaluations.

Next consider relaxing the budget balance constraint. If a central broker can charge less than it pays, it can charge each person who consumes evaluations the minimum value that the evaluations provide to any one of them, yet pay each evaluator the actual cost of providing the evaluation. The amount disbursed is likely to exceed the amount collected, creating a deficit.

One way to finance the broker's deficits is through dues that pay for one's participation in the evaluation of a large number of messages. Thus, a membership organization could collect dues and spend them by running a budget deficit on individual evaluation purchases.[4] Unfortunately, someone who expects to benefit by less than the dues would refuse to join such an organization, and would be denied access to the evaluations that the organization finances. Any system that does not share every evaluation with everyone who can benefit from it—and no break-even system with SASP and voluntary participation can assure this—is inefficient.

Another deficit financing strategy is government subsidy, an option that finances public goods such as roads and the military. However, government funding of evaluation provision would violate the voluntary participation provision, because it would compel all taxpayers to pay for the service, even though many are unlikely to benefit from it. In any case, the government is not likely to finance evaluations of bulletin board messages, although it does subsidize other informational efforts intended to improve the functioning of markets, such as the Securities and Exchange Commission.

Third, consider relaxing the SASP constraint. In this case, the central broker can charge each user of evaluations his or her full value for consumption, which would raise more than enough money to pay the evaluators. In order to balance the budget, the broker can evenly disburse the surplus among all the participants. Note that this scheme involves price discrimination among people who consume the same number of evaluations: The people who gain more utility from the evaluations pay more for them.

The allocation problem becomes much more difficult when individuals' preferences are not public knowledge at the outset, and may be reported strategically. Strategic reporting can take many forms; one person might report too high a cost of evaluating in order to get paid more when it is efficient for him to evaluate early, and a second person might report too low a benefit from evaluations in order to free-ride on the willingness of others to pay the early evaluators.

Several mechanisms could determine the correct set of evaluators, even when preferences are unknown and people can report them strategically (Avery et al., in preparation-b). The difficulty is that two of our desirable properties must be sacrificed. For example, Groves–Clarke levies (Clarke, 1971; Groves, 1973), which in essence charge each individual the cost his participation imposes on the rest of

4. Economists would label this *two-part pricing*, with the dues being a fixed cost that enables individual items to be priced closer to marginal cost.

the players, can achieve voluntary participation and induce honest reporting, at the cost of sacrificing both budget balance and SASP.

If there is some initial stage when people agree about the probability distributions of each others' preferences, and can contract before finding out their own preferences, then they can use an expected-externality mechanism (Pratt & Zeckhauser, 1987), which charges each individual the cost his participation will on average (although not in any particular instance) impose on the rest of the participants. Honest reporting and budget balance can be met, but only a weaker (ex ante) form of voluntary participation is secured and SASP is lost.[5]

When preferences are not public, which we expect to be the norm, any efficient system must give up SASP and either budget balance or voluntary participation. Given the theoretical limitations on even the most sophisticated possible system, even when there are no practical problems such as transmitting information or computing prices, achieving a reasonably effective exchange of material on the information superhighway will take all the help we can get. Objective, impartial electronic brokers have an important role to play.

INTEGRATION, COMPETITION, AND FEES

The degree to which brokerage services should be vertically integrated with information provision is a policy concern. Telecommunications policy generally distinguishes between and often separates information provision and carriage. For example, after the AT&T breakup, the Baby Bells (information carriers) were not initially permitted to provide information services such as pay-per-call stock quotes.[6] The separation of provision and carriage allows competition and innovation in one realm to proceed independent of developments in the other realm.

Vertical integration of two services may be desirable for either of two reasons: (a) to reduce production costs when there are technological economies returning to integration, and (b) to reduce deadweight losses when each of the two services operating in isolation would be monopolized or cartelized. (Two separate stages of monopolistic pricing—producing what is called double marginalization—generate higher deadweight losses than a single overall stage.[7])

The principal argument against integration is that market power at one stage will be transmitted to the other. Years ago, for example, DeBeers exploited its control

5. A weaker relative of SASP, in which bids are considered actions, is met by both Groves–Clarke and expected externalities approaches. Individuals who report the same preferences and take the same actions face the same price.

6. By the same reasoning, AT&T, as a long-distance carrier was prohibited from providing local telephone service. The vertical integration argument is also at the center of current cable debates. Should cable services operate merely as a common carrier, or should they also create and buy programs?

7. Consider one monopolist who sells to another, who then sells to a market. If the first raises its price, this reduces the profits of the second, and vice versa. But because the separate monopolists do not take into account these reductions, the ultimate price is greater than it would be if they integrated vertically. The additional deadweight loss due to double marginalization may occur in the form of lower quality, not just lower quantity and higher cost (Economides, 1994).

over diamond production to exert control of the distribution network, leading to greater total efficiency loss than if it merely controlled production.[8]

Given the miracles of electronic exchange, there do not appear to be substantial economies of vertical integration between information provision and brokering. Therefore, the argument for and against permitting integration must involve a trade-off between double marginalization concerns and the undesirable extension of market power from one stage to the other. How market power will evolve in information provision and brokering is difficult to predict. Nevertheless, we consider it unlikely that substantial market power will develop at both stages, hence our concern with double marginalization is not great. The possibility of market power at just one stage is considerably greater, with the unfortunate possibility of additional efficiency loss when it is transferred to the other. Moreover, if separation is required initially, and there turns out to be significant market concentration both for brokerage and for information provision, a policy move toward integration should not be difficult to achieve. The need for a policy shift in the other direction, from integration toward separation, would be harder to recognize, and implementing such a shift would be more disruptive.

Based on this analysis, we recommend that the provision of information should initially be separated from the value-added services that a broker might provide, such as matching providers and consumers, or conducting auctions to determine prices. For example, in a future where consumers can watch videos on demand, a brokering service might keep track of a consumer's preferences in order to suggest a video to watch. A second company would actually send movies to the consumer's home, with transmission provided by yet a third company. The services would appear integrated to consumers, just as today one is not necessarily aware of all the telephone companies that participate in each phone call. The underlying separation, however, would encourage innovation in all services and prevent biases from creeping into the suggestion service, as might well happen if the suggestion service were owned by, say, Disney[9].

There are technical barriers to the separation of brokering and information provision. The software run by brokers and information providers must work together. For

8. This set of issues has also arisen in an array of other arenas. For example, Microsoft has been accused of selling its operating systems at lower prices to computer manufacturers that agreed not to offer any other operating system. Government policy toward competition will be sound, we suspect, if appropriate analogies are drawn, say from the telephone or airline industries, about the success of existing methods. The market may also play a helpful role: In the computer industry in recent years, "open systems" strategies, in which downstream firms are encouraged to make complementary products, have fared better than proprietary strategies. By contrast, efforts to gain or maintain predominant control over upstream software resources have fared poorly. The disasters Sony and Matsushita have suffered in the Hollywood movie business are perhaps the best examples.

9. Similarly, information carriage could be broken into two pieces, actual carriage and value-added brokering services. For example, if accounting and bill collection were separate from carriage, consumers would likely have the choice of telephone bills with calls itemized at a somewhat higher charge than unitemized bills, and an even wider array of pricing packages than is currently available. Other brokerage services might identify the cheapest carrier for a particular phone call or even for a single data packet. Brokering of information carriage, however, is beyond the scope of this chapter.

example, the broker needs to know what information is available from each provider, and on what terms. If many independent vendors write software for brokers and information providers, the software must conform to compatibility specifications.[10]

A simple but important corollary of the argument for separation is that brokers should be permitted to charge fees, even if the information providers may not or do not. Much of the information now exchanged on the Internet is provided free of charge and a spirit of altruism pervades the Internet community. At first blush, it seems unfair that a broker should make a profit by identifying information that is available for free, and some Internet user groups would likely agitate for policies to prevent for-profit brokering. However, so long as the use of the brokering service is voluntary, it helps some information seekers without hurting any others. Anyone who does not wish to pay can still find the same information through other means, at no charge.

With brokerage services, there is a tension between the advantages of competition and those of monopoly provision. First, a competitive market with many brokers will permit the easy introduction of new innovations and the rapid spread of useful ones. Because of the rapid spread, however, the original innovator may gain little market advantage and so may have little reason to innovate in the first place. Patents or other methods of ensuring a period of exclusive use for innovations may be necessary. Second, some brokering services may be a natural monopoly. For example, Better Bit Bureaus work best if all evaluations are shared freely, so competing brokering services with private collections of evaluations would be inefficient. Similarly, auction and other pricing services may be most effective if all buyers and sellers participate in the same market.[11] One solution might be for all evaluations to be collected in one place, with brokers competing to sell different ways to aggregate them. More generally, some aspects of brokering may be best organized as monopolies; others should be competitive.

CONCLUSION

The information superhighway brings together millions of individuals who could exchange information with one another. Any conception of a traditional market for

10. Technical standards must be set with great care. Once a system is in place with even a moderate number of users, there can be considerable technological inertia. An outmoded interface in current use may win out over a much better one just being introduced because it will be very costly for users to switch to the new interface without the guarantee that others will be switching as well.

11. Experience with stock exchanges is instructive, although the data are unclear about the advantages and disadvantages of centralization. The New York Stock Exchange long argued for its monopoly on the theory that gathering all bids and asks in one place provided for the most liquid market. Parallel markets, however, are now well established, and are often quite innovative.

In the pure information exchange realm, there are now services that enable employers to compare medical costs against those of other employers on the system, standardized by such factors as age and occupation. The more employers on a single system, the richer and more informative will be the comparison data. The virtue of aggregating information provides one argument for a natural monopoly, assuming data would not be readily exchanged between services.

making beneficial exchanges, such as an agricultural market or trading pit, or any system where individuals respond to posted prices on a computer screen is woefully inadequate for the extremely large number of often complex trades that will be required.

Electronic brokers will be required to permit even reasonably efficient levels and patterns of exchanges. Their ability to handle complex, albeit mechanical, transactions, to process millions of bits of information per second, and to act in a demonstrably even-handed fashion will be critical as this information market develops. Electronic brokers can also run pricing systems, charging and crediting slight amounts to individual accounts as bits careen along the superhighway.

Such power does not come without dangers. Might privacy be hijacked by tomorrow's equivalent of the computer hacker, or sacrificed merely by providing anonymous but extensive statistical data? Can we be confident that we can avoid groupthink or censorship concerns? We must find electronic brokerage mechanisms that address these questions satisfactorily, for without brokers we will be stymied in our ability to use the information superhighway effectively.

ACKNOWLEDGMENTS

Support from the corporate sponsors of the MIT Center for Coordination Science (Resnick) and the Decision, Risk, and Management Science Division of the National Science Foundation (Zeckhauser and Avery) is gratefully acknowledged.

REFERENCES

Andrews, E. L. (1993, July 18). A chip that allows parents to censor TV sex and violence. *New York Times,* p. 14.

Avery, C., Resnick, P., & Zeckhauser, R. (in preparation-a). *The market for evaluations.* Unpublished manuscript.

Avery, C., Resnick, P., & Zeckhauser, R. (in preparation-b). *Price discrimination and the provision of public goods.* Unpublished manuscript.

Brynjolfsson, E., & Resnick, P. (1993, December 2). Electronic solution, Letter to the editor. *New York Times,* p. A14.

Clarke, E. (1971). multipart pricing of public goods. *Public Choice, 11,* 17–33.

Crawford, V., & Sobel, J. (1982). Strategic information transmission. *Econometrica, 50,* 1431–1452.

Economides, N. (1994). *Quality choice and vertical integration* (Rep. No. EC-94-22). New York: Stern School of Business, New York University.

Glazer, J., & Ma, C. (1989). Efficient allocation of a prize—King Solomon's dilemma. *Games and Economic Behavior, 1,* 222–233.

Goldberg, D., Nichols, D., Oki, B. M., & Terry, D. (1992). Using collaborative filtering to weave an information tapestry. *Communications of the ACM, 35*(12), 61–70.

Groves, T. (1973). Incentives in teams. *Econometrica, 41,* 617–631.

Hill, W., Stead, L., & Rosenstein, M. (1995). Recommending and evaluating choices in a virtual community of use. In *CHI 95: Conference on Human Factors in Computing Systems*. New York: ACM.

Malone, T. W., Grant, K. R., Turbak, F. A., Brobst, S. A., & Cohen, M. D. (1987). Intelligent information sharing systems. *Communications of the ACM, 30*(5), 390–402.

Maltz, D. A. (1994). Distributing information for collaborative filtering on Usenet net news. (Tech. Rep. TR-603). Cambridge, MA: MIT Laboratory for Computer Science.

Myerson, R., & Satterthwaite, M. (1983). Efficient mechanisms for bilateral trade. *Journal of Economic Theory, 28*, 265–281.

Nichols, P. M. (1994, August 26). Home video: What to rent? Ask a jury of your peers. *New York Times,* p. D15.

Pratt, J., & Zeckhauser, R. (1987). Incentive-based decentralization: Expected-externality payment induced efficient behavior in groups. In G. R. Feiwel (Ed.), *Arrow and the ascent of modern economic theory* (pp. 439–483). New York: New York University Press.

Resnick, P., Iacovou, N., Suchak, M., Bergstrom, P., & Riedl, J. (1994). GroupLens: An open architecture for collaborative filtering of netnews. In *CSCW 94: Conference on Computer Supported Cooperative Work* (pp. 175–186). New York: ACM.

Shardanand, U., & Maes, P. (1995). Social information filtering: Algorithms for automating "word of mouth." In *CHI 95: Conference of Human Factors in Computing Systems*. New York: ACM.

Varian, H. (1974). Equity, envy, and efficiency. *Journal of Economic Theory, 9*, 63–91.

VI

THE FIRST AMENDMENT AND CHANGING TECHNOLOGY

16

Game Theory and the First Amendment: Strategic Considerations and Freedom of the Press

Timothy J. Brennan
University of Maryland

THE INFORMATION AS COMMODITY APPROACH TO FIRST AMENDMENT ECONOMICS

That researchers and practitioners in constitutional law, journalism, communications, political theory, and philosophy would have abiding interests in First Amendment policy and jurisprudence is to be expected. Economists, however, have not had a great deal to say about the First Amendment. This, might be a bit more surprising than First Amendment scholars and practitioners from traditionally associated disciplines might think. Our intellectual era has fostered the work of Gordon Tullock, Gary Becker, Richard Posner, and the rise of "economics imperialism." The work of these writers and the scholars following in their wake has led to economic analyses of political institutions, marriage and the family, crime, jurisprudence, and sex. The relative neglect of First Amendment issues suggests that they are less important than these other areas or less amenable to economic analysis.

Two strains in the literature are exceptions to this generalization. The first, pioneered by Owen (1979), focuses on First Amendment considerations in determining access to technologies for delivering or receiving speech—newspapers, broadcast, telephones, and cable service. Owen's primary theme is to view the First Amendment as guaranteeing that those who want to use a communications medium should have to pay a price no greater than marginal cost to send or get a message. Prices higher than marginal cost would suppress speech for which speakers or listeners would pay its cost of transmission, but not the higher price. More controversially in some First Amendment circles, perhaps, is that prices below marginal cost would lead to excessive speech. The sense of excessive here is that transmission capability would be used at the margin by communicators unwilling to pay the actual cost of using that capability.

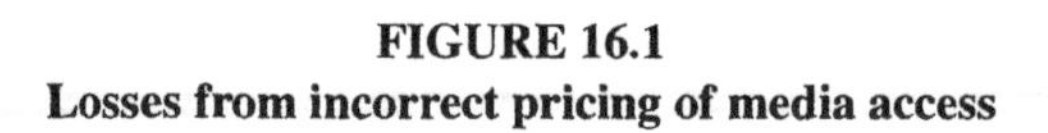

FIGURE 16.1
Losses from incorrect pricing of media access

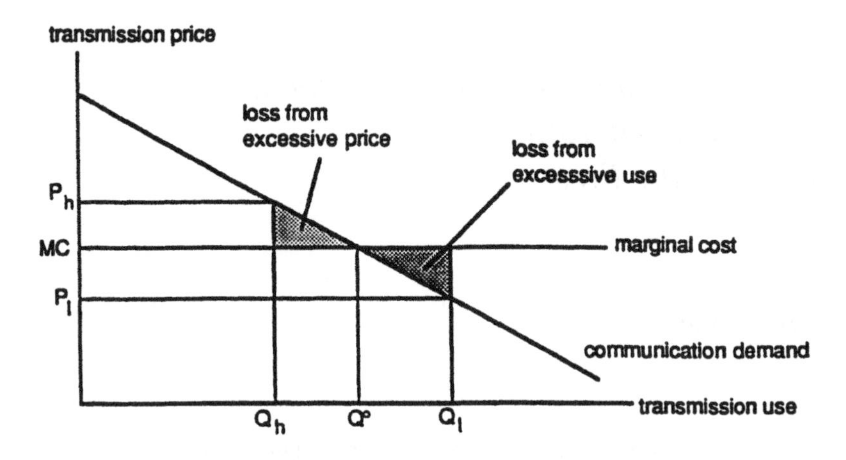

Both types of losses are diagrammed in Fig. 16.1. The downward sloping line for communication demand reflects the economist's conventional assumption that people want to do more of anything as its price falls. Whether a fall in price induces a lot more communicating or a little more—whether demand for media access is elastic or inelastic—is, as they say, an empirical question. *MC* is the marginal cost of media access; if the transmission price is set equal to *MC*, speakers and listeners will undertake the efficient amount of communicating. This is indicated by quantity $Q°$ of transmission use, measured in appropriate units such as minutes of telephone traffic, column-inches of newspaper space, seconds of commercial broadcast time, or bit rates in a digital data line. An excessive price $P_h > MC$ leads to too little use $Q_h < Q°$; a price P_l below *MC* leads to excessive use $Q_l > Q°$.

The lightly shaded area in Fig. 16.1 indicates the difference between the value of deterred communications and their cost, when a high price leads to too little communicating. The darker triangle indicates how much the cost of media use exceeds the value of added use when its price is too low. Applying economic policy norms to media access leads to the recommendation that the performance of communications industries and their regulators be judged by whether these losses are minimized, ideally through marginal cost pricing. The virtue of competition, apart from any content or ownership diversity value, is that the desire of competitors to undercut each other's price will drive overall industry price down to marginal cost. This provides support for limiting the number of stations a single person or firm can own in a local market, and for preventing newspaper or other media mergers among viable competitors.

At the other extreme, a media market may be a natural monopoly (Sharkey, 1981), because two competitors could not cover total operating costs at competitive prices. Local telephony has long been thought to be a natural monopoly. Considerable evidence and some theory supports similar claims regarding cable television. Monopolists, as a rule, will set price above marginal cost, to extract additional revenue from consumers in the absence of a competitive threat. This creates a loss of the sort indicated by the lighter triangle in Fig. 16.1, as the monopolist denies media service to consumers willing to pay marginal cost but not the higher monopoly price. Regulation can move price closer to marginal cost, limiting these losses,[1] but perhaps reducing innovation as well.[2] On the economic account, therefore, a First Amendment *qua* marginal cost analysis of regulation simply uses different rhetoric to reinforce the advice that regulation should be designed to produce efficient outcomes.[3]

Subsequent analyses have taken this marginal cost interpretation as a starting point. This approach has been applied to First Amendment issues raised by allowing transmission monopolists to diversify into the provision of content (Brennan, 1990). Claims by telephone companies that information service and cable television programming restrictions in the antitrust decree in U.S. v. AT&T and federal cable legislation are unconstitutional on First Amendment grounds motivated this work. Allowing a profit-maximizing transmission monopolist to provide content can thwart the effectiveness of regulation to reduce access prices from the monopoly level to one closer to marginal cost. On a marginal cost interpretation, therefore, allowing a telephone company to provide information services could limit the First Amendment rights of other information service providers and consumers.

This marginal cost interpretation has been extended in a number of directions. One is to integrate it with criteria for distributive justice and moral rights to analyze the philosophical merits of First Amendment-related policies. Earlier work in the Fairness Doctrine provides an example (Brennan, 1989b). Incorporation of ef-

1. Unless the monopolist or regulator can identify speakers and listeners only willing to pay a low media access price, perhaps because of low income, the regulated price will typically still be above marginal cost. Where media monopoly is "natural," in the sense that one firm can meet demand at a lower cost than two or more smaller firms, it will typically be the case that average cost falls with output. This condition mathematically implies that marginal cost, the cost of producing additional units, is less than the average cost of producing the total level of service. An access price equal to marginal cost will therefore not generate enough revenue to cover total cost. Absent a subsidy from outside the market, price must be above marginal cost to ensure any supply at all.

2. Regulation that actually sets the access price equal to average cost will discourage a communications transmission supplier from innovating or making other efforts that would reduce costs. If demand for transmission is inelastic, a relatively small level of inefficiency that results in only slightly higher than ideal costs will outweigh the economic gains from large reductions in price.

3. Some media markets may be neither perfectly competitive nor monopolized. Over-the-air television broadcasting, and cities with more than one daily general-interest newspaper are examples. In those markets, as in other oligopolies, direct price regulation is rare. The primary oligopoly policy in the U.S. is antitrust, directed at preventing unjustified market concentration and collusive practices. Interestingly, broadcasting and newspapers are not subject to exactly the same antitrust treatment as other industries. Broadcasting mergers are regulated by the FCC to ensure lower levels of local and national ownership concentration than what antitrust enforcement might ensure. Local newspaper mergers to monopoly do not need to pass the same "failing firm" burdens as similar mergers face in other industries, if the merging newspapers maintain separately published papers with separate editorial staffs. The jointly operated newspapers can sell advertising in common.

ficiency in a broader ethical framework for policy evaluation is not peculiar to media access; it can be applied to education, smoking, and other issues.

A policy question that might illustrate the special treatment for First Amendment considerations might be the extent to which owners of media in concentrated markets might go beyond simple increases in price to express or favor particular points of view. The narrow marginal cost First Amendment interpretation would generally not restrict such uses. That, say, the owner of the only newspaper in a city would use it exclusively to promote a particular political perspective may be a manifestation of the media owner's wealth. In effect, he or she is (perhaps) outbidding other advertisers, speakers, and readers by foregoing more profitable opportunities in order to speak his or her mind.

The general First Amendment protection of such uses suggests that the marginal cost perspective may underlie jurisprudence in that area. Doing so, however, raises the question of why have a First Amendment at all, instead of a constitutional preference for competitive markets or marginal cost pricing. Some differences between usual First Amendment perspectives and economic prescriptions arise when we turn from media access prices to speech itself. Treating information as a commodity like anything else, the economic perspective suggests that a speaker be liable for harms caused by his speech, just as he would be liable for harms caused through nuisance (under property law), tort, or pollution. A similar analogy may be used to minimize apparent contrasts between common property law and copyright (Brennan, 1993). These contrast with First Amendment presumptions against content-based restraints or penalties on speech.[4]

In the second, related vein, Schauer (1992) suggested that those harmed by speech be compensated, but from tax revenues rather than the speaker. If there is a public purpose in speech that harms, there is more equity and perhaps efficiency in having the public pay the cost rather than having it borne by the target of the speech. One might regard this as an application of the takings concept from constitutional and property law. The takings analogy is that if society wants a particular kind of land use, for example, banning construction of homes at the beach, it should compensate those who lose from the ban, rather than have them pay the full cost (Lucas v. South Carolina Coastal Council, 1992). The efficiency argument is that the public should only have its government undertake those activities for which the public as a whole is willing to pay. Uncompensated takings, of land, privacy, or reputation seem more likely to fail this test.

Economic perspectives—Owen's marginal cost approach to media access, extensions to regulated monopolies and broader philosophical frameworks, and Schauer's separation of liability from the duty to compensate—treat the objects of

4. The economic perspective sees little difference between *ex ante* restraints and *ex post* fines. A large enough fine can restrain speech, and a prior restraint is only as credible as the expected penalty for its violation. Posner (1992) asserted that the two differ because (a) *ex ante*, regulators are more likely to be procensorship than judges *ex post*, (b) *ex ante* burden-of-proof standards are likely to be lower than *ex post* criminal law standards, and (but?) (c) harms from disclosed information may be large and difficult to reverse. The point about the two regimes providing equivalent deterrence, in principle, remains.

the First Amendment like any other actions or property. The special character of information tends to fall by the wayside. This is ironic not just because of the sense many of us have that the First Amendment speaks to special social needs. This style of analysis focuses much more on rights of speech or access to communications media. More notably, this perspective as applied in these ways neglects much of the role information plays in prominent economic models used to analyze industry practices and government policy making. This role is most crucial in game theoretic models with imperfect information. In those models, different agents have to plan their interactions with each other on the basis of past learning, current uncertainty, and future credibility of communications.

A potentially important contribution of this strategic approach to First Amendment analysis is that it gives special meaning to freedom of the press. Neither the marginal cost perspective nor more socially oriented justifications of First Amendment protections treat freedom of the press as deriving from a more generic freedom of speech. As such, they say little if anything about justifying freedom of the press to investigate, protect sources, and the like. Viewing freedom of the press as a strategic device—namely, to allow government officials to make credible commitments to keep their activities open to the public—offers independent warrant for granting the press the right to uncover as well as the right to print. This rationale for freedom of the press differs from those based on the value in protecting unpopular speech. Game theoretic arguments might show that access to government information could be a First Amendment right, rather than something requiring separate statutory implementation. A corollary is that these arguments provide some justification for giving public figures less protection than private individuals against undesired press coverage.

The purpose of this chapter is to introduce game theoretic ideas and analytical tools into the description and justification of First Amendment protections. Two generic kinds of problems associated with imperfect information are adverse selection and moral hazard; I begin with some ways in which First Amendment freedoms might mitigate these problems when they surface in relations between the public and its government. I then turn to a series of examples illustrating more specific applications of game theoretic ideas to potential First Amendment situations. These include the Biblical story of Joseph as the Pharaoh's treasurer, a look at conflicts of interest, and the specific difficulty of monitoring politicians with multiple constituencies. First Amendment protections of the press need not be in a nation's interest when there are international conflicts; I review a model showing that possibility and important criteria for justifying these restrictions. I conclude with some observations regarding which First Amendment contexts appear, at this early stage of research, to be more empirically amenable than others to game theoretic analysis.

IMPERFECT INFORMATION MODELS AND THE FIRST AMENDMENT: FUNDAMENTALS AND APPLICATIONS

The essential characteristic of game theoretic approaches to economic problems is that each actor specifically recognizes the beliefs and actions of other actors in forming his or her own beliefs and taking his or her own actions (Fudenberg & Tirole, 1992; Tirole, 1989). Neither the competitive nor monopoly models underpinning the marginal cost approach to the First Amendment share this characteristic. In the competitive model, each actor takes the outside world as given, with all pertinent information revealed in the prices it faces for the goods and services it wishes to buy and sell. Competitive agents assume that they can buy and sell as much as they please without driving prices up or forcing them down. The monopolist is equally passive regarding its environment; the sole difference between it and a competitively acting agent is that it knows and takes as given the amounts its buyers would demand to purchase at each potential price. In neither case is there anything gamey or strategic about the actions of competitive agents or monopolists. Given the information available, they make purchases and sales to maximize profit or pleasure.

A conventional assumption in these models is *perfect information*, that everyone knows everything they might want to know about prices, product quality, and in the case of the monopolist, buyer demand. This is widely viewed as unrealistic; not everything of interest is currently or even prospectively knowable. However, *imperfect information*, in and of itself, does not force one to take a more strategic view of economic actors. A common form of imperfect information is subjective uncertainty, modeled as probability distributions over possible events. Where no one actor knows any more than any other actor, however, the competitive and monopoly models can still be used. Instead of basing a good's price on its known properties, it is based on the expected value, depending on the chances it has more or less desirable properties. Risk-averse agents will discount the value of the object further, because for them avoiding worse than average outcomes outweighs achieving equally better than average outcomes. Risk can become a commodity that more risk-averse agents pay less risk-neutral agents to absorb—this is the fundamental economic rationale for insurance (Dionne & Harrington, 1992).

The more crucial form of imperfect information is where some agents know more than others. In these asymmetric information situations, strategic considerations come into play, as each agent has to take actions based on their expectations of what others believe, and how their actions might change those beliefs and, consequently, those other agent's actions. Asymmetric information and uncertainty are not required for strategic considerations come into play. In classic minimax and Prisoner's Dilemma games, the players are fully informed as to the values and outcomes each other will get. In the former, a player maximizes expected benefits in a zero-sum game by randomly choosing among different tactics. The latter shows how parties who would be better off cooperating nevertheless all refuse to cooperate, because each is better off acting selfishly regardless of whether others act cooperatively. Nevertheless, asymmetric information effects are significant, widespread, and particularly amenable to First Amendment policy analysis.

There are two broad categories of asymmetric information models, classified on the basis of whether the information is generated before or after the parties elect to deal with each other (Hirshleifer & Riley, 1992; Rasmussen, 1991). In *adverse selection* models, the information is known to some of the parties prior to the dealing; in *moral hazard* models, the information concerns actions taken after the dealing is arranged.

Adverse Selection

Overview

The market failure created by asymmetric information in general and adverse selection in particular was first made clear in economics by Akerlof (1970). The setting for his story was a stereotypical market for used cars. The sellers of the cars knew the quality; the buyers did not. Because the buyers did not know the quality of the cars, they would base their payment of the cars on the average quality of the used cars they expected to be available. Sellers, knowing the quality of their cars, would sell only if this payment was worth more to them than the car. Sellers with worse than average cars would find the payment attractive, but sellers with better than average cars would not. When the sellers with better than average cars pull theirs off the market, however, the average quality of cars sold falls. This will depress the price ignorant buyers are willing to pay, leading sellers of the higher quality remaining cars to pull those off the market. The price continues to drop, until an equilibrium is reached where the average quality of cars actually sold equals the highest quality of cars actually sold. This happens only when the all the cars sold are of the lowest quality—the lemons.

A market subject to reduced sales, only at minimum quality levels, is regarded as having a "lemons" problem. The situation is not limited to used cars. In some credit markets (e.g., student loans), borrowers may know their future earning prospects if they get the loan; the bank does not. Thanks to bankruptcy laws, a borrower with low prospects would expect to not have to pay back the loan, and thus would be more willing to borrow than those with good prospects. As lenders raise rates to account for potential default, the fraction of borrowers who expect to default rises. This pushes up interest rates further, driving the remaining relatively less risky borrowers out of the market. In equilibrium, the only borrowers left could be those expecting not to pay back the loans, in which case the credit market would disappear altogether (Stiglitz & Weiss, 1981).

An apparent remedy for this adverse selection problem, and others, is to eliminate the asymmetry in the information. The problem is in making communications informative. Anyone can say that their used car is a peach or that they will get a good job and pay back the loan. For that reason, buyers and sellers in these markets want a kind of message that allows a receiver to tell whether the sender has the characteristic in question. The defining property of such messages, referred to as signals, is that those with the characteristic in question find it less expensive to send the message than do those without it (Spence, 1974).[5] Mere statements of

5. In some settings, a principal can design a compensation scheme that, at some cost, could lead an agent to disclose truthful information about itself (Baron & Myerson, 1982).

quality or diligence are not signals, because people lacking those characteristics can make those statements just as easily as those with them.

A couple of examples of signals may be useful. For cars, a warranty is a signal. Warranties may provide a direct value to consumers as a form of insurance, but this is not an entirely persuasive explanation; mere insurance could be provided by third parties. The reason sellers provide warranties is that they signal the quality of a car. Because a seller of a low-quality car will find it more costly on average to honor a warranty than will a seller of a high-quality car, issuing a warranty allows buyers to sort high-quality from low-quality cars (Carlton & Perloff, 1990). More familiar examples of signals are grades, test scores, and diplomas. Simple assertions that one is bright and diligent are not signals; the dim and lazy can make the same claims. However, if the latter group finds it more difficult to get good grades, score high on tests, and earn diplomas, those devices signal those attributes—even if they provide no training or otherwise evidence any relevance to posteducational careers.[6]

First Amendment Applications

Consider the set of potential candidates for a public office. Each of them knows, beforehand, his or her character traits, including how he or she balances the desire to serve the public against the desire for personal gain. The voters do not know which candidates are motivated to serve the public and which are motivated to serve themselves. Assertions that one is dedicated to public service are not signals, because anyone can make them. Suppose, however, that desire to serve the public interest is correlated with a sense of self-worth, and that those with such a desire are not inclined to be regarded as mouths at the public trough. Because the voters do not know which candidate is which, they regard them all equally. Those with the most pride drop out, raising the proportion of potential candidates who are prideless. Voter regard falls, repeating the now familiar cycle, until the only candidates left are those out for themselves. The prophecy becomes self-fulfilling.

Now, suppose candidates were to commit collectively *ex ante* to allow outsiders access to private information. One might imagine people agreeing behind Rawls's (1971) "veil of ignorance" to decide on the degree of access without knowing which of them would actually be candidates. Suppose further that those outsiders—the press—could profit by making accurate information available in a timely

6. An incidental speculation is that signaling theory may explain why students pay huge premiums for educations at expensive institutions, when it is not clear that the content of the education differs in great degree between institutions. One obvious answer is that the admissions committees of these institutions save future employers the expense of verifying an individual's characteristics. Moreover, were one unlikely to be financially successful in the future, one would not be able to recoup the expenses of an "elite" institution. If so, only those individuals who would be successful would be willing to signal that by burning money up front.

Lest this seem implausible, a similar argument is one of the leading explanations economists have for seemingly noninformative advertising. Unless consumers are likely to buy the product over and over, an advertising campaign will lose money. Thus, advertising's cost becomes a signal of a product's desirability, regardless of the advertising's content.

manner to those interested in the attributes of those candidates. Protecting the access, speech, and publication rights of the press would, on this account, reduce the asymmetry in information between candidate and voter, slowing if not halting the adverse selection process that leads only the sleaziest to run for office.[7]

When we turn to politicians, opportunities for direct signaling are unclear. Attributes such as law school degrees, military service, or families with children may be analogous to grades and test scores as signals of political worth—or they may not. Warranties are inviting in theory, but rarely does a candidate state that he will not run again if he fails to deliver on campaign promises, much less offer voters anything like their money back. If elected office involves a large sacrifice in future job opportunities, a candidate's willingness to run would signal expectation of reelection; candidates expecting to lose in subsequent elections might find campaigning not worthwhile.[8] In the multiple branch government system, with numerous legislators, it is difficult for a successful candidate, even a chief executive, to guarantee many outcomes.[9]

The *ex ante* aspect of the First Amendments open access commitment is crucial. Candidates *ex post* may not wish to have the press provide accurate information about themselves. Contracting on a candidate-by-candidate basis with newspapers would not be a credible remedy for adverse selection. Unless a candidate contracted with all in the press to allow access, speech, and publication rights, voters would be appropriately suspicious that these contracts were contingent on secret terms to bias reports in the candidate's favor. The contract has to be at a social level, that is, constitutional. The First Amendment's mandate for freedom of speech could prevent adverse selection in elections. The constitutionality of rights to find and publish information regarding candidates and their associates could be viewed with that goal in mind.

Moral Hazard

Overview

The second class of market failures associated with asymmetric information, moral hazard, stems from a combination of two factors. The first is that in many instances, after two parties have committed to an agreement, one or both may be able to take advantage of that commitment at the other's expense. The paradigmatic example is smoking in bed after buying fire insurance; the term *moral hazard*

7. The assumption that the press has the incentive to report accurately the information that it can obtain is crucial for these strategic justifications to be valid in practice. If, as many believe, either insufficient competition among news outlets or insufficient demand by its audience for accurate information limits this incentive, favoring the press through the Constitution is less justified on these grounds, if not others. The conclusion of this chapter discusses these considerations at more length.

8. This could be an argument against term limits, if it is empirically valid. My sense is that ex-politicians find lucrative employment, however, making this explanation only theoretical entertainment.

9. I leave others to investigate whether large contributors or political action committees elicit and enforce such guarantees.

arose in connection with *ex post* carelessness when insurance covers losses from risky behavior (Heimer, 1985). Similar problems may arise in principal/agent situations, where the latter may not look after the former's interests. Workers, once hired, may shirk; mechanics, once found, may falsely report car problems; management, once appointed, may indulge in perks rather than maximize stockholders' returns.[10]

The mere possibility of this kind of exploitation is not inherently problematic. In theory, if the insurer can observe the behavior of the insured, or if the principal can observe the behavior of the agent, contracts between them can include incentives to prevent *ex post* exploitation of the commitment. The insurance company could reduce payment or refuse to pay altogether for fire damage if caused by the insured's smoking in bed; an employer can cut pay or fire employees who shirk; voters can turn out of office a politician who does not do the job. Moreover, also in theory, the undesired behavior need not be observed directly. If the potentially exploited party could infer the potential exploiter's undesired conduct from the outcome, then a compensation schedule incorporating proper incentives could be based on the outcome rather than the behavior itself. For example, if managerial effort can be precisely inferred from reported profits, profit-based compensation contracts could induce the amount of effort that the stockholders desire.

Market failures due to moral hazard therefore require a number of conditions. One side, the agent, has to be able to take an action that the other side, the principal, can neither observe directly nor infer from other observables. Smoking in bed by the insured (the agent), which the insurance company (the principal) cannot observe, would be an example, if other accidental causes for house fires cannot be ruled out. These actions are problematic only if they both benefit the agent and harm the principal. Smoking in bed, when insurance covers losses from fire, would be worrisome if the agent likes a cigarette before turning in; the insurance company loses in having a higher than expected payout.

Insurance companies will not simply absorb these losses, of course. Premiums will rise to compensate for the fact that moral hazard is predictable but unpreventable. If customers of fire insurance generally had sufficiently strong desires to smoke, the outcome would not be troublesome. Even if smoking were observable, they would be more willing to pay higher premiums for fire insurance contingent on smoking than to give up smoking. Moral hazard becomes interesting, however, in situations where (a) the agents would rather receive compensation for acting in the principal's interest than take advantage of the agency contract, but (b) the agents have no way of guaranteeing to the principal that they would live up to that contract.

Insurance policies specifically, and agency contracts generally, have to build in moral hazard costs that both sides would prefer to avoid but neither can commit to

---

10. A topical instance of moral hazard is the alleged overpurchasing of medical care when the insurance company, rather than the individual patient or doctor, bears the marginal cost of that care. In theory, we might all prefer to commit *ex ante* to lower levels of care, but such commitments are not credible after the insurance has been purchased. A cogent incorporation of such ideas into the ethical analysis of health care policy is in Dworkin (1994).

avoid. At best, this raises the costs of writing and enforcing contracts, and may lead to arrangements that are less than ideal, for example, by leaving agents with less independent discretion than they would have if they could convincingly demonstrate that their actions were in fact on the principal's behalf.[11] A cascading problem akin to the "lemons" adverse selection effect may occur. The least risky clients may prefer to take their chances rather than pay an insurance premium made higher by moral hazard. This raises the percentage of insured parties who are likely to engage in moral hazard, raising premiums further, causing the least likely of those to drop out. As in the lemons model, the only insured parties, if any, are those most likely to engage in moral hazard; the better risks are driven out of the market.[12]

First Amendment Applications

The key problem with moral hazard, pertinent to the First Amendment, involves these agency guarantees in political contexts. Politicians, once elected, may act to benefit themselves, possibly through favors to special interests, rather than acting on behalf of their constituents. Certainly citizens as principals, and most likely public officials as their agents, would prefer that those entrusted with managing the government could in fact be trusted. Were it possible for politicians and bureaucrats to commit to act on the electorate's behalf rather than their own, they might enjoy more autonomy at their workplace and respect beyond it. Those who prefer being trusted to being suspected might be more likely to seek public service. If so, moral hazard may create an effect similar to the "lemons" problem. Inability to convince the electorate that one would not abuse the public trust once in office would keep the most trustworthy from seeking such office. Offices are left to the least trustworthy, making the expectation of moral hazard rational, that is, that public officials unfortunately do behave as the electorate expects.

Public service seems susceptible to this moral hazard. Direct information on the efforts of public servants is difficult to obtain. Agents (officials) typically control the information on their activities, and thus could keep principals (citizens) from accurately observing what they were doing. Consequently, the citizens would skeptically discount the glowing reports they would hear, stripping the communication between them and the public officials of much of its potential content. Inferring postelection or postappointment effort from outcomes is also difficult. As noted earlier, there are factors beyond the control of public officials that determine the degree to which government could achieve its goals, making inferences of

11. Managed care health plans, in which insurance companies restrict the treatments doctors can choose, are responses to this kind of moral hazard, and are subject to the criticism that they may prevent doctors from undertaking ideal treatments in some circumstances, particularly where verifying ex post that such treatments were necessary is difficult.

12. An illustrative digression would be to note that this model may explain why there is privately provided life insurance, home insurance, even auto insurance, but no privately provided insurance against job loss, when unemployment is likely to be at least as probable and as devastating as some of the other disasters against which one can insure.

effort from outcome problematic. Moreover, information on outcomes as well as effort may also be controlled by the agents themselves.[13]

Inability to observe behavior leading to moral hazard in political contexts need not be an either/or phenomenon. If shirking or self-aggrandizement can be observed imperfectly or with some positive probability, contingent compensation schemes can give public officials some incentive to act in the electorate's interest. One such scheme is to pay public officials more than they could earn in other lines of work, firing them if they are caught in a moral hazard situation. If government pay scales exceed alternative incomes, a public official stands to lose substantially if undesired conduct is uncovered.[14] The lower the likelihood of detection, the higher the government premium needs to be to provide a significant incentive to work on the electorate's behalf. To the extent that government performance indicates that an official's effort is on the public's behalf, one could tie compensation to performance. For reasons already noted, however, this tie is unlikely to be strong enough to make performance-based compensation effective, even if it were otherwise feasible.[15]

The First Amendment provides a potential solution that goes to the heart of the problem rather than circumventing it. Following the adverse selection model, suppose that public officials could commit to let the electorate know what they do after they get into office. Constitutional guarantees of a free press could help ensure that this commitment is credible. The constitutionality of the free press guarantee prevents officials from reneging on the deal after their election or appointment. With such a commitment in place, candidates for office could put performance terms into the formal or informal contracts they have with their constituents. Not only would the citizens be better off, but those who would be inclined toward public service might then find it worthwhile to pursue government careers, knowing that they can credibly commit to official models of conduct they privately find more appealing.

The commitment is only as good as the credibility of the press own commitment to remain independent from influence by public officials. A full justification of

13. Even absent these problems, public officials may be unlikely to act in the general public interest. Mancur Olson, George Stigler, Russell Hardin, and Charles Wolf among others have noted that the incentive and ability of groups to influence governance need not be proportional to the net benefits they receive from public decisions. The key observation is that a diffuse group has disproportionate costs of organizing, whereas a small group less subject to free rider problems can act effectively on its behalf. Policy A may benefit the former group more than Policy B benefits the latter, but these organizing costs may lead policymakers to choose B over A.

14. A related suggestion is that politicians be overpaid by requiring that the press praise them enthusiastically unless they are caught. This works, however, only if the press praise generates higher regard among the public. A conscious policy to overpraise politicians would be self-defeating, because the public would discount the praise as being mandated by policy rather than genuinely earned.

15. In a *Saturday Night Live* skit, comedian Dana Carvey, in the character of Ross Perot, offered to become president at no pay, unless economic growth was at least 3%. If he met the 3% goal, he would take $1 billion, $20 billion with 4% growth, and $50 billion if growth were 5%. Because each percentage point of growth represented an addition of $40–50 billion to the economy, both Perot and the public would benefit from his efforts. A difference between economists and normal people may be that former would find the Carvey/Perot offer at least superficially appealing.

First Amendment guarantees may require that the press rewards depend on the impression readers have of its political independence. The modern stereotype of the Sam Donaldson-style nagging reporter, and zealous pursuit of public officials' warts, could be necessary to signal such independence, even if impolite, biased, or destructive compared to a theoretical ideal.

ILLUSTRATIONS OF ASYMMETRIC INFORMATION IN FIRST AMENDMENT CONTEXTS

The Greatest Bureaucratic Story Ever Told

An early and instructive example of the strategic effects of asymmetric information on bureaucratic behavior comes from the story of Joseph in the Book of Genesis.[16] Thanks to his psychoanalytic skills, Joseph interprets the Pharaoh's dreams to learn that Egypt will have 7 years of rich harvests, followed by 7 years of famine (Gen. 41.1–32). He suggests to the Pharaoh that the government impose a 20% in-kind tax on grain production and appoint someone to manage the stockpile. The Pharaoh, recognizing Joseph's prescience, gives him the job, essentially moving him from the dungeon to the Department of Agriculture (Gen. 41.14,33–41).[17] The Egyptian public, however, is apparently not told about the length of the forthcoming famine. Initially, Joseph extracts all of the Egyptian's money for grain. When the famine persists, they then turn over their livestock. In the following year, with nothing else left, the Egyptians turn over their land to Joseph to give to the Pharaoh, and sell themselves into government slavery (Gen. 47.13–25).

As an appointee of a Pharaoh, Joseph may not have been under the explicit obligations that an appointee of a democratically elected official might have in his role as a civil servant. Nonetheless, Joseph's success at extracting wealth from the Egyptians illustrates the power of an informational advantage. Assume that the objective for the Pharaoh had been to maximize the amount he could get from the Egyptians through sales of grain. With that goal in mind, the Pharaoh, through Joseph's efforts, would want to increase the amount of stored grain Egyptians would wish to purchase at any given price.

As illustrated in Fig. 16.2, this is equivalent to increasing the amount Egyptians would be willing to pay for any given amount of grain the Pharaoh wished to sell, leading to higher profits. Q_l and Q_h refer to sales of grain that maximize the Pharaohs profits respectively when demand is low and high; P_l and P_h are the corresponding prices. MC is the marginal cost to the Pharaoh of keeping grain available; it may be zero. The lightly shaded area is the Pharaoh's profits when demand is low. The darker area represents the increase in the Pharaoh's profit if demand is boosted.[18]

---

16. Citations to the Book of Genesis are from May and Metzger (1965).

17. Perhaps foretelling present impressions of the lives led by those who end up with political power, Joseph had landed in prison because a woman, who happened to be the wife of the captain of the Pharaoh's guard, spitefully accused Joseph of refusing to sleep with her just because he refused to do so (Gen. 39.7–20).

18. That profits increase with increasing demand follows from the envelope theorem, in Dixit (1976).

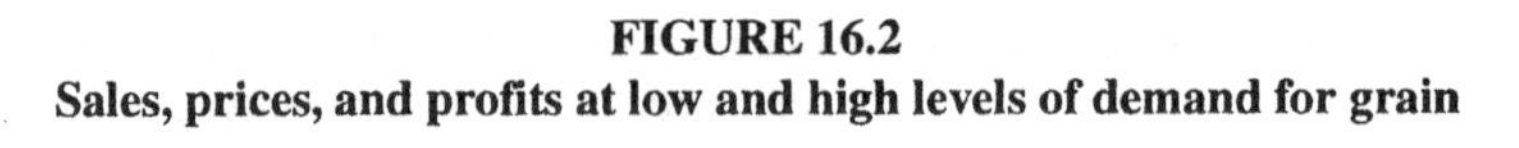

FIGURE 16.2
Sales, prices, and profits at low and high levels of demand for grain

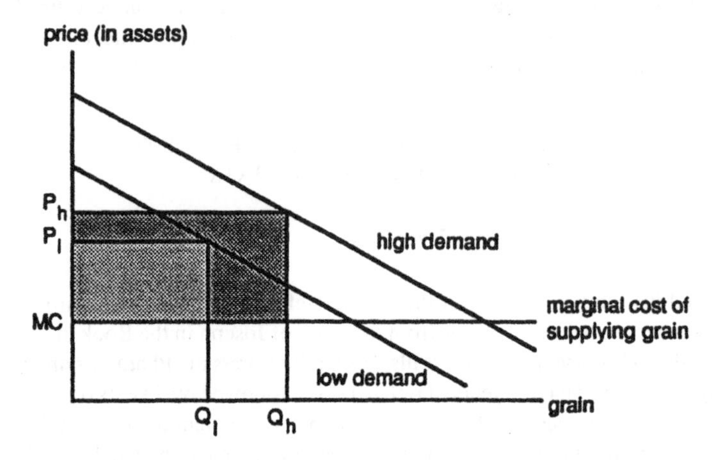

Under normal conditions, the greater the Egyptian populace's expectation that a famine would extend into later years, the fewer assets they would have been willing to exchange for grain in the early years.[19] In these circumstances, Joseph and the Pharaoh have no incentive to pass along their inside information regarding the length of the famine. Getting the populace to cede, voluntarily, all of their property over to the government will be easier to achieve if the populace lacks the information it would need to want to conserve. Refusal to disclose information in the story of Joseph illustrates the value for giving politicians a way to make credible commitments to disclose information they have or obtain regarding the economy's future performance. A free press, with First Amendment protections, could ensure that kind of disclosure.

The story of Joseph is valuable for First Amendment policy analysis also because it is incomplete. It assumes that the Egyptian public cannot infer from Joseph's conduct that famine will last longer than otherwise expected. If false disclosures are possible, the government would always make optimistic grain forecasts, none of which would be credible, leaving the public's estimation of fu-

19. The formal analysis is a bit messy, but the intuition is fairly straightforward. Suppose Egyptians regarded grain and these other assets as substitutes, in that the more grain one has, the less one benefits at the margin from additional units of assets. Because one's assets in the future depend on how much of them are used up in the present, this reduction in the marginal benefit of such assets in the future implies a greater willingness to consume them in the current period or to part with them for other benefits, such as having more grain in the current period. Consequently, the expectation of more grain in the future means that they would be more willing to trade assets for grain in the initial period of the famine. The expectation of a continued famine, conversely, would lead them to conserve assets for the future rather than spend them in the present on grain.

If assets and grain are complements in consumption rather than substitutes, this effect would run in reverse. A larger amount of expected grain in the future makes assets more valuable in the future, reducing willingness to part with them in the present for grain.

ture famines unchanged. If false disclosures are impossible, the fact of nondisclosure would indicate poor agricultural prospects. Another indicator could be a disclosure that the government is hoarding grain, because this would indicate its expectation that it would be able to extract large amounts of the populace's wealth over an otherwise unanticipated long-term famine.

The First Amendment advocate should note that inferential solutions to the information asymmetry between the government and the populace themselves require a protected, ambitious free press. To make nondisclosure a signal, press efforts would be necessary to ensure that false disclosures are too costly to be an effective strategy for the government. Grain hoarding by Joseph and the Pharaoh can warn the public of a long-term famine only if accurate information about the public's grain stockpiles, possibly including government grain imports from other countries, is available to the public. First Amendment protections may be necessary to meet those informational needs as well. Even when a more complete characterization reveals that exploiting asymmetric information may not be simple, the complexities themselves reveal more subtle justifications for First Amendment protections.

Serving Many Masters

Conflicts of Interest

Joseph's success in extracting wealth from the Egyptian populace for the Pharaoh illustrates the advantages to the public of a commitment by its agents to disclose information. Even if information on the official expectation of the length of the famine is not available, press access could provide stockpiling information or insurance against false disclosures that would indirectly limit the vulnerability of the governed to the governors. These indirect types of information illustrate the importance not just of the public knowing what the government knows, but knowing what it does.

A topical and significant class of conduct with strategic implications involves conflicts of interest (Brennan, 1989a). Rules limiting outside activities can limit or prevent moral hazard by an agent, when the agent would benefit from efforts expended in other directions, such efforts reduce the agent's effectiveness in acting on the principal's behalf, and the principal can neither observe the agent's effort on his behalf nor infer such effort from eventual outcomes.[20] As a theoretical matter, the principal would not want to take away all of an agent's outside opportunities, because allowing the agent some chance to benefit reduces the amount of compensation the principal has to pay. Nevertheless, an absolute conflict of inter-

20. There are two reasons to suspect that outside activities would reduce effort on the principal's behalf. The first, a distraction effect, occurs if efforts in outside activities reduce the marginal benefit to the agent of working for the principal. The second, an income effect, occurs if income earned in outside activities reduces the agent's marginal interest in expending effort on the principal's behalf (Brennan, 1989a). It may be noteworthy for some readers that the impetus for this work came from observing a faculty senate debate on rules limiting professors to 1 day per week of outside consulting.

est prohibition may be easier to enforce, eliminating distractions that might reduce the benefits the agent generates for the principal.[21]

Applications in governance abound. If a public official or civil servant maintains other employment, the electorate that employs him or her will typically have a difficult time observing the official's effort. As noted earlier, inference of effort from accomplishment will also be difficult. Moreover, even if such effort is observable at relatively low cost, the benefits to any one constituent of acquiring such information will likely exceed its cost of acquisition. Observing whether the official is expending any effort on behalf of principals other than the electorate at large should be much easier. For example, determining whether one's representative in the legislature is speaking on his or her constituency's behalf may be hard. It would be considerably easier to determine whether that representative had paid outside employment that would predictably lead to a redirection of effort away from the electorate's interest.[22]

Of course, voluntary disclosure of conflicts of interest is not likely, and candidates once in office are likely to want to pursue outside activities more than their constituents would prefer. Consequently, promises to disclose outside activity are not credible. A press protected from government control may be able to provide evidence of such conflicts. A press free of such control not only may limit conflicts postelection or postappointment. It makes it possible for candidates for election or appointment to commit in advance to avoid conflicts. This notice would help voters to judge which candidates would be more likely to work harder on behalf of the electorate.

Multiple Constituencies

The foregoing analysis assumed that the electorate had a single goal in mind, and the only conflict facing public officials involved outside activities. Typically, a government official will face multiple constituencies, with demands for public services lying along different dimensions. In such cases, asymmetric information between the government and these constituencies could bring about the exploitation of the constituencies by the officials for their own purpose. The root cause of the exploitation is that the official can argue before Constituency A that outcomes less than optimal from Constituency A's viewpoint arise from the government's need to respond to Constituency B. The government can similarly invoke Constit-

---

21. Brennan (1989a). Credit for this observation belongs to Marius Schwartz.

22. This rationale for proscribing conflicts of interest does not depend on the alternate activity being actually counter to the interests of the constituents. The central idea is simply that outside activity *per se* reduces efforts on the constituents behalf. The paid outside employment could be the equivalent of moonlighting in an occupation unrelated to legislative issues; it need not be, say, pursuing investments with an official of a savings and loan currently under investigation by banking regulators.

Another implication is that towns so small that public governance positions are part-time jobs may be particularly susceptible to moral hazard on the part of their elected officials.

uency A to justify outcomes that seem adverse to Constituency B. It benefits by taking advantage of each constituency's ignorance of the circumstances associated with serving the other.

A simple model can illustrate the possibility. Suppose we have two constituencies, A and B, served by the government. Constituency A might be rural voters; B might include urban voters. The costs to the government of serving A and B are C_A and C_B respectively; C_A might be the cost of agricultural price supports, whereas C_B is the cost of mass transit subsidies. The government collects taxes T from A and B, with the goal of maximizing its net return $T - C_A - C_B$. If the government officials can not keep the extra revenue directly, they could spend it on assorted perquisites of office. The government would prefer to let T be infinite, but it is presumably constrained in some way in its ability to collect funds in excess of the costs of governing. Finally, assume that the effectiveness of this constraint depends on the knowledge of the constituencies of those costs.

To incorporate asymmetric information into the model, assume that A knows T and C_A, but not C_B; similarly assume that B knows T and C_B, but not C_A. Each constituency knows what the government says it costs to serve the other constituency; A can compute $T - C_A$, and B can compute $T - C_B$. The constraint on the government's ability to bilk the constituencies depends on As estimate of the probability P^A that $C_B \leq T - C_A$, and Bs estimate of the probability P^B that $C_A \leq T - C_B$. One could regard P^A and P^B as the estimates by the respective constituencies of the chance that the government's cost announcements are exaggerations. To close the model, assume that the cost H the government bears in setting T above actual cost $C_A + C_B$ depends on the voter discontent V reflected in these probabilities, notated as

$$H(T; C_A, C_B) = V(P^A(C_B \leq T - C_A), P^B(C_A \leq T - C_B)).$$

Given C_A and C_B, increasing T reduces increases the constituency's estimation of the likelihood that the government is exaggerating in claiming that T is necessary to cover aggregate costs of governing $(C_A + C_B)$. Both P^A and P^B rise, increasing voter discontent V,[23] and H, limiting the tax that government officials find most rewarding to collect.

A constitutionally protected press that can ferret out and distribute more accurate information on the costs of providing government service can decrease A's and B's subjective estimates of the probability that the costs of serving the other constituency are as large as the government says. These changes in the probability estimates will lead the government to reduce its level of taxation, if more precision in the estimates increases the amount additional taxes contribute to voter discontent.[24] The role of a free press may be more crucial once we take a longer run per-

23. Formally, $\frac{\partial V}{\partial P^A} > 0$; $\frac{\partial V}{\partial P^B} > 0$, implying that $\frac{\partial H}{\partial T} > 0$.

24. Formally, let l be the parameter reflecting the precision of the estimated probabilities, so that $\frac{\partial H}{\partial \lambda} > 0$. The condition for increased precision leading to decreased taxation is $\frac{\partial^2 H}{\partial T \partial \lambda} > 0$.

spective incorporating how constituencies might infer estimates of the likelihood that taxation exceeds cost, knowing the strategic incentives of the government to misrepresent costs. The probability that the constituencies assign to excessive taxation above costs has to be consistent with the tax level they would predict that the government would set. In a deterministic situation, this probability of exaggeration can be either zero or one. If the probability is zero, the government would take advantage of the constituencies' expectations of good faith to tax excessively. If the probability of exaggeration is one, that is, exaggeration is certain, then so too is voter discontent, forcing the government to reduce taxes.[25] If both constituencies know both C_A and C_B, they can judge the government's performance directly without having to rely on inherently nonself-fulfilling predictions. A constitutional commitment to make this kind of information discoverable and publishable may not only improve government performance in the short run, it may allow it to stabilize in the long run.

GAME THEORETIC LIMITS ON FIRST AMENDMENT PRIVILEGE

The analysis so far has found support in strategic, asymmetric information theories for First Amendment rights in a free press and access to information on public figures and their activities. It would be remiss not to point out that such theories can support limitations on freedom of the press. Certainly, there are well-known contexts for considering limitations on freedom of the press, such as libel, invasion of privacy, protection of trade secrets, and limiting disclosure to an enemy of military strength and tactics. The rationales for these considerations are either fairly obvious or involve controversies that are not peculiarly strategic or game theoretic.

The interesting strategic rationales for limiting the freedom of the press would be contexts in which the electorate *qua* principal specifically does not want to know what its government *qua* agent knows or is doing. The key idea is that by keeping itself in the dark, the public may find it easier to make some kinds of commitments credible. One potential area would be behavioral commitments an individual wants to make to himself, and which could be thwarted by press reports of government knowledge or activity.[26] One possible if not fanciful instance would be where citizens believe that their resolve not to fall victim to urban paranoia requires that they remain ignorant of anecdotal crime reports to which they believe they would overreact.

A more plausible and more consequential context for strategic rational ignorance involves suppression of the press during military confrontations (Brennan, 1992). The underlying idea is that the credibility of a strong military response may be necessary to prevent aggression. The credibility of the threat would ensure that the aggression not take place, hence that the strong response never would be made. That threat might not be credible if a strong response would be opposed by the public as being either too costly or morally unwarranted by the size of the aggres-

25. A similar observation regarding a seller's product quality choice under rational expectations when customers are unsure of the level of quality is Faulhaber and Allen (1988).

26. The work that initiated recent attention to problems of self-commitment was Elster (1979); see also Elster (1985).

sion itself. If the potentially aggressive opponent knew that the public would either halt a strong response or punish military leaders who undertook such aggression, it would know that the threat was not credible, and hence might undertake the aggression. To make the threat credible, and thus to deter aggression, the public might want to commit to remain ignorant of the military response.

The obvious method for achieving this ignorance would be to ban press reports of military activities or inflicted damages.[27] The possibility that the public might infer a strong military response, however, means that a wider range of suppressed information may be necessary to ensure strategically valuable ignorance. If citizens could observe the enemy's effort, they might be able to infer that the military had acted in an objectionably intense manner. A second possibility is that the public may be uncertain as to whether an intense military response was necessary to deter aggression. For example, perhaps a light response would work under some weather conditions or political circumstances inside the enemy's boundaries, but that a heavy response would be necessary in other circumstances. If the *ex ante* probability that only high effort is necessary is sufficiently small, and if the citizens remain ignorant of the military's effort or whether the circumstances necessary for great effort actually held, then a military threat to use strong or excessive force could be credible.[28] Contrary to intuition, an enemy might find it in its interest to attempt to get information on its responses to the public. The press might find itself under pressure not to report such information, if doing so would render the objectionable response incredible.

Distinctions between theory, practice, and duty ought not go unmentioned here. Self-commitment and military strategy might rationalize ignorance. Those rationalizations also invite preemption of liberal democratic principles and abdication of moral responsibility for actions taken on one's behalf in time of war. Whether courts, legislators, policymakers, or the public could be trusted to recognize whether the narrow circumstances that could justify First Amendment limitations actually apply may be doubted. A more grave question is whether the expected gains from possible self-control or deterrence of aggression could outweigh the moral considerations underlying our obligations to liberty and control over our government.

CONCLUDING REMARKS

Game theoretic economic models of the First Amendment illustrate how constitutional commitments to freedom of the press could mitigate the harms arising from adverse selection and moral hazard in politics. These harms include a tendency toward self-fulfilling prophecies that public spirit would be little seen in public of-

27. The argument here might extend to the power of military equipment, and perhaps in some situations to espionage. It perhaps could go without saying that control over disclosure extends to topics beyond those directly associated with military tactics or damages. I do not intend to suggest that this argument could be plausibly extended to the range of information currently classified.

28. Brennan (1992b).

fice. The most inherently dedicated might choose not to run, and officeholders might act in the self-interested way their constituents expect. These bad effects would be less likely to ensue if candidates and officeholders could make credible, binding commitments to disclose their conduct in office and the information available to them. A protected free press may be an institutional device for making such commitments feasible, allowing potential officeholders to be tested on the basis of their willingness and ability to serve the electorate.

A virtue of this framework is that it focuses on freedom of the press rather than general freedom of speech. It may shed some light on what otherwise seems a mysterious addition to the First Amendment: If free speech is already guaranteed, what is the point of having a free press included? The focus on political figures in these examples here is also instructive. Adverse selection and moral hazard come about primarily when some person or group wishes another person or group to carry out tasks on its behalf. Politicians clearly fit that mold; private parties, in general, do not. The distinction between those who hold themselves out as agents for the public *qua* principal and those who do not could explain why press intrusion into the lives of public figures deserves greater protection than intrusion into private lives, suggesting criteria for drawing the jurisprudential line between private and public.[29] Although this distinction does not tell us whether the gambling activities or sex lives of athletes, entertainers, or business persons should be open to the press, it does offer the prospect of an intuitively plausible legal and policy test.[30]

These illustrations here are little more than suggestive. Future research in this area may expand the range of situations for which strategic information considerations could inform First Amendment analyses. It also may deepen as well as broaden the contribution of economics to these analyses. The models here do not consider sequential games, learning, agency problems in the press, and strategic interactions within the government and within the electorate. If the state of economic specialties where game theory has contributed provides any indication, these considerations will open up numerous possible explanations—perhaps beyond the point where our empirical abilities will allow us to tell one from the other (Peltzman, 1991).

On that note, we might examine whether the game theoretic justifications for the First Amendment are borne out empirically. This question might seem obvious, but it figures all too rarely in most speculations on justifications for the First Amendment. Does freedom of speech, to the degree achieved in the United States, actually increase either self-perceived autonomy or actual independence of thought and action among its residents? Does the truth corner the marketplace of

29. One can imagine, however, that private principal–agent relationships might incorporate broad *ex ante* discovery rights to ensure that the principals can verify the quality of the agents efforts.

30. A similar distinction exists in defining the limits of information that may be disclosed under the Freedom of Information Act (FDIA; 5 U.S.C. 552). Essentially, information the government happens to have on private persons should not be disclosed. Only that information pertaining to the performance of the government itself should be available (Cate, Fields, & McBain, 1994). A corollary of the analysis in this chapter suggests that the FOIA's guarantees of availability of information regarding government conduct could be implicit in the First Amendment itself.

ideas or, following Gresham, does bad currency drive out the good? The burden of proof in the debate perhaps ought to fall on the skeptics, if only because the harms of falling very far down the slippery slope of suppression appear so large. Nonetheless, the frequency with which unsupported psychological and political claims are used to buttress philosophical arguments is disturbing.

I will not buck this trend here, but some thoughts on the data might be of note. In this case, the question would be whether the First Amendment's free press guarantee permits politicians and bureaucrats to make credible disclosure commitments. This credibility would allow claims of trustworthiness to be valid, raising the quality of public service and its reputation among the public served. Unfortunately, a recent Washington Post poll does not offer much support. Whereas 50% of the respondents said that President Clinton generally tells the truth, the figure falls to 29% for members of Congress and 14% for politicians in general (Morin, 1994).[31] A hint as to why these percentages seem so low, shrinking as the constituency of the politician becomes more local, is that newspaper reporters fell in the truth-telling category for only 40% of the respondents.[32] Those who argue that the performance of the press does not merit its Constitutional protections may have some ammunition. In any event, game theoretic justifications for the First Amendment suggest that further research into the causes and remedies for press disrepute may be more than an idle academic pursuit. It could be important for ensuring the responsive performance of a democratic system of governance.

ACKNOWLEDGMENTS

Discussions with Sandra Braman, Mancur Olson, and Cindy Alexander helped clarify some of the ideas here. Special thanks go to Fred Schauer and seminar participants at the Joan Barone Shorenstein Center for Policy and the Press and the Kennedy School of Government, Harvard University. None of these bear any responsibility for errors and omissions.

REFERENCES

Akerlof, G. (1970). The market for lemons: Qualitative uncertainty and the market mechanism, *Quarterly Journal of Economics, 84*, 488–500.

Baron, D., & Myerson, R. (1982). Regulating a monopolist with unknown costs, *Econometrica, 50*, 911–30.

31. The survey covered about 1,000 adults; reported sampling error is 5%. The average person was regarded as a truth-teller by 71% of those surveyed.

32. Television news reporters did better, with a 60% favorable response. I suspect that the relative balance of television and newspapers as sources of information on public officials tilts toward the latter as one moves from national through state and Congressional district to the county, city, and precinct level.

Brennan, T. (1989a). Exclusive dealing, limiting outside activity, and conflict of interest, *Southern Economic Journal, 56*, 323–35.

Brennan, T. (1989b). The fairness doctrine as public policy, *Journal of Broadcasting and Electronic Media, 33*, 419–40.

Brennan, T. (1990). Vertical integration, monopoly, and the First Amendment, *Journal of Media Economics, 3*, 57–76.

Brennan, T. (1992). Rational ignorance: The strategic economics of military censorship, *Southern Economic Journal, 58*, 966–74.

Brennan, T. (1993). Copyright, property, and the right to deny, *Chicago-Kent Law Review, 68*, 675–714.

Carlton, D., & Perloff, J. (1990). Modern industrial organization. New York: HarperCollins.

Cate, F. H., Fields, D. A., & McBain, J. K. (1994). The right to privacy and the public's right to know: The central purpose of the Freedom of Information Act, *Administrative Law Review, 46*, 41–74.

Dionne, G., & Harrington, S. E. (Eds.). (1992). Foundations of insurance economics: Readings in economics and finance. Norwell, MA: Kluwer Academic.

Dixit, A. (1976). Optimization in economic theory. Oxford, UK: Oxford University Press.

Dworkin, R. (1994, January). Is Clinton's health care plan fair? *New York Review of Books, 41*,20–25.

Elster, J. (1979). Ulysses and the sirens. Cambridge, UK: Cambridge University Press.

Elster, J. (1985). Weakness of will and the free-rider problem, *Economics and Philosophy, 1*, 231–65.

Faulhaber, G., & Allen, F. (1988). Optimism invites deception, *Quarterly Journal of Economics, 103*, 397–407.

Fudenberg D., & Tirole, J. (1992). Game theory. Cambridge, MA: MIT Press.

Heimer, C. A. (1985). Reactive risk and rational action: Managing moral hazard in insurance contracts. Berkeley: University of California Press.

Hirshleifer, J., & Riley, J. G. (1992). The analytics of uncertainty and information. Cambridge, UK: Cambridge University Press.

Lucas v. South Carolina Coastal Council, 112 S. Ct. 2886 (1992)

May, H. G., & Metzger, B. M. (Eds.). (1965). The Oxford annotated Bible with the apocrypha, Rev. Standard Ed. Oxford, UK: Oxford University Press.

Morin, R. (1994, February 20). To tell the truth, Washington Post, p. C5.

Owen, B. (1979). Economics and freedom of expression. New York: Ballinger.

Posner, R. (1992). Economic analysis of law. Boston: Little Brown.

Peltzman, S. (1991). The handbook of industrial organization: A review article, *Journal of Political Economy, 99*, 201–17.

Rasmussen, E. (1991). Games and information: An introduction to game theory. Oxford, UK: Basil Blackwell.

Rawls, J. (1971). The theory of justice. Cambridge, MA: Harvard University Press.

Schauer, F. (1992). Uncoupling free speech, *Columbia Law Review, 92*, 1321–61.

Sharkey, W. (1981). The theory of natural monopoly. Cambridge, UK: Cambridge University Press.

Spence, A. M. (1974). Market signaling. Cambridge, MA: Harvard University Press.

Stiglitz, J., & Weiss, A. (1981). Credit rationing in markets with imperfect information, *American Economic Review*, *71*, 393–410.

Tirole, J. (1989). The theory of industrial organization. Cambridge, MA: MIT Press.

17

Lost on the Infobahn Without a Map: The Need for a Coherent First Amendment Approach

Robert Corn-Revere
Hogan & Hartson, Washington, DC

In addition to triggering an avalanche of bad metaphors, the popularization of the expression *electronic superhighway*[1] has perpetuated the typical confusion that arises in the First Amendment analysis of new communications technologies. Development of the "Infobahn" promises to bring about new broadband networks and a convergence of media, but it also is leading to a First Amendment identity crisis. Consider the following scenario:

A federal regulator walks into a room and and is confronted with five television sets, each displaying the same program. The show features a steamy sex scene between a man and a woman, complete with nudity, adult language, and lots of sweat. Although transparent to the viewer, each TV is fed via a different transmission source. The first television is receiving a terrestrial broadcast transmission, the second obtains the images by coaxial cable, the third is connected to a fiber optic common carrier network, the fourth is hooked to a VCR, and the fifth TV is receiving a direct broadcast satellite (DBS) feed. Leaving aside any questions of federal versus local jurisdiction, and assuming that the images are not obscene, what is the regulator's constitutional authority to control these images?

The answer is, well, it depends.

1. A monthly technology supplement to the *Washington Post* has launched a contest to search for a better term. *See Giving the Information Superhighway a Good Name*, *Fast Forward*, September 1994 at 4 ('Bad enough the phrase resonates with the oaken earnestness of Vice President Al Gore, the guy responsible for introducing it to the public. Far worse are the dozens of winky 'highway' metaphors, all too cute by half, that have thumbed a ride: 'Toll booth.' 'Traffic cops.' 'Road kill."). Based on a reader poll, the term 'CyFiWay' was chosen as a replacement. *See We Did It CyFiWay*, *Fast Forward*, December 1994 at 4. However, the poll result does more to reveal the flaws of democracy than to provide a realistic alternative. A better option—the 'E-Net'—came in at number two.

For the broadcast transmission, it depends on whether the images are sufficiently salacious to be considered "patently offensive" based on "contemporary community standards for the broadcast medium."[2] It also depends on whether the telecast is at a time of day when there is a "reasonable risk that children may be in the audience"—a concept that is far from settled.[3] Assuming these conditions are met, the government may require that the telecast be restricted to the appropriate time of day. With respect to the cable connection, the government's ability to regulate is far more limited. Various courts have held that indecency regulations are invalid when applied to cable television.[4] As with broadcast television, however, the law remains a work in progress. With respect to the third television, a court would likely apply the cases relating to "dial-a-porn" to the fiber optic common carrier network.[5]

The answer is even more murky with respect to televisions 4 and 5. Although there is much logic and some case law to suggest that the VCR-originated images would receive the same constitutional protection as the print media,[6] the issue has never been formally resolved by the courts.[7] The appropriate First Amendment standard for DBS transmissions is even further from resolution. To the extent satellite programmers operate as broadcasters, making their transmissions freely available to all receivers, they would be subject to the same statutory requirements

---

2. *Pacifica Foundation*, 56 F.C.C.2d 94, 98 (1975), quoted in *FCC v. Pacifica Foundation*, 438 U.S. 726, 731–732 (1978). This notion of community standards specific to "the broadcast medium" may well be a moving target following the success of *NYPD Blue*.

3. Thus far, the United States Court of Appeals for the D.C. Circuit has rejected the FCC's reasoning regarding the times within which there is a "reasonable risk" that children will be watching or listening. *See Action for Children's Television v. FCC*, 932 F.2d 1504 (D.C. Cir. 1991), *cert. denied*, 112 S. Ct. 1281 (1992) ("*ACT II*") (24-hour indecency ban rejected); *Action for Children's Television v. FCC*, 852 F.2d 1332 (D.C. Cir. 1988) ("*ACT I*") (12 a.m. to 6 a.m. "safe harbor" period rejected).

4. *Community Television of Utah, Inc. v. Wilkinson*, 611 F. Supp. 1099 (D.C. Utah 1985), *aff'd sub nom. Jones v. Wilkinson*, 800 F.2d 989 (10th Cir. 1986), *aff'd mem.* 480 U.S. 926 (1987); *Cruz v. Ferre*, 755 F.2d 1415 (11th Cir. 1985); *Daniels Cablevision, Inc. v. United States*, 835 F. Supp. 1, 9-10 (D.D.C. 1993); *Community Television, Inc. v. Roy City*, 555 F. Supp. 1164 (D. Utah 1982); *Home Box Office, Inc. v. Wilkinson*, 531 F. Supp. 987 (D. Utah 1982). *See generally* Note, *Content Regulation of Cable Television: "Indecency" Statutes and the First Amendment*, 11 *Rutgers Computer & Technology L.J.*, 141 (1985).

5. After dial-a-porn regulations were twice struck down as being too restrictive of adults' rights to receive sexually oriented telephone messages, *Carlin Communications, Inc. v. FCC*, 749 F.2d 113 (2d Cir. 1984) (*Carlin I*); *Carlin Communications, Inc. v. FCC*, 787 F.2d 846 (2d Cir. 1985) (*Carlin II*), the FCC fashioned judicially acceptable regulations. *Carlin Communications, Inc. v. FCC*, 837 F.2d 546 (2d Cir. 1987), *cert. denied*, 488 U.S. 924 (1988) (*Carlin III*). *See also Information Providers' Coalition For Defense of the First Amendment v. FCC*, 928 F.2d 866 (9th Cir. 1991); *Dial Information Corp. of New York v. Thornburgh*, 938 F.2d 1535 (2d Cir. 1991), *cert. denied*, 112 S. Ct. 966 (1992). Congress attempted to ban all interstate dial-a-porn calls, including calls that were indecent but not obscene, but this measure was invalidated as a First Amendment infringement. *Sable Communications of California, Inc. v. FCC*, 492 U.S. 115 (1989).

6. See *Video Software Dealer's Association v. Webster*, 968 F.2d 684 (8th Cir. 1992).

7. Groskaufmanis, *What Films We May Watch: Videotape Distribution and the First Amendment*, 136 U. Pa. L. Rev. 1263, 1284 (1988).

as terrestrial TV stations.[8] Additionally, Congress has determined that DBS operators shall be subject to many of the same "public interest" obligations as traditional broadcasters.[9] However, as a constitutional matter, the spectrum scarcity that has served to justify less First Amendment protection for broadcasters appears inapplicable to DBS operators who may be able to provide hundreds of video channels

It is likely to take years for our hypothetical federal regulator to know the constitutional limits of his authority with respect to the five televisions. If case law develops as it has in the past, it is entirely possible—if not probable—that the five transmissions would be governed by distinct First Amendment standards. Moreover, by the time those legal standards are in place, there is likely to be a sixth television, fed by computer-generated images, and perhaps even a seventh, receiving transmissions from some other source such as terrestrial microwaves.[10]

WHAT IS WRONG WITH THIS PICTURE?

Certainly there are differences between the various transmission media beyond what may be readily apparent to the casual viewer. Broadcast signals come to the home free of charge and can be received by any television within range of the transmission, cable television and common carrier fiber optic links require a physical connection and are (or will be) provided to customers by subscription, videotapes must be obtained from some external source and require additional hardware for playback, and DBS requires specialized receiving equipment and will be provided, for the most part, by subscription. But do these differences support constitutional distinctions between the various media?

Printed material comes in many forms and is distributed in a wide variety of economic arrangements. Leaflets, handbills, and some newspapers are distributed without charge and are made available to all within the range of the publisher. In addition to such free distribution, newspapers are sent through the government mail system and sold on public rights of way. The same can be said of magazines. Some printed material can be read only with the aid of specialized equipment, such as a microfilm or microfiche reader, or—increasingly—an electronic book or personal computer. Text may be transmitted to the computer screen from a floppy disk or CD-ROM format, from an online service or, it is anticipated, from some over-the-air source. Despite these differences, the print media all are subject to the

8. 18 U.S.C. § 1464 provides: "Whoever utters any obscene, indecent, or profane language by means of radio communication shall be fined not more than $10,000 or imprisoned not more than two years, or both."

9. Cable Television Consumer Protection and Competition Act of 1992, Pub. L. No. 102-385, 106 Stat. 1460 (West Supp. 1993) ("1992 Cable Act")

10. Various microwave delivery systems already exist, including satellite master antenna television (SMATV), multichannel multipoint distribution service (MMDS) and local multipoint distribution service (LMDS). Each provides customers with programming otherwise available on cable television systems.

same First Amendment protections (although this proposition is largely untested with respect to electronic texts).[11]

This chapter explores whether the First Amendment standards applicable to various communications media are consistent with settled constitutional principles, and whether such a multifaceted approach can be sustained in light of rapid technological change. It examines the history of constitutional treatment of new technologies, and how the First Amendment status of communications media generally has corresponded to the regulatory classification scheme established by the government. Initially, new technologies are given little or no First Amendment protection, but as each medium gains cultural penetration and becomes more mainstream, courts are increasingly willing to recognize its First Amendment status.

This evolutionary process has become more difficult and less reliable as the pace of technological change has accelerated and as regulatory distinctions among media have blurred. Accordingly, courts have been left with little guidance for developing new standards, as demonstrated by the search for the First Amendment status of broadcasting, cable television, and common carriage. The search for a coherent approach has been impeded by the use of such metaphors as the "electronic superhighway," which tends to focus on the method of transmission. A regulatory and judicial preoccupation with the various modes of communication—as opposed to the fact of communication—has perpetuated the current confusion over constitutional standards. This chapter concludes that a more unified First Amendment approach would reduce confusion and more reliably preserve constitutional values.

THE CONSTITUTIONAL STATUS OF CABLE TELEVISION

The Cable Communications and Consumer Protection Act of 1992 focused attention as never before on the First Amendment status of cable television operators and programmers. As part of a sweeping bid to reregulate the cable industry, the 1992 Act imposed a wide variety of new obligations on cable operators, including must-carry for commercial and noncommercial broadcasters,[12] retransmission consent,[13] leased-access channel rate regulation,[14] indecency restrictions on both leased-access and public access channels,[15] notice requirements for previews of unsolicited R-rated movies on premium channels,[16] and vertical and horizontal ownership limits.[17] Not surprisingly, a First Amendment challenge was filed the

11. But see *Cubby, Inc. v. Compuserve*, 776 F. Supp. 135 (SDNY 1991) ("A computerized database is the functional equivalent of a more traditional news vendor.").

12. 47 U.S.C.A. §§ 534-35 (West Supp. 1993).

13. 47 U.S.C.A. § 325(b) (West Supp. 1993).

14. 47 U.S.C.A. § 532(c) (West Supp. 1993).

15. 47 U.S.C.A. §§ 531, 532(h) (West Supp. 1993)

16. 47 U.S.C.A. § 544(d) (West Supp. 1993)

17. 47 U.S.C.A. § 533(f) (West Supp. 1993).

same day the must-carry requirements became law.[18] A broadly based First Amendment attack also was brought against most other provisions of the Act.[19]

As these cases have progressed, the central question has been the selection of a First Amendment standard for cable television. In *Turner Broadcasting System v. FCC*, the government argued that must-carry rules trigger only minimal scrutiny as "a reasonable attempt to correct ... market dysfunction" that restricts the transmission of broadcast signals.[20] Although acknowledging that "cable television is not affected by the scarcity of the broadcast spectrum," the government asserted that cable should be governed by a constitutional standard "comparable" to that applied to broadcasting.[21] Alternatively, the government argued that must-carry rules could be upheld under what it described as the "more exacting standard" of *United States* v. *O'Brien*, which is applicable to content-neutral regulations that have incidental effects on speech.[22] The District Court had upheld the must-carry rules using this constitutional standard.[23]

The cable industry, in sharp contrast, argued that First Amendment "strict scrutiny," the standard applicable to printed communications, should be used to analyze the must-carry rules. The rules, according to the industry briefs, are content based because they compel carriage on the grounds that local broadcast signals convey information important to the public interest.[24] On a more general level, however, the industry argued that none of the particular characteristics of cable communications justified a lower level of constitutional scrutiny. Cable television operators do not have power to distort the market for television signals, according to the industry, and such economic power does not justify a different constitutional approach.[25] Nor do such purported factors as a scarcity of physical space to place cables or the receipt of a government benefit via franchise rights support a lower First Amendment standard.[26]

The Supreme Court in *Turner Broadcasting System* stopped short of resolving this dispute. Although the Court emphasized that "[t]here can be no disagreement" that "[c]able programmers and cable operators engage in and transmit speech, and they are entitled to the protection of the speech and press provisions of the First Amendment," it did not articulate a standard for evaluating these rights.[27] It agreed with the cable industry that the less rigorous scrutiny applicable to broad-

18. *Turner Broadcasting System v. FCC,* 819 F. Supp. 32, 37 (D.D.C. 1993).

19. *Daniels Cablevision, Inc. v. United States*, 835 F. Supp. 1 (D.D.C. 1993).

20. Brief for the Federal Appellees, *Turner Broadcasting System v. FCC*, No. 93-44, at 13.

21. *Id.* at 14, 32–36.

22. *Id.* at 37–47.

23. *Turner Broadcasting System v. FCC,* 819 F. Supp. at 41.

24. E.g., *Brief for Appellant National Cable Television Association, Inc., Turner Broadcasting System v. FCC*, No. 93-44, at 16–23; *Brief for Appellant Time Warner Entertainment Company, L.P., Turner Broadcasting System v. FCC*, No. 93-44, at 14–21.

25. *Brief for Appellant Time Warner Entertainment Company, L.P., Turner Broadcasting System v. FCC*, No. 93-44, at 33–36.

26. *Brief for Appellant Time Warner Entertainment Company, L.P., Turner Broadcasting System v. FCC*, No. 93-44, at 32–33, 36–38.

27. *Turner Broadcasting System v. FCC*, 114 S. Ct. 2445 (1994).

casting "does not apply in the context of cable regulation" because of "fundamental technological differences."[28] It also rejected the government's assertion that market dysfunction justified "industry-specific antitrust legislation" in the form of must-carry rules, subject only to rational basis scrutiny.[29] Accordingly, the Court found that "at least some degree of heightened First Amendment scrutiny is demanded."[30]

However, at the same time, the Court did not accept industry arguments that strict scrutiny must be applied to must-carry rules. It rejected the notion that broadcast carriage obligations are content based, either in purpose or effect, and concluded that burdens were imposed on cable operators "only because they control access to the cable conduit."[31] As with constitutional questions relating to other technologies, the Court predicated its findings on "the unique physical characteristics of cable transmission" but foreshadowed the time when such characteristics may be less pivotal. Indeed, the majority noted that "given the rapid advances in fiber optics and digital compression technology, soon there may be no limitation on the number of speakers who may use the cable medium."[32]

Even before such advancements are available, however, the Court was unwilling to find that must-carry rules are constitutional. Thus, it remanded the case to determine whether the economic health of broadcasters actually was at risk and whether the must-carry rules are an appropriately tailored means of addressing the problem. In particular—and despite "unusually detailed statutory findings"—the Court found no evidence that broadcast stations had been harmed by cable operators, no findings regarding the adverse effects of must-carry on cable programming services, and no judicial findings on less restrictive measures that might be available.[33] In short, the decision means that the debate over the First Amendment status of cable television, as well as the constitutionality of must-carry rules, will continue.

The dispute outside the courtroom has been no less intense. Some have argued that cable television systems are like newspapers, and should be accorded full First Amendment status.[34] Others have focused on some of the particular characteristics of cable technology, or on various public policy goals, and have argued

---

28. *Id.*

29. *Id.*

30. *Id.*

31. *Id.* The Court also pointed to "an important technological difference between newspapers and cable television.... [T]he cable network gives the cable operator bottleneck, or gatekeeper, control over most (if not all) of the television programming that is channeled into the subscriber's home."

32. *Id.* See also *NCTA v. FCC*, (D.C. Cir. 33 F.3d 66, 1994).

33. *Turner Broadcasting System v. FCC*, 114 S. Ct. 2445.

34. E.g., J. Emord, *Freedom, Technology, and the First Amendment* (1991); L. Powe, *American Broadcasting and the First Amendment* 216–47 (1987); G. Shapiro, P. Kurland, & J. Mercurio, *Cablespeech* (1983); Cole, *The Cable Television "Press" and the Protection of the First Amendment—A Not So "Vexing Question,"* 28 *Cal. Western L.Rev.*, 347 (1991); Saylor, *Municipal Ripoff: The Unconstitutionality of Cable Television Franchise Fees and Access Support Payments*, 35 *Cath. U. L. Rev.*, 671 (Spring 1986); Lee, *Cable Franchising and the First Amendment*, 36 *Vand. L. Rev.*, 867 (1983).

that cable television should be subject to a less demanding constitutional regime.[35] One experienced observer of developments in telecommunications law noted snidely that the courts have begun to recognize "yet another First Amendment right: the right to string wires on poles."[36]

But this is a somewhat strange statement in the First Amendment context. Logically extending its emphasis on the method of transmission, it suggests that "freedom of the press" is nothing more than the right to spread ink on paper. Lofty statements about a free press being the bulwark of human liberty tend to lose their punch when one is focused on the messy and technical elements of the printer's art. It is strange as well in light of Vice President Gore's insistence that new communications media on the wired network will "entertain as well as inform. [T]hey will educate, promote democracy, and save lives." The Vice President based his vision on the understanding that the various media are converging.[37] If the National Information Infrastructure is to provide this rebirth of free expression and democratic ideals, how could it be that the means of providing it is less worthy of constitutional protection?

The idea of denigrating the First Amendment status of cable by virtue of its means of delivery underscores an essential point that often is obscured in the debate: The outcome of the current controversy will determine not just the First Amendment rights of cable operators, but of all electronic publishers.[38] The constitutional standard for cable television likely will shape the rules of the road for the electronic superhighway. Thus, choices made today regarding the right to speak electronically will determine the vitality of the First Amendment in the next century.

THE REAL ISSUE: APPLYING THE FIRST AMENDMENT TO NEW TECHNOLOGIES

Transition to the Multimedia Age

It is entirely likely that by the time the Supreme Court finally takes a definitive stand on cable television's First Amendment status, cable could be a relic of the past. Some people suggest this is already happening with broadcasting. Although broadcasting continues to be a healthy industry, its demise has been predicted with

35. E.g., Price, *Congress, Free Speech, and Cable Legislation: An Introduction*, 8 *Cardozo Arts & Ent. L. J.*, *225* (1990); Barron, *On Understanding the First Amendment Status of Cable: Some Obstacles in the Way*, 57 *Geo. Wash. L. Rev.*, 1495 (1989); Brenner, *Cable Television and Freedom of Expression*, 1988 *Duke L.J.*, 329; Miller & Beals, *Regulating Cable Television*, 57 *Wash. L. Rev.*, 85 (1981).

36. Robinson, *Implications of the Court's Video Programming Decision: Telcos Will Enter the Cable Industry in a Big Way, The Cable-Telco Report* (September 13, 1993) at 13.

37. Speech of Vice President Al Gore at UCLA, Los Angeles, California, January 11, 1994.

38. See, e.g., *Chesapeake & Potomac Telephone Co. v. United States*, 830 F. Supp. 909 (E.D. Va. 1993); *Daniels Cablevision, Inc. v. United States*, 835 F. Supp. 1, 10-11 (D.D.C. 1993) (First Amendment status of DBS operators considered).

increasing frequency,[39] and broadcasters have become uneasy about being left behind on our nation's trip down the electronic superhighway.[40] At the same time, although many courts have questioned the continuing validity of the scarcity doctrine, they have not been able to bring themselves to revisit a First Amendment standard predicated on the particular technical and market characteristics of the

39. See Federal Communications Commission, *Through The Looking Glass: Integrated Broadband Networks, Regulatory Policies, and Institutional Change* (OPP Working Paper No. 24, November 1988) ("*Through The Looking Glass*"); K. Auletta, *Three Blind Mice* (1990). Some advertising executives have suggested that within the foreseeable future the networks' "days as a mass medium will be over." Levine, *The Last Gasp of Mass Media? Forbes*, September 17, 1990 at 9. See also Zoglin, Goodbye to the Mass Audience, *Time*, November 19, 1990 ("The era of the mass TV audience may be ending .. . "); Carter, *Little Improvement in Sight As Networks End Bad Year, New York Times*, December 24, 1990 ("a senior network executive, who insisted on remaining anonymous [stated,] 'We're presiding over networks as they head out of business.'"); Werts, *Look Who's Watching, Newsday*, December 23, 1990 ("the networks are dying, and single-interest cable channels are premiering monthly"); Mahoney, *Network Woes Are Barter's Gain, Electronic Media*, March 25, 1991 at 16 (According to Tim Duncan, executive director of the Advertiser Syndicated Television Association, the networks' ability "to deliver 99 percent of the nation at the flip of a switch ... isn't the case in many network dayparts anymore. That doesn't exist outside of prime time and shortly will not exist in prime time."); Shales, *The Endangered NBC Peacock, Washington Post*, March 29, 1991 at B2; Shales, *The FCC and the Threat to Free TV, Washington Post*, April 8, 1991 at C2. However, reports of broadcasting's demise may be just a bit premature. See Foisie, *TV Finances Fare Mostly Better*, Broadcasting & Cable, February 7, 1994 at 42.

40. See Stern & McAvoy, *Broadcasters Claim Stake on Superhighway, Broadcasting & Cable*, February 7, 1994 at 48.

broadcasting industry circa 1969.[41] Certainly by the time the Supreme Court reconsiders *Red Lion*, traditional broadcasting will be an even less potent force in the context of the rise of other new communications technologies, and the same fate may await cable television.

Vice President Gore has proclaimed that, like our Universe:

> Current communications industries—cable, local telephone, long distance telephone, television, film, computers, and others—seem to be headed for a Big Crunch/Big Bang of their own. The space between these diverse functions is rapidly shrinking—between computers and televisions, for example, or inter-active communication and video. But after the next Big Bang, in the ensuing expansion of the information business, the new marketplace will no longer be divided along current sectoral lines. There may not be cable companies or phone companies or computer companies, as such. Everyone will be in the *bit* business. The *functions* provided will define the marketplace.[42]

This is not a new insight, but it is an especially important point. Over a decade ago Ithiel de Sola Pool described the "convergence of modes" that is "blurring the lines between media." He noted that "[a] single physical means—be it wires, cables, or airwaves—may carry services that in the past were provided in separate ways. Conversely, a service that was provided in the past by any one medium—be it broadcasting, the press, or telephony—can now be provided in several different physical ways.'[43] The Congressional Office of Technology Assessment similarly found that "technology is ushering in a convergence of forms of press publishing that were once partitioned by technology: print publishing, mail, broadcasting and

---

41. The constitutionality of broadcast regulation is not an immutable fact; it is based on "'the present state of commercially acceptable technology' as of 1969." *News America Publishing, Inc. v. FCC*, 844 F.2d 800, 811 (D.C. Cir. 1988), quoting *Red Lion Broadcasting Co. v. FCC*, 395 U.S. 367, 389–90 (1969). See *Meredith Corp. v. FCC*, 809 F.2d 863, 867 (D.C. Cir. 1987). The Supreme Court has noted that "because the broadcast industry is dynamic in terms of technological change[,] solutions adequate a decade ago are not necessarily so now, and those acceptable today may well be outmoded 10 years hence." *CBS, Inc. v. Democratic National Committee*, 412 U.S. 94, 102 (1973). Both courts and commentators have questioned the continuing validity of the scarcity rationale for the constitutionality of regulating broadcast content. E.g., *FCC v. League of Women Voters of California*, 468 U.S. 364, 376–377 n.11 (1984); *Arkansas AFL-CIO v. FCC*, 11 F.3d 1430 (8th Cir. 1993) (*en banc*) (Arnold, C.J., concurring); *News America Publishing, Inc.*, 844 F.2d at 811 ("The Supreme Court ... has recognized that technology may render the [scarcity] doctrine obsolete—indeed, may have already done so."); *Telecommunications Research and Action Center v. FCC*, 801 F.2d 501, 506-09 (D.C. Cir. 1986), *cert. denied*, 482 U.S. 919 (1987); *Loveday v. FCC*, 707 F.2d 1443, 1459 (D.C. Cir. 1983), *cert. denied*, 464 U.S. 1008 (1984). See L. Tribe, *American Constitutional Law* 1005–1006 (2d ed. 1988) ("reconsideration [of the scarcity argument for broadcast regulation] seems long overdue").

42. Speech of Vice President Al Gore at UCLA, Los Angeles, California, January 11, 1994.

43. I. Pool, Technologies of Freedom 23 (Cambridge, MA: Harvard University Press, 1983). See also Nadel, *A Unified Theory of the First Amendment: Divorcing the Medium From the Message*, 11 *Fordham Urban L. J.*, 163, 166 (1982); *Media Blur Predicted By Turn of Century, Multichannel News* (May 2, 1988).

telephone."[44] This change in the media environment has seriously complicated the once-simple task of regulatory classification.[45] It also is causing extensive changes in communications networks, with trends toward reduced costs, declining sensitivity to distance, faster communications, increasing information traffic, greater channel diversity, increasing interactivity, increasing flexibility and expandability, and increasing interconnectivity.[46]

More relevant for purposes of this chapter, however, is the fact that such evolution is changing the very concept of a cable system. FCC Commissioner Andrew Barrett has pointed out that "to pursue the multimedia future, cable companies must replace their existing one-way, coaxial-based networks with optic-fiber based interactive information superhighways."[47] In fact, this transformation is already underway. Since 1989, the use of fiber optics by cable operators has increased by 675% and is expected to grow by 25% annually through the next decade.[48] Most cable plant installed since 1987 has interactive capability, and certain projects, such as Time Warner's "Full Service Network" in Orlando, are using digital switching to provide such services as two-way video, video on demand, interactive full-motion video educational services, and interactive video games.[49]

Moreover, it is important to keep in mind that the changes affecting the cable television industry are not occurring in a vacuum. Technology is evolving along various lines, each of which promises to expand individual citizens' access to information. Thus, in addition to the technical advancements in the cable industry, telephone companies are aggressively pursuing the development of video and data networks, both with cable partners and independently;[50] the FCC has authorized PCS service, which promises to add wireless two-way data transmission capabilities;[51] satellites for DBS service were recently launched, opening the door to a competitive method of digital video delivery;[52] interactive information for personal computers and other consumer devices is increasingly available on CD-

44. U.S. Congress, Office of Technology Assessment, *Science, Technology and the First Amendment*, OTA-CIT-369 (Washington, DC: U.S. Government Printing Office, January 1988) at 27.

45. See generally Stern, Krasnow, & Senkowski, *The New Video Marketplace and the Search for a Coherent Regulatory Philosophy*, 32 *Cath. U. L. Rev.*, 529, 571–76 (1983).

46. W. R. Neuman, *The Technological Convergence: Television Networks and Telephone Networks*, printed in *Television for the 21st Century: The Next Wave* 3–17 (Aspen Institute, C. Firestone, ed. 1993). See also W.P. Dizard, *Old Media/New Media* 38–56 (New York: Addison-Wesley/Longman 1993); G. Gilder, *Life After Television* (New York: W.W. Norton & Co., 1992).

47. *Id.* at 49.

48. National Cable Television Association, *Twenty First Century Television*, 9 (1993).

49. *Id.* at 15, 33.

50. Barrett, *Shifting Foundations: The Regulation of Telecommunications in an Era of Change*, 46 *Fed. Comm. L.J.*, 39, 40 (December 1993).

51. *Amendment of the Commission's Rules to Establish New Personal Communications Services*, Gen. Docket No. 90-314, FCC 93-451 (released October 22, 1993).

52. See, e.g., *Prices for DBS Programming Launched, Broadcasting & Cable*, January 3, 1994 at 47 (DBS service to customers is scheduled to begin in April 1994).

ROM;[53] and online computer services are becoming increasingly popular, delivering, among a great many services, interactive newspapers and magazines.[54]

These developments are quite unlike previous transformations of media technology. For example, the introduction of steam-powered presses and inexpensive pulp paper in the mid-19th century made possible book and newspaper publication on a mass scale, but it was essentially an enhancement of an existing method of communication.[55] The second transformation, brought about by the introduction of broadcasting, introduced a totally new means of conveying information.[56] However, it was a means that was not a complete substitute for print or for speech. Each medium continued to play a fairly distinct role in the information marketplace. The current transformation of the media, however, is of a distinctly different nature. Multiple methods of delivering video images are evolving including multimedia forms that combine video and print. Print can be delivered electronically, and, with interactive capability, assumes the attributes of speech. Many examples of convergence can be described, but the point is, the current transformation is not conducive to analyzing new media forms in terms of their particular characteristics.[57]

Historical Treatment of New Technologies

Cycles of Repression

In many ways, censorship is the bastard child of technology. Before the printing press, government suppression of expression was largely unnecessary and seldom practiced. There was no central authority over scribes, nor was there any need for one. They worked in isolation on individual manuscripts that largely were incapable of causing a major controversy.[58] But the advent of the printing press changed all of that. Commonly cited examples of censorship of the 16th and 17th centuries were direct reactions to "a new communications environment in which dissatisfied individuals possessed a capacity for finding allies or reaching others in ways that had not existed previously."[59] Accordingly, it "is no accident that shortly after Gutenberg invented the printing press, official authorities invented the first cen-

53. Dizard, *supra*, at 26–29. See Langberg, *Gabriel Brings CD-ROM Into the Rock Revolution, Washington Business*, January 24, 1994 at 21.

54. E.g., Shaw, *Inventing the 'Newspaper' of Tomorrow, Los Angeles Times*, June 3, 1991 at A1; Shaw, *'Electronic' Newspaper Emerging After Slow Start, Los Angeles Times*, June 3, 1991 at A1.

55. Dizard, *supra* at p. 19.

56. *Id.*

57. See, e.g., Corn-Revere, *Multimedia and the Future of the First Amendment, QuickTime Forum*, September/October 1993 at 20, 22.

58. Pool, *supra* at 14–15. *See also* M. Katsh, *The Electronic Media and the Transformation of Law*, 136 (New York: Oxford University Press, 1989) ("Writing itself was mainly a means of acquiring and exercising power but was not a threat to power. Those in power did not worry about it or have to censor it.").

59. Pool, *supra* at 15–16; Katsh, *The First Amendment and Technological Change: The New Media Have a Message*, 57 *Geo. Wash. L. Rev.*, 1459, 1467 (1989); *See* R. Smolla, *Free Speech in an Open Society* 337–338 (New York: Alfred A. Knopf, 1992); L.R. Sussman, *Power, The Press and the Technology of Freedom*, 10–12 (New York: Freedom House,1989); *The First Amendment—The Challenge of New Technology*, 9–11 (S. Mickelson, E. Mier, Y. & Teran, Eds., New York: Praeger, 1989).

sorship bureau."[60] As M. Ethan Katsh explained in his book, *The Electronic Media and the Transformation of Law*:

> The spread of printing in the last half of the 15th century created a new communications environment that undermined the authority of powerful institutions. Those whose power derived from their ability to control the written word were threatened by a reduced ability to control the new medium of print. As a result, many censorship laws were enacted, trials held, and punishments meted out. By the late 16th century, "censorship of the printed word had become the universal practice of the lay and church authorities throughout Europe."[61]

Thus, governments employed censorship because of an acute awareness that the authority of the state waned as the power of the press ascended. In particular, press licensing laws were "an attempt to foster only books that promoted the values or interests of the authorities, something the scribal system did automatically."[62]

Yet even as the new technology of print increased the government's need to censor, it thwarted the accomplishment of this state objective. The ability of the press to mass produce books and other works negated most efforts to exert control.[63] In Britain, for example, the government successively attempted the creation of state monopolies, press licensing, taxation, and criminal libel as methods of restricting the press.[64] In the end, however, such attempts at control were abandoned, not "due to any philosophical conclusion concerning the advisability of a free press but primarily to an inability to devise an enforceable system of regulation capable of achieving the results desired."[65]

Consequently, the rise and fall of government regulation over the press has tended to be cyclical. New technologies tend to increase pressure for government control by challenging established state policies and by threatening to undermine official authority. The government responds by enacting measures to reassert its authority and to otherwise regulate the press. Such efforts ultimately fail, however, because of the power of a given technology or because of technological expansion of the means of communication. This evolutionary process reinforces movement toward a system of free expression.

---

60. Smolla, *supra* at 338.

61. M. Katsh, *supra* at 136, quoting S. H. Steinberg, *500 Years of Printing* 260 (3d ed. Baltimore, 1974). See also Pool, *supra* at 14–15; Loevinger, *Earl F. Nelson Lecture: Law, Technology and Liberty*, 49 *Missouri L. Rev.* 767, 777 (1984).

62. Katsh, *supra* at 142; Pool, *supra* at 15–16.

63. Katsh, *supra* note 57 at 1469–70. It has been suggested that as new electronic communications technologies become universal, "censors will be overwhelmed, and finally made superfluous." Sussman, *supra* at 12.

64. Pool, *supra* at 15–16.

65. M. Katsh, supra at 145 (citation omitted). See also *id.* at 146–165.

The American Experience With New Technologies

By adopting the First Amendment, the United States became the first nation to embrace the new technology as an essential component of its political system. This choice evolved not only from the colonists' experience with suppression but from the framers' appreciation for "the highly active and uninhibited communications environment" that print made possible.[66] It has been suggested that the nature of the technology and the actual practices of publishers of the period may be a better guide to understanding the First Amendment than attempts to divine the intentions of the framers by dissecting their words or reading contemporary common law. Thus, whereas "[t]he particular words chosen for the First Amendment may have been fortuitous or accidental ... the evolution of a law that was more protective of expression than anything that existed pre-Gutenberg was not."[67]

Although the new technology of the printing press was "born free" in the United States, this break with tradition was not sufficient to end the cycle of repression. As new technologies have been introduced, courts and other policymakers have been slow to recognize their First Amendment status. Professor Laurence Tribe has noted that the decisions "reveal a curious judicial blindness, as if the Constitution had to be reinvented with the birth of each new technology."[68] Thus, contrary to the First Amendment tradition, and particularly with the rise of the regulatory state, new technologies now are born in captivity.[69]

Examples are not hard to find. In 1915, film was too new a medium to qualify for constitutional protection as "speech." The Supreme Court in a trilogy of cases[70] upheld the authority of state censorship boards to subject moving pictures to prior restraint. Analyzing the regulatory scheme in terms of state constitutional protections for freedom of speech, the Court found, as a matter of "common sense," that the constitution was inapplicable to cinema.[71] The Court said that the technology of film poses a special danger that "a prurient interest may be excited and appealed to," and noted that "there are some things which should not have pictoral representation in public places and to all audiences."[72] It concluded that "the exhibition of moving pictures is a business, pure and simple, originated and con-

66. *Id.* at 1470.

67. *Id.*, at 148. *See generally*, L. Levy, *Emergence of a Free Press* (New York: Oxford University Press, 1985)

68. Tribe, *The Constitution in Cyberspace: Law and Liberty Beyond the Electronic Frontier*, Keynote Address at the First Conference on Computers, Freedom & Privacy (San Diego, California, March 26, 1991).

69. See generally Lively, *Fear and the Media: A First Amendment Horror Show*, 69 *Minn. L. Rev.*, 1071 (1985). See also Groskaufmanis, *What Films We May Watch: Videotape Distribution and the First Amendment*, 136 *U. Pa. L. Rev.*, 1263, 1284 (1988).

70. *Mutual Film Corp. v. Industrial Comm'n of Ohio*, 236 U.S. 230 (1915); *Mutual Film Co. v. Industrial Comm'n of Ohio*, 236 U.S. 247 (1915); *Mutual Film Corp. of Missouri v. Hodges*, 236 U.S. 248 (1915).

71. *Mutual Film Corp. v. Industrial Comm'n of Ohio*, 236 U.S. at 244.

72. *Id.* at 242.

ducted for profit, like other spectacles, not to be regarded ... as part of the press of the country or as organs of public opinion."[73]

Courts first confronted the First Amendment status of broadcasting in 1932 and again were reluctant to extend constitutional protection to a new medium of expression. In *Trinity Methodist Church, South v. Federal Radio Comm'n*,[74] the U.S. Court of Appeals for the District of Columbia Circuit upheld against constitutional attack a Federal Radio Commission (FRC) decision to revoke a radio station license.[75] The FRC argued in its brief to the court that broadcasting is not protected speech under the First Amendment.[76] Although the court did not exclude radio from constitutional protection in the same stark terms used by the Supreme Court in reference to film 17 years earlier, the result was the same. It described radio as a mere "instrumentality of commerce," and upheld the license revocation as simply "application of the regulatory power of Congress in a field within the scope of its legislative authority."[77]

The "application of regulatory power" at issue was the denial of a license renewal because of a licensee's intemperate attacks on public officials and for broadcasts that were "sensational rather than instructive."[78] The Supreme Court declined to review the holding, even though it had struck down a Minnesota press law a year earlier on srikingly similar facts.[79] When it finally did consider the First Amendment rights of broadcasters, the Court recognized some application of constitutional protections, but at a lower level than to "traditional" media.[80]

Over time, a growing number of courts have questioned the factual predicates underlying the constitutional status of broadcasting.[81] But for more than six decades, the law has allowed greater government intrusion into the editorial processes of broadcasters than traditional media. Courts have continued to be exceedingly reluctant to revisit the First Amendment standard for broadcasting, despite overwhelming evidence that the conditions supporting the weaker constitutional pro-

---

73. *Id.*

74. 62 F.2d 850 (D.C. Cir. 1932), *cert. denied*, 288 U.S. 599 (1933).

75. The FRC was the predecessor agency to the Federal Communications Commission.

76. *See* Powe, *supra* at 16. When placed in historical context, the FRC's position may seem less extreme. See Price, *Congress, Free Speech, and Cable Legislation: An Introduction*, 8 *Card. Arts & Ent. L. J.*, 225, 230 (1990) ("At the outset, radio was perceived primarily not as a medium for speech, but as a device to aid ships at sea... . No substantial body of thought conceived of radio or television in their infancy, as a new form of newspaper.").

77. *Trinity Methodist Church, South*, 62 F.2d at 850–51.

78. *Id.*

79. 288 U.S. 599 (1933). *Compare Near v. Minnesota*, 283 U.S. 697 (1931) (scandalous attacks on public officials by newspaper protected from prior restraint). For an insightful comparison of the two cases, *see* Powe, *supra* at 13–21.

80. E.g., *NBC v. United States*, 319 U.S. 192 (1943); *Red Lion Broadcasting Co. v. FCC*, 395 U.S. 367 (1969).

81. *FCC v. League of Women Voters of California*, 468 U.S. 364, 377 n.11, 379 n.12 (1984); *Arkansas AFL-CIO v. FCC*, 11 F.3d 1430 (8th Cir. 1993) (Arnold, C.J., concurring); *Loveday v. FCC*, 707 F.2d 1443, 1459 (D.C. Cir. 1983), *cert. denied*, 464 U.S. 1008 (1984).

tections have changed.[82] As noted earlier, courts have shown a similar ambivalence about applying First Amendment protections to cable television.

The practice of extending First Amendment rights incrementally has been supported rhetorically by treating different communications delivery methods as being constitutionally distinct. As Justice Robert Jackson wrote in his concurring opinion in *Kovacs v. Cooper*, "The moving picture screen, the radio, the newspaper, the handbill, the sound truck and the street corner orator have differing natures, values, abuses and dangers. Each ... is a law unto itself... ."[83]

This oft-repeated maxim of First Amendment jurisprudence that "differences in the characteristics of new media justify differences in the First Amendment standards applied to them"[84] has been institutionalized through the intermediary of a regulatory classification scheme. Differences in the characteristics of new media first result in some type of categorization and each category is accorded different treatment in constitutional inquiries. Or, as the FCC's Office of Plans and Policy noted, "[t]he regulatory/legal world is ruled by definitions."[85]

The Regulatory State and Freedom of Expression

When different methods of communication were created and put to commercial use, the government classified the media according to the types of services provided and subjected them to various levels of regulation. The Communications Act of 1934 set out the basic regulatory models: private radio, broadcasting, common carrier and, with the addition of the Cable Communications Act of 1984, cable television.

Broadcasting, typified by over-the-air radio or television, is defined as "the dissemination of radio communications intended by the public" and broadcast licensees are charged with certain "public trustee" obligations.[86] These include requirements that licensees serve their community needs and interests,[87] that licensees provide reasonable amounts of air time to candidates for federal elective office[88] and "equal opportunities" to appear on air to candidates at all levels whose

82. *Syracuse Peace Council v. FCC*, 867 F.2d 654 (D.C. Cir. 1989), *cert. denied*, 493 U.S. 1019 (1990); *Branch v. FCC*, 824 F.2d 37 (D.C. Cir. 1987), *cert. denied*, 485 U.S. 959 (1988); *Telecommunications Research & Action Center v. FCC*, 801 F.2d 501, *rehearing en banc denied*, 806 F.2d 1115 (D.C. Cir. 1986), *cert. denied*, 482 F.2d 919 (1987).

83. 336 U.S. 77, 97 (1949) (Jackson, J., concurring).

84. *Red Lion Broadcasting Co.*, 395 U.S. at 386.

85. *See* Federal Communications Commission, *Through The Looking Glass: Integrated Broadband Networks, Regulatory Policies, and Institutional Change* 21 (OPP Working Paper No. 24, November 1988) ("*Through The Looking Glass*").

86. 47 U.S.C. § 153(o).

87. *Id.* § 307(b); *United States v. Southwestern Cable Co.*, 392 U.S. 157, 174 (1968); *Malrite TV of New York v. FCC*, 652 F.2d 1140, 1144 (2d Cir. 1981).

88. 47 U.S.C. § 312(a)(7).

opponents have appeared,[89] that licensees not transmit obscene or indecent programming,[90] and that licensees announce sponsored programming.[91]

Common carriers, typified at least originally by standard telephone service providers, are defined as "any person engaged as a common carrier for hire, in interstate or foreign communication by wire or radio or in interstate or foreign radio transmission of energy."[92] The Act's definition also stresses that "a person engaged in radio broadcasting shall not ... be deemed a common carrier."[93] Title II of the Communications Act requires carriers to provide service on reasonable request therefore, at reasonable rates and without discrimination between customers.[94] Also, unlike broadcasters, common carriers generally have no editorial control over communications, but rather, transmit intelligence of a customer's design and choosing.[95]

Private radio generally covers all users of the radio spectrum that are not involved in broadcasting or common carrier service. The regulatory category is a catch-all for "nationwide and international uses of radio by persons, businesses, state and local governments, and other organizations licensed to operate their own communications systems for their own use as an adjunct of their primary business or other activity."[96] Eligibility for private carrier status is generally limited to certain well-defined categories, such as public safety radio services, special emergency radio services, industrial radio services, and land transportation radio services.[97] But service within a designated category is no guarantee of regulatory treatment as a private operator. If a licensee acts as a common carrier within its range of permissible service, for example, by holding itself out indiscriminately to serve all those who may benefit from its particular offering, it will be treated as a common carrier.[98] The practical consequence of qualifying for private carrier status is exemption from broadcast or common carrier regulations.[99] This is not to say that private radio operators are entirely unregulated. Rather, the Commission

89. *Id.* at § 315(a)(1).

90. *Id.* at § 312(a)(6); 18 U.S.C. § 1464.

91. 47 U.S.C. § 317.

92. *Id.* at § 153(h).

93. *Id.*

94. *Id.* §§ 201–202. See *National Ass'n of Regulatory Utility Commissioners v. FCC*, 525 F.2d 630, 640–42 (D.C. Cir.), *cert. denied*, 425 U.S. 992 (1976).

95. *National Ass'n of Broadcasters v. FCC*, 740 F.2d 1190, 1203 (D.C. Cir. 1984) ("the sine qua non of a common carrier is the obligation to accept applicants on a non-content oriented basis"); *Frontier Broadcasting Co.*, 24 F.C.C. 251, 253–55 (1958). See Note, *Common Carriers Under the Communications Act*, 48 *U. Chicago L. Rev.*, 409, 428 (1981). But see *Carlin Communications, Inc. v. Southern Bell Telephone & Telegraph Co.*, 802 F.2d 1352 (11th Cir. 1986); *Carlin Communications, Inc. v. The Mountain States Telephone and Telegraph Co.* , 827 F.2d 1291 (9th Cir. 1987).

96. 47 C.F.R. § 0.131 (1987).

97. See 47 C.F.R. Part 90 (1987).

98. *NARUC I*, 525 F.2d at 642–44.

99. *Id.* at 645.

imposes various technical and procedural rules to allocate radio spectrum and to ensure its orderly use.[100]

Cable television initially defied classification as either broadcasting or common carriage.[101] To resolve this confusion, Congress created a new but complex regulatory definition in the Cable Communications Policy Act of 1984, defining a "cable system" as: "a facility, consisting of a set of closed transmission paths and associated signal generation, reception, and control equipment that is designed to provide cable service which includes video programming and which is provided to multiple subscribers within a community ... ."[102] The Act provides guidelines for cable regulation through local franchising, with the proviso that cable systems "shall not be subject to regulation as a common carrier or utility,"[103] but it is not entirely clear what this statement means. Although the Act avoids imposing certain indicia of common carrier status it treats cable operators as common carriers in other respects. For example, operators are required to set aside channel capacity under reasonable price, terms, and conditions for "leased access" by unaffiliated entities, rates in most communities are regulated according to complex formulas, and operators are prohibited from exerting any editorial control over the leased access programming.[104]

By the same token, the FCC has adopted rules to enable common carriers to compete with cable operators through the provision of "video dial tone" service.[105] Although video dial tone involves the offering of many of the same video entertainment services to subscribers as are available via cable, the FCC determined that it did not entail "transmission" of video service "directly to subscribers" as defined by law. Accordingly, video dial tone providers are not subject to Cable Act regulations or franchising requirements. The D.C. Circuit upheld this regulatory distinction because of the common carrier status of telephone companies.[106] The court noted that a telephone company that provides video dial tone

100. See 47 C.F.R. Part 90 (1987). See generally *Direct Broadcast Satellite Report and Order*, 90 F.C.C.2d 676 (1982).

101. See *United States v. Southwestern Cable Co.*, 392 U.S. at 164 (according to the FCC, cable systems are "neither common carriers nor broadcasters, and therefore are within neither of the principal regulatory categories created by the Communications Act").

102. 47 U.S.C. § 522(6). The Cable Act expressly excludes from the definition (a) a facility that serves only to transmit the signals of one or more television stations; (b) a facility that serves only subscribers in multiple dwelling units under common ownership, management or control (so long as no public rights-of-way are used); (c) common carrier facilities regulated under Title II of the Communications Act unless video programming is transmitted directly to subscribers; and (c) facilities of electric utilities when used solely for operating utility systems. *Id.*

103. *Id.* at § 541(c).

104. *Id.* at § 532. See generally Cable Communications Consumer Protection and Competition Act of 1992, 47 U.S.C.A. §§ 532(c) (West Supp. 1993).

105. *Further Notice of Proposed Rulemaking, First Report and Order and Second Further Notice of Inquiry*, 7 FCC Rcd. 300 (1991); *Memorandum Opinion and Order on Reconsideration In the Matter of Telephone Company-Cable Television Cross-Ownership Rules, Sections 63.54-63.58*, 7 FCC Rcd. 5069 (1992), *aff'd sub nom. National Cable Television Assn. v. FCC*, 33 F.3d 66 (D.C. Cir., 1994).

106. *National Cable Television Assn. v. FCC*, 33 F.3d at 71–75.

service lacks control over programming content, and acts merely as "a transparent conduit" to subscribers. Consequently, an entirely different regulatory structure applies to the business, even though the end product may be indistinguishable to the end user.

Particularly with the traditional classifications, designation of a particular regulatory pigeonhole for a given medium has a profound effect on determining the relevant constitutional standard to be applied. For example, courts have recognized sharply different First Amendment rights for broadcasters compared to common carriers. But dictum that "of all the forms of communication, it is broadcasting that has received the most limited First Amendment protection"[107] is somewhat misleading; common carriers are given less protection in terms of operators' editorial control.[108]

This difference was highlighted in the Supreme Court's opinion in *Columbia Broadcasting System v. Democratic National Committee.*[109] The Court held that a broadcast licensee could have a blanket policy of refusing to air paid editorial announcements without running afoul of the public interest mandate of the Communications Act. In reaching this conclusion, the Court examined the legislative history of the Radio Act of 1927 and the Communications Act of 1934 and found that Congress considered and "firmly ... rejected the argument that the broadcast facilities should be open on a nonselective basis to all persons wishing to talk about public issues."[110] After all, the Court reasoned, the Communications Act specifies that a person "engaged in radio broadcasting shall not ... be deemed a common carrier."[111] Common carriers, then, would appear to receive the lowest level of First Amendment protection, for they do not have a recognized right to speak on their own and are denied editorial control over their communication traffic.[112]

Broadcasting, at least in some important respects, has been accorded less First Amendment deference than its closest video competitor, cable television. Even without a fully articulated constitutional standard for cable, courts have held that

107. *FCC v. Pacifica Foundation*, 438 U.S. 726, 748 (1978).

108. *But see Chesapeake & Potomac Telephone Co. v. United States*, 830 F. Supp. 909 (E.D. Va. 1993) (cross ownership restrictions on common carriers struck down using First Amendment rationale).

109. 412 U.S. 94, 105–14 (1973).

110. *Id.* at 105–08.

111. *Id.* at 108–09, quoting 47 U.S.C. § 152(h).

112. *See United States v. Western Electric Co.*, 552 F. Supp. 131, 189–90 (D.D.C. 1982), *aff'd sub nom. Maryland v. United States*, 460 U.S. 1001 (1983); *United States v. Western Electric*, 673 F. Supp. 525, 586 n.273 (D.D.C. 1987). But see *Chesapeake & Potomac Telephone Co. v. United States*, 830 F. Supp. 909 (E.D. Va. 1993); *Carlin Communications, Inc. v. Southern Bell Telephone & Telegraph Co.*, 802 F.2d 1352 (11th Cir. 1986); *Carlin Communications, Inc. v. The Mountain States Telephone and Telegraph Co.* , 827 F.2d 1291 (9th Cir. 1987).

certain regulations permissible for broadcasters could not be applied to cable.[113] This is perhaps best illustrated by the Federal Communications Commission's policy restricting indecent radio and television programming.[114] In 1988, the FCC issued a notice of forfeiture to a Kansas City television station for a prime-time broadcast of the uncut film *Private Lessons*, holding that the movie was "indecent" within the meaning of the U.S. Criminal Code. The film depicted the seduction of a teenage boy by his governess and contained some nudity. Although the notice was later withdrawn on other grounds, it was the first time the Commission stated its intention to fine a television licensee for indecent broadcasts.[115]

But *Private Lessons* and other films of the "teen sex comedy" genre have been staples of many premium movie channels on cable television and are routinely transmitted with impunity. Certain local jurisdictions have attempted to ban "indecent" cable programming using almost identical language as that chosen by the FCC to regulate broadcasting, but courts have invalidated these laws.[116] The courts found that "fundamental differences between the broadcast medium and cable television require that [government power to regulate indecency] not be extended to cable television."[117] Of course, in certain other respects described earlier, such as with FCC signal carriage rules[118] or local access requirements,[119] cable operators have enjoyed less First Amendment protection than broadcasters.

One natural consequence of the role assigned regulatory classifications is that courts must determine the correct category before addressing the substance of First Amendment claims. In *City of Chicago v. Day*,[120] for example, the Circuit Court of Cook County, Illinois was asked to decide whether a satellite master antenna television system (SMATV), should be classified as a cable television system and subjected to franchising requirements under the federal cable act. The court refused to consider First Amendment defenses, saying "[f]or [defendant's]

113. E.g., *Preferred Communications, Inc. v. City of Los Angeles*, 754 F.2d 1396, 1403-04 (9th Cir. 1985), *aff'd on narrower grounds*, 476 U.S. 488 (1986); *Quincy Cable TV, Inc.*, 768 F.2d at 1450; *Century Federal, Inc.*, 648 F. Supp. at 1470–75. However, early cases, decided before the cable industry developed as a serious competitor to broadcasting, treated the two technologies as constitutionally indistinguishable. *Black Hills Video Corp. v. FCC*, 399 F.2d 65, 69 (8th Cir. 1968). See *Quincy Cable TV, Inc.*, 768 F.2d at 1443–44 (citing cases).

114. The FCC's constitutional authority to regulate indecent broadcast programming was recognized in *FCC v. Pacifica Foundation*, 438 U.S. 726 (1978). However, the Supreme Court has emphasized the narrowness of its *Pacifica* holding. *See Bolger v. Young's Drug Products Corp.*, 463 U.S. 60 (1983).

115. *KZKC Television, Inc.*, 4 FCC 6706 (1989). Virtually all forfietures for broadcast indecency have involved radio stations.

116. *Community Television of Utah, Inc. v. Wilkinson*, 611 F. Supp. 1099 (D.C. Utah 1985), *aff'd sub nom. Jones v. Wilkinson*, 800 F.2d 989 (10th Cir. 1986), *aff'd mem.* 480 U.S. 926 (1987); *Cruz v. Ferre*, 755 F.2d 1415 (11th Cir. 1985); *Community Television, Inc. v. Roy City*, 555 F. Supp. 1164 (D. Utah 1982); *Home Box Office, Inc. v. Wilkinson*, 531 F. Supp. 987 (D. Utah 1982).

117. *Community Television of Utah, Inc. v. Wilkinson*, 611 F. Supp. at 1109–10.

118. *Turner Broadcasting System v. FCC*, 819 F. Supp. 32, 41 (D.D.C. 1993).

119. *Chicago Cable Communications, Inc. v. Chicago Cable Commission*, 678 F. Supp. 734 (N.D. Ill. 1988), *aff'd*, 879 F.2d 1540 (7th Cir. 1989), *cert. denied*, 110 S. Ct. 839 (1990).

120. No. 88-MC-313994 (Circuit Ct., Cook Co., IL, May 21, 1990).

argument to have any legal merit, it would have to prove that it is a SMATV system. However, this it has failed to do."[121] The court reasoned the the defendant would first be required to submit to regulations appropriate to its regulatory category before sorting out its constitutional status.

Given the overriding importance of regulatory classification to the constitutional analysis, the question necessarily arises as to the level of scrutiny courts should bring to the government's classifications. The Supreme Court addressed this question, although not on First Amendment grounds, in *FCC v. Beach Communications, Inc.*[122] The case involved the same basic question as *City of Chicago v. Day*: whether a SMATV system could be subjected to franchising requirements under the Cable Act. The Court of Appeals had struck down a statutory distinction that exempted SMATV systems from franchising requirements where such systems connected commonly owned or managed buildings (and to the extent no public rights of way were crossed) but subjecting to regulation identical SMATV systems that connected buildings not commonly owned or managed.[123] The court held that the statutory definition violated the implied equal protection guarantee of the Due Process Clause of the Fifth Amendment in that it was "unable to imagine" any conceivable basis for the distinction.[124]

The Supreme Court reversed the D.C. Circuit, holding that the Cable Act's definition of a cable system that excluded certain SMATV systems but included others, was entitled to the presumption of having a rational basis. "In establishing the franchise requirement," the Court noted, "Congress had to draw the line somewhere; it had to choose which facilities to franchise. This necessity renders the precise coordinates of the resulting legislative judgment virtually unreviewable, since the legislature must be allowed leeway to approach a perceived problem incrementally."[125] Like the D.C. Circuit, however, the Court emphasized that it was limiting its review to "the question presented" of whether the regulatory classification is "rationally related to a legitimate government purpose under the Due Process Clause." Whether heightened First Amendment scrutiny should apply—particularly in light of the "burdens imposed on franchised cable systems under the newly enacted [Cable Act of 1992]"—was a question left open for the Court of Appeals to decide on remand.[126]

As the regulatory divisions among the various media become less distinct, it may well be that the government will face an increasing obligation to justify its media classification scheme. Increasingly, broadcasters, common carriers, and cable operators are providing the same or similar services. As this occurs, each segment of the industry will gain a more sound basis for arguing that it has been

---

121. *Id.*, slip op. at 8.

122. 113 S. Ct. 2096 (1993).

123. *Beach Communications, Inc. v. FCC*, 959 F.2d 975 (D.C. Cir.), *aff'd following remand*, 965 F.2d 1103 (D.C. Cir. 1992).

124. *Id.* at 987. The court expressly declined to address the SMATV operators' First Amendment claims, holding them to be "unripe." *Id.* at 984–85.

125. 113 S. Ct. at 2102.

126. *Id.* at n.6.

unreasonably singled out for burdensome treatment based on regulatory classifications that have little to do with real-world distinctions among the media. In short, convergence of the media will undermine the system of regulatory classifications, which, in turn, will undercut the rationale for different constitutional treatment of various communications technologies.

The Cycle Continues

Just as precolonial regulatory schemes faded as it became evident that they were no match for the technology they attempted to control, regulation of new media forms in the United States tends to relax over time. Courts always have seemed somewhat uneasy about the "law unto itself" approach to First Amendment analysis. Perhaps for that reason, once a communication technology is no longer novel they have honored that dictum more in the breach than in the observance. Media that were pegged with one of the traditional regulatory classifications—particularly broadcasting—have been most susceptible to separate treatment, but even that tendency may be changing.

Courts' ambivalence toward the command to treat each medium differently has been underscored by their recognition of traditional First Amendment values. For example, 37 years after the Supreme Court held that cinema was not "speech," it expressly overruled *Mutual Film Corp. v. Industrial Comm'n of Ohio* and found that "expression by means of motion pictures is included within the free speech and free press guaranty of the First and Fourteenth Amendments."[127] Although the Court felt compelled to observe that "[e]ach method [of communication] tends to present its own peculiar problems," it more importantly found that "the basic principles of freedom of speech and the press, like the First Amendment's command, do not vary. Those principles, as they have frequently been enunciated by this Court, make freedom of expression the rule."[128]

Thus, as a particular medium becomes more commonplace, the recognition of "core values" tends to outweigh the rhetoric regarding its "peculiar problems." At the same time, however, dictum about the uniqueness of each communications medium lives on long after courts have chosen to apply traditional First Amendment doctrine. The Court's rejection in *Joseph Burstyn, Inc. v. Wilson*, of an almost four-decades'-old precedent that excluded film from constitutional protection is a clear example of this phenomenon. Rather than create a new First Amendment theory tailored to the medium, it relied on established First Amendment prohibitions against prior restraint and discriminatory taxation of the traditional press.[129] Although the Court suggested that the constitution does not necessarily require "ab-

127. *Joseph Burstyn, Inc. v. Wilson*, 343 U.S. 495, 502–503 (1952). See also *United States v. Paramount Pictures, Inc.*, 334 U.S. 131, 166 (1948) ("We have no doubt that moving pictures, like newspapers and radio, are included in the press whose freedom is guaranteed by the First Amendment.").

128. *Joseph Burstyn, Inc.*, 343 U.S. at 503.

129. The Court expressly relied on *Near v. Minnesota*, 283 U.S. 697 (1931) and *Grosjean v. American Press Co.*, 297 U.S. 233 (1936). *See Joseph Burstyn, Inc.*, 343 U.S. at 503 n.14.

solute freedom to exhibit every motion picture of every kind at all times and all places," the exceptions it recognized to First Amendment protection were well settled at the time for established media.[130]

Even though subsequent decisions suggested that requiring predistribution submission of films to "censorship boards" is not necessarily unconstitutional, closer examination belies the notion that films were accorded lesser protection than "traditional media." In *Freedman v. Maryland*, for example, the Court struck down a Maryland film censorship statute as providing inadequate procedural safeguards.[131] In doing so, the Court applied "the settled rule of our cases" and suggested as a model "a New York injunctive procedure designed to prevent the sale of obscene books."[132] In short, the Court removed any basis for treating films differently from print media.[133] It also repudiated precedent that suggested otherwise.[134] Accordingly, the Maryland legislature disbanded the state film licensing board in 1981 after 65 years of operation. Censorship boards in all other states had

130. *Id.* at 502–503 & n.13. The Court cited *Feiner v. New York*, 340 U.S. 315 (1951) (threat of violent crowd reaction may justify restricting speech); *Kovacs v. Cooper*, 336 U.S. 77 (1949) (government may regulate decibel level of sound amplification devices); *Chaplinsky v. New Hampshire*, 315 U.S. 568 (1942) ("fighting words" not constitutionally protected); *Cox v. New Hampshire*, 312 U.S. 569 (1941) (government may require parade permits).

131. 380 U.S. 51 (1965). The Court held that certain procedures must be followed where the government seeks to halt distribution of a film. First, the government bears the burden of instituting judicial proceedings and proving that the material is unprotected. Second, any restraint prior to court proceedings is strictly limited to a brief, specified period solely in order to maintain the status quo. Third, rapid judicial determination must be guaranteed. *Id.* at 58–59.

132. *Id.* at 58, 60. Virtually all of the precedent cited in *Freedman* related to the traditional media.

133. The one exception the Court allowed was in the time limits prescribed for review of films as contrasted with that for books. It found that the long lead times generally associated with film exhibition may lead to a different standard for what constitutes "prompt judicial determination" of the status of a film as compared to a book. But the Court laid down no "rigid time limits or procedures" and made no concrete findings other than "the statute would have to require adjudication considerably more prompt than has been the case under the Maryland statute." *Id.* at 60–61.

134. An earlier case, *Times Film Corp. v. City of Chicago*, 365 U.S. 43 (1961) had suggested that the government may require submission of motion pictures in advance of exhibition. But in *Freedman*, the Court limited the holding in *Times Film* to the narrow and very abstract proposition that "a prior restraint was [not] necessarily unconstitutional *under all circumstances*." 380 U.S. at 53–54 (emphasis in original). Indeed, the Court disavowed the notion that *Times Film* had upheld "the specific features of the Chicago censorship ordinance." *Id.* at 54. As Justice Douglas pointed out in his concurring opinion, "the Chicago censorship system, upheld by the narrowest of margins in *Times Film Corp. v. Chicago*, 365 U.S. 43, could not survive under today's standards." *Id.* at 62.

been abandoned by the mid-1960s.[135] By 1982, the Court was willing to describe film (at least in dictum) as "one of the traditional forms of expression such as books" that are protected as "pure speech."[136]

The procedural safeguards applied in *Freedman* have been used interchangeably among various media ever since. The Court has required the same protections in cases involving censorship of mail[137] and seizure of imported material by U.S. customs agents.[138] In *Southeastern Promotions, Ltd. v. Conrad*, the Court applied the same procedural requirements to theatrical performances.[139] Despite the application of traditional First Amendment doctrine, the Court nevertheless repeated dictum that "[e]ach medium of expression ... must be assessed ... by standards suited to it, for each may present its own problems."[140] However, it reasoned that theater generally involves the acting or singing out of the written word and found "no reason to hold theatre subject to a drastically different standard."[141] In short, the rhetoric regarding "peculiar problems" has little effect on the result.

This coincidence of traditional First Amendment protection and cultural penetration of a given medium is more problematic when applied to newer communications technologies. The classification of new media for regulatory purposes tends to institutionalize, and thereby prolong, distinct constitutional treatment. Government control over broadcasting is the clearest example of this. Commercial television existed for more than 40 years and had long become a dominant social force before courts began to reconsider their constitutional approach. Most observers have concluded that the original justification for different treatment of

135. E. DeGrazia & R. Newman, *Banned Films*, 147 (1982). By 1992, Dallas, Texas was the only city in the United States that continued to have a film review board, and it was eliminated the following year. *See* Kastor, *It's a Wrap: Dallas Kills Film Board, Washington Post*, Aug. 13, 1993 at D1; *'Kuffs' Compromise, USA Today*, January 8, 1992 at D1 col.1. It is interesting to note that mainstream films began to include more realistic depictions of reality—particularly sexual relations—after the demise of licensing boards. To respond to this trend, the film industry in 1968 established a voluntary rating system (on a scale of G to X) to provide guidance to prospective audience members. *See* Hinson, *The 20-Year Rating Game, Washington Post*, G1 col. 4, Nov. 6, 1988. Although the rating system does not necessarily pose a First Amendment problem, it has been subject to increasing criticism. *See Miramax Films, Inc. v. MPAA*, 560 N.Y.S.3d 730 (1990) (rating system is "an effective form of censorship"); Masters, *Judge Blasts Movie Rating System, Wash. Post*, July 20, 1990 at A1 col. 2.

136. *See New York v. Ferber*, 458 U.S. 747 (1982).

137. *Blount v. Rizzi*, 400 U.S. 410, 419–421 (1971).

138. *United States v. Thirty-Seven Photographs*, 402 U.S. 363, 367 (1971).

139. 420 U.S. 546, 559–560 (1975).

140. *Id.* at 557.

141. *Id.* at 557–558.

broadcasting—the purported scarcity of frequencies—has for years been nothing more than a legal fiction.[142]

Still, the increasing tensions that have taken the luster off the "public trustee" model for broadcasting seem to have persuaded a number of courts to be more concerned about the First Amendment concerns inherent in regulation. Accordingly, they have begun to analyze free speech claims of broadcasters by giving less weight to—or by not relying on at all—the "special" nature of the medium. In *FCC v. League of Women Voters of California*, for example, the Supreme Court invalidated a statutory prohibition on editorializing by public broadcasting stations that received funds from the Corporation for Public Broadcasting.[143] Although the Court expressly upheld the public trustee concept of constitutional analysis over strict scrutiny, it subjected the government's asserted interests to a far more rigorous analysis than ever before, and questioned the continuing validity of the scarcity rationale.[144] The Court conducted a thorough review of the purposes of public broadcasting and the legislative objectives, and found that the ban on editorializing was not narrowly tailored and did not serve the asserted governmental interests.[145] As in other cases of the "law unto itself" genre, the Court continued to pay lip service to the "public trustee" concept, but it emphasized that "the broadcasting industry is indisputably a part [of the press]," and supported its ultimate conclusions with precedents involving traditional media.[146] As the Court stressed most recently in *Turner Broadcasting System v. FCC*, "the FCC's oversight responsibilities do not grant it the power to ordain any particular type of programming that must be offered by broadcast stations" and "the Commission may not impose upon them its private notions of what the public ought to hear."[147]

142. See, e.g., H. Geller, *Fiber Optics: An Opportunity for a New Policy?*, 15 (1991) ("the broadcast regulatory model is a failed concept" and "the public trustee scheme . . . is a joke"); L. Bollinger, *Images of a Free Press,* 88–90 (Chicago: University of Chicago Press, 1991) (describing the rationale of *Red Lion* as having "devastating—even embarrassing—deficienc[ies]," as "illogical," and as being based on "the simple-minded and erroneous assertion that public regulation is the only allocation scheme that can avoid chaos in broadcasting"); Lively, *supra*, at 1085; Fowler & Brenner, *A Marketplace Approach to Broadcast Regulation*, 60 *Texas L. Rev.*, 207, 221–226 (1982).

143. 468 U.S. 364 (1984).

144. *Id.* at 374–381. See also *id.* at 377 n.11 (Noting criticisms of scarcity rationale, the Court indicated that it would be willing "to reconsider our longstanding approach" if given "some signal from Congress or the FCC that technological developments have advanced so far that some revision of the system of broadcast regulation may be required."); *id.* at 378–379 n.12 ("were it to be shown by the Commission that the fairness doctrine '[has] the net effect of reducing rather than enhancing' speech, we would then be forced to reconsider the constitutional basis of our decision in [*Red Lion*]").

145. *Id.* at 384–399.

146. *Id.* at 382, citing *United States v. Paramount Pictures, Inc.*, 334 U.S. 131, 166 (1948), in which the Court for the first time stated that "[w]e have no doubt that moving pictures, like newspapers and radio, are included in the press whose freedom is guaranteed by the First Amendment." See also *id.* at 382–399, citing, *inter alia*, *Bolger v. Youngs Drug Products Corp.*, 463 U.S. 60 (1983); *Carey v. Brown*, 447 U.S. 455 (1980); *First National Bank of Boston v. Bellotti*, 435 U.S. 765 (1978); *Wooley v. Maynard*, 430 U.S. 705 (1977); *Buckley v. Valeo*, 424 U.S. 1 (1976) (per curiam); and *New York Times v. Sullivan*, 376 U.S. 254 (1964).

147. 114 S. Ct. 2445.

Lower courts have been more willing to dispense with the public trustee doctrine or to apply its First Amendment precepts in a far stricter way. In *Community Service Broadcasting of Mid-America, Inc. v. FCC*, the D.C. Circuit, sitting *en banc*, emphatically noted that "spectrum scarcity cannot be invoked to support a government attempt to penalize or suppress speech, based on its general content, by some, but not all, broadcast licensees."[148] The court's plurality opinion stated that under either strict scrutiny, or the *O'Brien* test for incidental speech restrictions, a requirement that public broadcast stations make and retain recordings of programs "in which any issue of public importance is discussed," violated the First Amendment.[149] Similarly, in *News America Publishing Co. v. FCC*, the D.C. Circuit demanded "a better fit between the law and its asserted legitimate purposes" than was evident in a congressional restriction on the FCC's ability to grant waivers of the newspaper–television cross-ownership rule.[150] The court pointedly outlined the weakness of separate constitutional treatment based on spectrum scarcity, but noted that even under the public trustee doctrine, regulations must be narrowly tailored to further a substantial government interest.[151] It found the cross-ownership limit at issue to be "astonishingly underinclusive," and therefore unconstitutional.[152]

Even where the courts have rejected First Amendment claims, they have begun to do so without reference to scarcity. In *United States v. Edge Broadcasting Co.*, for example, the Supreme Court upheld against a First Amendment challenge a prohibition on the broadcast of lottery advertisements in states that did not have a government lottery. Although the decision is a highly fragmented one in which 7 Justices supported the outcome for various reasons, not a single one relied on the rationale of *Red Lion*. Indeed, none even cited it.[153] The Court's decision ultimately rested on the commercial speech doctrine as articulated in *Central Hudson Gas & Electric Corp. v. Public Service Commission of New York*.[154]

The change in judicial attitudes toward broadcasting is shown even by the language courts use in framing their constitutional analyses. As Dean Lee Bollinger has observed, the *Red Lion* Court "never referred to the broadcast media as the press nor to broadcasters as editors or journalists; they were consistently described as licensees and fiduciaries."[155] But a different view has emerged in later cases. The Supreme Court has stated that "broadcasters are engaged in a vital and inde-

148. 593 F.2d 1102, 1111 n.21 (D.C. Cir. 1978) (*en banc*).

149. *Id.* at 1111–1122. Only four judges (Wright, Bazelon, McGowan, and Wilkey) endorsed this section of the opinion. Judge Robinson found that the regulation could not survive even minimal scrutiny, and found it unnecessary to apply a stricter test. *Id.* at 1127 (Robinson, J., concurring in part and concurring in the judgment).

150. 844 F.2d 800, 805 (D.C. Cir. 1988).

151. *Id.* at 810–812.

152. *Id.* at 814.

153. 113 S. Ct. 2696 (1993). See also *Valley Broadcasting Co. v. United States*, 820 F. Supp. 519 (D. Nev. 1993), in which the court invalidated a prohibition of broadcast advertising of casino gambling using the *Central Hudson* test. The court did not mention *Red Lion* or the public trustee standard.

154. 447 U.S. 557, 566 (1980).

155. Bollinger, *supra* at 91.

pendent form of communicative activity" and that "the First Amendment must inform and give shape to the manner in which Congress exercises its regulatory authority in this area."[156] The recognition of broadcasters as being an essential element of the press has been even more direct among the lower courts. As the D.C. Circuit noted in *Community Service Broadcasting of Mid-America, Inc.*, public affairs programming on broadcast stations "lies at the core of the First Amendment's protections."[157] This rhetorical shift is significant.

As they did with film a generation ago, courts appear to be distancing themselves from the historic justifications for separate constitutional treatment of broadcasting. In certain cases, this has meant increased First Amendment scrutiny of regulations, even as courts continue to recite some of the time-worn dictum about the "special characteristics" of broadcasting. In other cases, courts have directly eschewed reliance on prior justifications. To the extent courts have avoided taking the next logical step of reconsidering *Red Lion*, perhaps it is because they seek to avoid creating legal uncertainty in an increasingly confusing media marketplace. On the other hand, one observer has suggested that courts will refuse to take this step until they have devised a new theory that would continue to permit government control of the media.[158] Whatever the explanation, it seems evident that courts will have to move beyond their present approach.

THIS IS NO WAY TO RUN AN ELECTRONIC SUPERHIGHWAY

All signs suggest that we are on the brink of a major shift in First Amendment doctrine. At the same time, federal policymakers are focused on the development of the National Information Infrastructure. The evolution of such a "network of networks" will have profound implications for future constitutional analyses. Will notions of scarcity continue to play a role, or will the physical characteristics of the network be the most important factor? To what extent will regulatory classifications circumscribe the First Amendment treatment of new technologies? Each of these questions will have to be addressed over time. But if history can teach, the lesson should be this: The gradual evolution of constitutional rights based on regulatory classifications is utterly unsuited to the new media environment.

156. *League of Women Voters of California*, 468 U.S. at 378. See also *CBS, Inc. v. FCC*, 453 U.S. 367, 395 (1981) (broadcasters are "entitled under the First Amendment to exercise the widest journalistic freedom consistent with [their public duties]").

157. 593 F.2d at 1110 ("noncommercial licensees are fully protected by the First Amendment"). *See also Syracuse Peace Council v. FCC*, 867 F.2d at 654; *News America Publishing, Inc.*, 844 F.2d at 812–813; *Johnson v. FCC*, 829 F.2d 157, 161–163 (D.C. Cir. 1987).

158. See, e.g., Lively, *supra*, at 1085 ("What may be evinced is a long-standing mind-set, traceable to *Mutual Film*, that the risk of abandoning control premises, no matter how unpersuasive or irreconcilable with the first amendment, is unacceptable.").

The Glacial Pace of Doctrinal Change

There is a wide and growing chasm between the rate of technological change and that of legal development. The case-by-case legal process by which courts seek to define the appropriate constitutional standard for a given medium typically takes decades, but the communications industry is evolving far more quickly. If anything, the disparity between the two is growing, as the nation moves steadily toward creation of broadband, digital, interactive networks as courts and policymakers continue to debate the constitutionality of the fairness doctrine.[159] As Professor Rodney Smolla has perceptively pointed out, "[s]cientists move more quickly than lawyers."[160]

This is especially true when the lawyers are judges. The fact that courts are reluctant to resolve the difficult questions raised by new technologies is not a new phenomenon. The Supreme Court delayed taking up cases on the status of radio, perhaps because it found the new medium too intimidating. Chief Justice Taft is reported to have explained his lack of eagerness as follows: "[I]nterpreting the law on this subject is something like trying to interpret the law of the occult. It seems like dealing with something supernatural. I want to put it off as long as possible in the hope that it becomes more understandable before the court passes on the questions involved."[161] If any of these feelings are shared by members of the modern judiciary, who may not have grown up with computers, then the prolonged search for a new constitutional standard becomes more understandable.

The shifting legal status of electronic eavesdropping under both constitutional and statutory law further illustrates the problem of evolving technology. In 1928, the Supreme Court considered whether warrantless wiretapping violated the Fourth Amendment prohibition against unreasonable searches and seizures. The Court found no constitutional violation because the surveillance was accomplished without intruding on the physical property of the defendant.[162] By failing to acknowledge that technology permitted the government to intrude on communications in a way that previously was impossible, Chief Justice Taft (still no futurist) was able to conclude that "[t]here was no searching [and there] was no seizure." The Fourth Amendment "does not forbid what was done here" because "[t]he United States takes no such care of telegraph or telephone messages as of mailed sealed letters."[163]

Justice Brandeis, whose views ultimately prevailed, argued in dissent that constitutional principles were undermined to the extent the Court focused excessively on the method chosen for communication. He argued forcefully that constitutions must be interpreted with technological advancements in mind to preserve funda-

159. L. Tribe, *American Constitutional Law*, 1007 (2d ed. 1988); Katsh, *supra* at 1493; Pool, *supra* at 7.

160. Smolla, *supra* at 321.

161. Coase, *The Federal Communications Commission*, 2 *J. Law & Econ.*, 1, 40 (1959), quoting C.C. Dill, *Radio Law*, 1–2 (1938).

162. *Olmstead v. United States*, 277 U.S. 438, 464 (1928).

163. *Id.*

mental rights. In particular, Justice Brandeis wrote, constitutions must be designed "to approach immortality" and "our contemplation cannot only be what has been but of what may be."[164] Anticipating the rise of a computer-based society, he warned that:

> Discovery and invention have made it possible for the Government, by means far more effective than stretching upon the rack, to obtain disclosure in court of what is whispered in the closet.
>
> The progress of science in furnishing the Government with means of espionage is not likely to stop with wire-tapping. Ways may some day be developed by which the Government, without removing papers from secret drawers, can reproduce them in court, and by which it will be enabled to expose to a jury the most intimate occurrences of the home. Advances in the psychic and related sciences may bring means of exploring unexpressed beliefs, thoughts and emotions.
>
> Can it be that the Constitution affords no protection against such invasions of individual security?

Justice Brandeis concluded that if the courts did not adapt to new realities, then constitutional principles would be "converted by precedent into impotent and lifeless formulas" and that "[r]ights declared in words might be lost in reality."[165]

The Court eventually adopted the same view toward wiretapping, but it took nearly 40 years to do so. In *Katz v. United States*, the Court declared that the Fourth Amendment "protects people, not places" and held that wiretapping is allowable only after a valid warrant is issued—the same as for any other search.[166] Congress enacted legislation to codify the law as set out in *Katz*, but it soon became outdated and had to be rewritten.[167] The advent of fiber optic communications networks has created pressure for further legal change.[168]

The very nature of law, with its emphasis for creating certainty, makes keeping up with rapid technological development difficult if not impossible. Even Justice Brandeis, the champion of a dynamic constitution in *Olmstead*, wrote that "in most matters it is more important that the applicable rule of law be settled, than

---

164. *Id.* at 472–473 (Brandeis, J., dissenting).

165. *Id.* at 473–474 (internal quotations omitted).

166. 389 U.S. 347 (1967).

167. Omnibus Crime Control and Safe Streets Act of 1968, Pub. L. No. 90-351, §§ 801–804, 82 Stat. 197, 211–225; Electronic Communications Privacy Act of 1986; See generally, Fein, *Regulating the Interception and Disclosure of Wire, Radio, and Oral Communications: A Case Study of Federal Statutory Antiquation*, 22 *Harv. J. on Legislation*, 47 (1985).

168. See Mintz & Schwartz, *Clinton Backs Security Agencies on Computer Eavesdropping*, *Wash. Post*, February 5, 1994 at A1; *NII Task Force Searches For Privacy Guidelines*, *Communications Daily*, February 3, 1994 at 3–4.

that it be settled right."[169] Consequently, the nature of constitutional adjudication makes it easy to understand why it took 37 years for the Supreme Court to change its First Amendment approach to cinema, and why it has continued to spend decades debating the appropriate standards for broadcasting and cable television.

This time-consuming quest for a stable legal standard creates a special dissonance in the dynamic field of electronic communications. Congress created the Federal Communications Commission precisely because the field is rapidly changing; it recognized that legislative changes could not keep up with advancements in radio communication. The Communications Act of 1934 was envisioned as a flexible regulatory system "because the broadcast industry is dynamic in terms of technological change." The administrative approach it created is predicated on the assumption that "solutions adequate a decade ago are not necessarily so now, and those acceptable today may well be outmoded 10 years hence."[170] Yet even in a regulatory system based on this premise, it is difficult for the administrative agency to keep up with changes and adjust its rules accordingly. Consequently, burdensome regulations may live on long after their reason for existence has vanished.[171]

So we are left with an evident paradox. On one hand, the law is criticized for failing to keep up with innovations. On the other, it seems that the purpose of the law is undermined if it changes too quickly.[172] The dilemma is magnified to the extent that the speed with which innovations are introduced is accelerating.

But this is a false paradox. It exists only to the extent that a new legal standard is expected to spring into being with each new transmission technology. Where First Amendment principles are not dependent on the specific communications technology, there will be a greatly diminished perception that the law has not kept pace, for there will be no expectation of a major doctrinal shift with each new invention. Such an approach would also preserve the law's function of promoting certainty. Current conflicts about the appropriate First Amendment standard for broadcasting and cable television have done more to create instability—both from the perspective of the regulator and the regulated industries—than perhaps any other single factor in the law.

Breakdown of the Classification Scheme

The doctrine that "each [communications medium] is a law unto itself"[173] makes constitutional analysis exceedingly complex in a world of burgeoning technology and proliferating classifications. Courts, policymakers, and legal scholars simultaneously are being presented with an expansion and contraction of regulatory op-

169. *Burnett v. Colorado Oil and Gas Co.*, 285 U.S. 393 (1932).

170. *CBS, Inc. v. Democratic National Committee*, 412 U.S. at 102.

171. See, e.g., *Evaluation of the Syndication and Financial Interest Rules (Report and Order)*, 6 FCC Rcd. 3094 (1991), *rev'd and remanded, Schurz Communications v. FCC*, 982 F.2d 1043 (7th Cir. 1992). *See also Quincy Cable TV, Inc. v. FCC*, 768 F.2d at 1455–1457.

172. Katsh, *supra* at 17–19.

173. *Kovacs v. Cooper*, 336 U.S. at 97 (Jackson, J., concurring).

tions. In the first instance, the variety of delivery systems and media services has multiplied, as have the number of regulatory classifications. This raises the possibility that a separate First Amendment test must be applied to each medium and a new standard developed with each technical innovation. Second, with convergence, the discrete functions of the various media are coming together. For example, with the advent of fiber optics it is conceivable that a single transmission medium could become the conduit for newspapers, electronic mail, local and network broadcasting, video rentals, cable television, and a host of other information services.[174] The synthesis of form and function vastly complicates segregating the different media for separate constitutional treatment.

Courts and legislators generally attempt to fill gaps in legal doctrine by analogy rather than by developing new concepts.[175] Just as notions of "common carriage" and "public interest, convenience and necessity" in the Communications Act were drawn from 19th-century concepts of transportation law,[176] courts usually have borrowed the constitutional analysis articulated for established media for application to new technologies. But Ithiel de Sola Pool has cited the weakness of this approach. He explained that "[a] long series of precedents, each based on the last and treating clumsy new technologies in their early forms as specialized business machines, has led to a scholastic set of distinctions that no longer correspond to reality. As new technologies have acquired the functions of the press, they have not acquired the rights of the press."[177]

Courts did create a genuinely new First Amendment standard for broadcasting, but they have failed so far to do the same for cable television or other new video delivery systems. In the search for a new standard, the debate in most cases comes down to whether the new technology in question has more characteristics in common with broadcasting than with print. If the medium is deemed more like over-the-air television, a standard more forgiving of government intrusion is applied; if

174. See generally *Through The Looking Glass, supra*; Geller, *supra* n.137.

175. E.g., Kalven, *Broadcasting, Public Policy and the First Amendment*, 10 *J. Law & Econ.*, 15, 38 (1967) ("Law, it has been said, is determined by a choice between competing analogies."). Policymakers have been forced to develop new analogies as technology and the communications marketplace have evolved. Former FCC Commissioner Patricia Diaz Dennis half facetiously suggested that broadcasters should be regulated as if they were in a "game preserve," as opposed to the unregulated "jungle" advocated by opponents of government control or the paternalistic "zoo" favored by proponents of public intervention. Speech of Patricia Diaz Dennis before the Broadcast Financial Management Association (April 18, 1988). See *Trying a New Policy On For Size*, Broadcasting, April 25, 1988 at 41. Coming up with new concepts to accommodate the rapidly changing communications landscape is no easy task, and Commissioner Dennis was forced to admit, "I am no closer to solving this problem than scientists are to coming up with a unified theory to explain how the universe operates." *Id.*

176. See *Pensacola Tel. Co. v. Western Union Tel. Co.*, 96 U.S. 1, 8–9 (1878); *National Ass'n of Regulatory Utility Commissioners v. FCC*, 525 F.2d 630, 640–642 (D.C. Cir.), *cert. denied*, 425 U.S. 992 (1976); S. Mickelson, J. Opfer, & S. Whalen, *The Common Carrier Principle*, 3–5 (1989); Comment, *The Diversity Principle and the MFJ Information Services Restriction: Applying Time-Worn First Amendment Assumptions to New Technologies*, 38 *Cath. U. L. Rev.*, 471, 496–497 (Winter 1989); Note, *Common Carriers Under the Communications Act*, 48 *U. Chicago L. Rev.*, 409 (1981).

177. Pool, *supra* at 250.

it is considered more akin to traditional publishing, full First Amendment rights attach.

A basic problem with this approach is that it lacks a principled or consistent method of application. The similarities among media are in the eye of the beholder, and the resulting answers have been mixed. Another problem is this: What if the real answer is "all of the above?" Multimedia, for example, is "like" newspapers because it transmits text; it is "like" books when presented over a personal player on CD-ROM. But it is also "like" broadcasting or cable because it may transmit video. And it may be "like" common carriers when transmitted over the telephone network. The philosophy that "differences in the characteristics of new media justify differences in the First Amendment standards applied to them" is illogical in the case of multimedia.[178]

One response at least to the administrative law problem has been the development of "flexible" classification systems. The number of video sources has proliferated in recent years, forcing the FCC to develop various methods of classification. In addition to over-the-air broadcasting, the FCC issues licenses for such services direct broadcast satellites (DBS), multichannel multipoint distribution service (MMDS), and other microwave video services including operational fixed service (OFS) and instructional television fixed service (ITFS). Not only are each of these services classified differently despite a similarity of function, but an operator may choose among various options to determine its regulatory status.

The FCC first groped for new ways to classify video services in its 1982 order authorizing DBS service.[179] The Commission adopted what it called a "flexible regulatory approach," wherein the service could be regulated either as broadcasting or common carriage. An operator that retains control over the content of transmissions and provides service directly to homes was treated as a broadcaster; an operator that leases transponder capacity on a first-come, first-served basis and relinquishes editorial control was treated as a common carrier. "Customer-programmers" who leased satellite capacity were essentially unregulated. The D.C. Circuit rejected this approach and held that the Communications Act definition of broadcasting encompasses most DBS applications.[180] It remanded the issue of regulatory classification for further consideration.

In response, the FCC initiated a proceeding "to determine what criteria may be used by the Commission to determine whether a communications service should be treated as 'broadcasting' under the Communications Act."[181] The Commission determined that subscription video services should be classified as nonbroadcast services and freed from broadcast regulation. The appropriate classification hinges on the operator's intent: the service is not broadcasting if the licensee does not

178. Corn-Revere, *supra* at 22.

179. *DBS Report and Order*, 90 F.C.C.2d 676 (1982), *aff'd in part, vacated in part sub nom. National Ass'n of Broadcasters v. FCC*, 740 F.2d 1190 (D.C. Cir. 1984).

180. *National Ass'n of Broadcasters*, 740 F.2d at 1205. See also *Telecommunications Research and Action Center v. FCC*, 836 F.2d 1349 (D.C. Cir. 1988) (remanding FCC decision to classify nonsubscription use of ITFS capacity as nonbroadcasting).

181. *Subscription Video, Report and Order*, 2 FCC Rcd 1001, 1003 (1987).

intend to serve the public generally.[182] Based on the *Subscription Video* rules, a DBS operator could opt for regulatory treatment as a broadcaster, a nonbroadcaster, or a common carrier. The D.C. Circuit affirmed.[183]

Resolving the administrative classification problem, however, does not resolve the constitutional puzzle. For example, Clinton administration proposals regarding the National Information Infrastructure sought to "ensur[e] flexibility so that the newly-adopted regulatory framework can keep pace with the rapid technological and market changes that pervade the telecommunications and information industries."[184] It would have permitted service providers to opt for regulatory classification under a new section of the Communications Act. Yet even under this new section, new media would qualify for constitutional treatment roughly equivalent to that accorded cable operators and common carriers. The proposal authorized the FCC to adopt rules to "address public interest concerns" such as the transmission of indecent or obscene communications.[185] It also sought to "[e]nsure that delivery of video programming directly to subscribers over broadband facilities is consistent with certain principles now applicable to cable services ... dealing with: retransmission consent; public, educational, and governmental access; must carry; and protection of subscriber privacy."[186] The administration's proposal ultimately was not incorporated into telecommunications legislation, but it provided some insight into developing approaches to resolve the problem of regulatory classification in an era of convergence.

Regulation and First Amendment Traditions

To begin construction of the electronic superhighway with the assumption that access issues and content will be regulated in the same way as previous "new" technologies begs an important question. Proponents of a regulatory approach assume that the justifications that supported a different First Amendment standard for other media, such as broadcasting, can be transplanted and applied to broadband digital networks. Another possible assumption is that additional or new justifications support a different constitutional approach, a question that is explored in more detail in the next section of this chapter. Whatever theory may be used, it is vital to recognize that an important choice is being made, and that fundamental differences exist between a First Amendment model based on press autonomy and one based on regulation.

In comparing the competing visions of the First Amendment, Dean Bollinger has noted how Supreme Court decisions with respect to broadcasting amount to a

182. *Id.* at 1006. As indicia of intent, the Commission focused on whether the customer needs a special encoder to receive the transmission, the information is encrypted, and the operator and subscriber are in a contractual relationship.

183. *National Ass'n For Better Broadcasting v. FCC*, 849 F.2d 665 (D.C. Cir. 1988).

184. Administration White Paper on Communications Act Reforms, at 1.

185. *Id.* at 10. The administration specifically cites Section 223 of the Communications Act, which has been used to regulate so-called "dial-a-porn" services.

186. *Id.*

"virtual celebration of public regulation."[187] This, he concludes, is "[n]othing less ... than a complete conceptual reordering of the relationships between the government, the press, and the public that was established with *New York Times v. Sullivan*."[188] To read cases like *Red Lion* is to "step into another world," where the press itself represents the greatest threat to First Amendment values and government intervention in editorial choices is the preferred method of salvation.[189] It is a vision of the First Amendment, in the words of William O. Douglas, "that is agreeable to the traditions of nations that have never known freedom of the press."[190]

In this regard it is essential to keep in mind the fact that convergence of the media has significant implications far beyond its effect on the integrity of existing regulatory classifications. Constitutional analysis of electronic media has been tied to the means of transmission. Consequently, regulatory justifications for a lower level of constitutional protection for one medium may well be communicable as traditional media move toward new means of delivery.

Former FCC Commissioner Lee Loevinger predicted that "the computer and the electronic screen will become the printing presses of the next century."[191] Perhaps this was a safe prediction, but it is well on its way to being fulfilled a decade early. Newspapers and magazines across the country have embarked on a wide variety of projects to create electronic publications.[192] It is also significant that traditional presses now use electronic production methods, including computer terminals and local area networks to support writing, editing, and production as well as satellite links to transmit copy between remote plants.[193] As new technologies become the

187. Bollinger, *supra* at 71.

188. *Id.* at 66.

189. *Id.* at 72.

190. *Columbia Broadcasting System v. Democratic National Committee,*412 U.S. at 163 (Douglas, J., concurring). See Katsh, *supra* at 138–139 ("It may be ... that the greatest insight into what will occur over the next 10 to 20 years can be derived from looking back even further, at the period after printing was introduced. This was the era when the modern struggle between individual expression and state control over expression was forming. It may be that more can be discerned about the future of free expression, about attempts by government to control a new medium, and even about the likely direction of Supreme Court decisions by looking at the spread of printing and by comparing the qualities of print and the qualities of the new media than by analyzing either the trend of court decisions or the thoughts or intentions of the framers."). See also Baeza, *Safeguarding the First Amendment in the Telecommunications Era*, 97 *Harv. L. Rev.*, 584, 590 (1983) ("today we find ourselves at another crossroads as we face a choice between increased freedom and increased repression of speech"); Bazelon, *The First Amendment and the "New Media"—New Directions in Regulating Telecommunications*, 31 *Fed. Comm. L. J.*, 201, 212 (1979); Goldberg & Couzens, *"Peculiar Characteristics": An Analysis of the First Amendment Implications of Broadcast Regulation*, 31 *Fed. Comm. L. J.*, 1, 40–41 (1979).

191. Loevinger, *supra* at 776. See also Pool, *supra* at 224 ("Networked computers will be the printing presses of the twenty-first century.").

192. See, e.g., *Telcos and Newspapers Fill in New Relationships, Communications Daily*, February 17, 1994 at 2; Carmody, *Time's Readers to Talk Back, on Computers*, *New York Times*, July 26, 1993 at D6; Markoff, *A Media Pioneer's Quest: Portable Electronic Newspapers*, *New York Times*, June 28, 1992 at F11; Shaw, *Inventing the "Newspaper" of Tomorrow*, *Los Angeles Times*, June 3, 1991 at A1; Shaw, *"Electronic" Newspaper Emerging After Slow Start*, *Los Angeles Times*, June 3, 1991 at A1.

193. Loevinger, *supra* at 776.

predominant forms of communication and distribution of ideas, the overall level of First Amendment protection in society may be diminished, even among traditionally protected media.[194]

This issue was presented squarely in *Telecommunications Research and Action Center v. FCC*[195] (*TRAC*), in which the D.C. Circuit held that broadcast content controls apply to teletext transmissions. Teletext is a means of transmitting textual and graphic material to television screens of home viewers, using an otherwise unused portion of the broadcast signal.[196] In its *Report and Order* authorizing teletext service, the FCC declined to apply political broadcasting controls "primarily [because of] a recognition that teletext's unique blending of the print medium with radio technology fundamentally distinguishes it from traditional broadcast programming."[197] The Court of Appeals reversed the Commission's decision, focusing on the means of *delivering* the printed word. "The dispositive fact is that teletext is transmitted over broadcast frequencies that the Supreme Court has ruled scarce and this makes teletext's content regulable," the Court reasoned. "Teletext, whatever its similarities to the print media, uses broadcast frequencies, and that, given *Red Lion*, would seem to be that."[198]

TRAC suggests that newspapers delivered by electronic means have less constitutional protection than when the exact same stories written by the exact same reporters and edited by the same editors are delivered on paper. Consistent with this reasoning, one writer has advocated applying political broadcasting regulations to online computer services such as Prodigy, CompuServe and America Online.[199] To avoid the risk that online services might discriminate between candidates, Congress could require such services to provide "reasonable access" to candidates, "equal time" in the event an opponent uses the service and limit prices to the "lowest unit charge." This could be accomplished constitutionally, according to the article, by assuming that broadcasting provides the appropriate regulatory and constitutional metaphor.[200]

---

194. See Lively, *supra*, at 1074.

195. 801 F.2d 501 (D.C. Cir.), *reh'g en banc denied*, 806 F.2d 1115 (D.C. Cir. 1986), *cert. denied*, 482 U.S. 919 (1987).

196. *Id.* at 503. Teletext or videotext may also be transmitted by way of cable or telephone, but the *Telecommunications Research and Action Center* decision dealt only with over-the-air teletext transmissions.

197. 53 R.R.2d 1309 (1983).

198. 801 F.2d at 508, 509. Nevertheless, the court ruled that the FCC could refrain from enforcing the fairness doctrine for teletext transmissions because the doctrine was an FCC policy and not a statutory requirement. *Id.* at 516–518.

199. Campbell, *Political Campaigning in the Information Age: A Proposal for Protecting Political Candidates' Use of On-Line Computer Services*, 38 *Villanova L. Rev.*, 517 (1993).

200. *Id.* at 519 & n.9, 521, 542–545 ("The assimilation of computer-based communications is remarkably similar to the process by which radio became an accepted medium of communication."). *See also* Computer Professionals for Social Responsibility, *Serving the Community: A Public Interest Vision of the National Information Infrastructure*, 22–23 (1993) (advocating government policies to promote diversity in content markets).

Such a theoretical approach poses an interesting logical question. If traditional media are properly subject to a different constitutional standard when the link between the publisher and the reader is electronic, what is the appropriate standard when the electronic link is between the writer and the publisher? In other words, so long as electronic methods are used at some stage in the production process, doesn't the government have jurisdiction to regulate the content of the publication, just as with broadcasting?

In a 1987 Senate hearing on the fairness doctrine, Professor Robert Shayon of the Annenberg School of Communications appeared to suggest that any use of spectrum in the production process would justify content regulation of the press. Shayon asserted that content controls might constitutionally be imposed on the *New York Times* or the *Wall Street Journal* because they transmit their copy via satellite to printing plants across the country. "I think that the spectrum is limited," Shayon observed, "And if the big users shut out the small users, then the government should act to make fairness the ruling guideline... . The government is not only a repressive factor, it represents the total community and sometimes can be used constructively."[201] In other words, based on the choice of distribution media, the "total community" may gain the ability to tell the *New York Times* and the *Wall Street Journal* what is "fair" and to enforce any such determination. Former FCC Chairman Charles Ferris, a staunch supporter of the fairness doctrine, has raised a similar question.[202]

In short, the choice of a First Amendment standard for new communications technologies is not a simple decision that will determine how a particular new technology will be regulated. Because the media are converging, the constitutional approach selected now could well determine the nature of the First Amendment for the 21st century for all media. The choice might represent a fundamental shift in the relationship between the government and the press. Some would regard such a change as a welcome event because it would allow far greater flexibility in the realm of public policy. Others view it as a threat that would undermine the central purpose of the First Amendment—to free the press from government oversight.

CONCLUSION

The existing approach that determines the level of constitutional protection for a communications technology based on its regulatory classification cannot be sustained. Technology changes too quickly for the law to keep up, and the result is a complex web of artificial and ill-suited distinctions. Moreover, the convergence of new media functions undermines many of the attempts at classification. The regulatory approach also tends to erode traditional free speech protections. As the media converge, this erosion threatens to extend to "old" media as well as the

201. *Fairness in Broadcasting Act of 1987*, S. Hrg. 100–148, 100th Cong. 1st Sess., 73–74 (March 18, 1987). See *Licensing Broadcasters: Just What the Framers Feared?, Cato Policy Report* 6, 8–9 (January/February 1988).

202. Pool, *supra* at 1.

new. Such problems can be reduced by viewing these issues in historical context and recognizing the the First Amendment is predicated on the protection of new technologies.

ACKNOWLEDGMENTS

This chapter is excerpted and adapted from Chapter 2 of a forthcoming treatise, *Modern Communications Law,* to be published by West Publishing Company (publication anticipated 1996). An expanded version of this chapter was presented at the Columbia Institute for Tele-Information Conference on the 1992 Cable TV Act: Freedom of Expression Issues, Columbia University School of Business, February 25, 1994 and was published under the title *New Technology and the First Amendment: Breaking the Cycle of Repression*, 17 *Hastings Comm/Ent L.J.*, 301 (1994). The views expressed in this chapter are those of the author alone and do not necessarily reflect those of his clients or other parties.

Author Index

*Numbers in parentheses designate footnotes in text that contain reference information.

V

W

Y, Z

Subject Index

J

K, L

T

For Product Safety Concerns and Information please contact our EU representative GPSR@taylorandfrancis.com Taylor & Francis Verlag GmbH, Kaufingerstraße 24, 80331 München, Germany

Printed and bound by CPI Group (UK) Ltd, Croydon, CR0 4YY

01/05/2025

01858394-0001